# Current Progress in Botany Research

# Current Progress in Botany Research

Editor: Jerry Miller

R CALLISTO REFERENCE

www.callistoreference.com

**Callisto Reference,**
118-35 Queens Blvd., Suite 400,
Forest Hills, NY 11375, USA

Visit us on the World Wide Web at:
www.callistoreference.com

ISBN: 978-1-64116-106-0 (Hardback)

**Cataloging-in-Publication Data**

Current progress in botany research / edited by Jerry Miller.
    p. cm.
Includes bibliographical references and index.
ISBN 978-1-64116-106-0
1. Botany. 2. Botany--Research. I. Miller, Jerry.
QK45.2 .C87 2019
580--dc23

# Table of Contents

Preface.................................................................................................................................VII

Chapter 1    **Activity Regulation by Heteromerization of Arabidopsis Allene Oxide Cyclase Family Members**...........................................................................................1
Markus Otto, Christin Naumann, Wolfgang Brandt, Claus Wasternack and Bettina Hautse

Chapter 2    **High Constitutive Overexpression of Glycosyl Hydrolase Family 17 Delays Floral Transition by Enhancing FLC Expression in Transgenic *Arabidopsis***...........................12
Shinichi Enoki, Nozomi Fujimori, Chiho Yamaguchi, Tomoki Hattori and Shunji Suzuki

Chapter 3    **The Occurrence of Flavonoids and Related Compounds in Flower Sections of *Papaver nudicaule***.....................................................................................19
Bettina Dudek, Anne-Christin Warskulat and Bernd Schneider

Chapter 4    **Caffeoylquinic Acids from the Aerial Parts of *Chrysanthemum coronarium* L.**........................27
Chunpeng Wan, Shanshan Li, Lin Liu, Chuying Chen and Shuying Fan

Chapter 5    ***FLOWERING LOCUS T* Triggers Early and Fertile Flowering in Glasshouse Cassava (*Manihot esculenta* Crantz)**...............................................................34
Simon E. Bull, Adrian Alder, Cristina Barsan, Mathias Kohler, Lars Hennig, Wilhelm Gruissem and Hervé Vanderschuren

Chapter 6    **Paired Hierarchical Organization of 13-Lipoxygenases in *Arabidopsis***........................42
Adeline Chauvin, Aurore Lenglet, Jean-Luc Wolfender and Edward E. Farmer

Chapter 7    **Differential Mechanisms of Photosynthetic Acclimation to Light and Low Temperature in *Arabidopsis* and the Extremophile *Eutrema salsugineum***.....................55
Nityananda Khanal, Geoffrey E. Bray, Anna Grisnich, Barbara A. Moffatt and Gordon R. Gray

Chapter 8    **First Report on the Ethnopharmacological uses of Medicinal Plants by Monpa Tribe from the Zemithang Region of Arunachal Pradesh, Eastern Himalayas, India**........................................................................84
Tamalika Chakraborty, Somidh Saha and Narendra S. Bisht

Chapter 9    **Selecting Lentil Accessions for Global Selenium Biofortification**........................96
Dil Thavarajah, Alex Abare, Indika Mapa, Clarice J. Coyne, Pushparajah Thavarajah and Shiv Kumar

Chapter 10    **Influence of Nitrogen Availability on Growth of Two Transgenic Birch Species Carrying the Pine GS1a Gene**....................................................................107
Vadim G. Lebedev, Nina P. Kovalenko and Konstantin A. Shestibratov

Chapter 11   **Effect of Drought on Herbivore-Induced Plant Gene Expression: Population Comparison for Range Limit Inferences**.................................................................................**116**
Gunbharpur Singh Gill, Riston Haugen, Steven L. Matzner,
Abdelali Barakat and David H. Siemens

Chapter 12   **Litter Accumulation and Nutrient Content of Roadside Plant Communities in Sichuan Basin, China**.........................................................................................**137**
Huiqin He and Thomas Monaco

Chapter 13   **An Improved Syringe Agroinfiltration Protocol to Enhance Transformation Efficiency by Combinative use of 5-Azacytidine, Ascorbate Acid and Tween-20**...............................**149**
Huimin Zhao, Zilong Tan, Xuejing Wen and Yucheng Wang

Chapter 14   **Patterns of Growth Costs and Nitrogen Acquisition in *Cytisus striatus* (Hill) Rothm. and *Cytisus balansae* (Boiss.) Ball are Mediated by Sources of Inorganic N** .................................................................................**159**
María Pérez-Fernández, Elena Calvo-Magro, Irene Ramírez-Rojas,
Laura Moreno-Gallardo and Valentine Alexander

Chapter 15   **Comparative Phenotypical and Molecular Analyses of Arabidopsis Grown under Fluorescent and LED Light**...............................................................................**170**
Franka Seiler, Jürgen Soll and Bettina Bölter

Chapter 16   **Quantification of *Plasmodiophora brassicae* using a DNA-Based Soil Test Facilitates Sustainable Oilseed Rape Production** ...................................................**185**
Ann-Charlotte Wallenhammar, Albin Gunnarson, Fredrik Hansson
and Anders Jonsson

Chapter 17   **Antibacterial Properties of Flavonoids from Kino of the Eucalypt Tree, *Corymbia torelliana*** ...................................................................................**191**
Motahareh Nobakht, Stephen J. Trueman, Helen M. Wallace, Peter R. Brooks,
Klrissa J. Streeter and Mohammad Katouli

**Permissions**

**List of Contributors**

**Index**

# Preface

The main aim of this book is to educate learners and enhance their research focus by presenting diverse topics covering this vast field. This is an advanced book which compiles significant studies by distinguished experts in the area of analysis. This book addresses successive solutions to the challenges arising in the area of application, along with it; the book provides scope for future developments.

Botany is the study of plant morphology, anatomy, genetics and ecology. It has applications across agriculture, horticulture, medicine, forestry and environmental management. Botanical research explores studies of plant structure, growth, diseases, evolutionary history, taxonomy, etc. It is also engaged in the study and provision of materials like timber, rubber, oil and drugs. Dominant aspects in the study of modern botany are the study of molecular genetics and epigenetics of plants. The topics included in this book on botany are of utmost significance and bound to provide incredible insights to readers. It traces the progress of this field and highlights some of its key concepts and applications. Students, botanists, researchers and other experts associated with this field will immensely benefit from the analyses and information provided in this book.

It was a great honour to edit this book, though there were challenges, as it involved a lot of communication and networking between me and the editorial team. However, the end result was this all-inclusive book covering diverse themes in the field.

Finally, it is important to acknowledge the efforts of the contributors for their excellent chapters, through which a wide variety of issues have been addressed. I would also like to thank my colleagues for their valuable feedback during the making of this book.

**Editor**

# Activity Regulation by Heteromerization of Arabidopsis Allene Oxide Cyclase Family Members

Markus Otto [1,2], Christin Naumann [1], Wolfgang Brandt [3], Claus Wasternack [2,4] and Bettina Hause [1,*]

Academic Editor: Debora Gasperini

[1] Department of Cell and Metabolic Biology, Leibniz Institute of Plant Biochemistry, Weinberg 3, D-06120 Halle (Saale), Germany; mrmarkus.otto@googlemail.com (M.O.); christin.naumann@ipb-halle.de (C.N.)
[2] Department of Molecular Signal Processing, Leibniz Institute of Plant Biochemistry, Weinberg 3, D-06120 Halle (Saale), Germany; cwastern@ipb-halle.de
[3] Department of Natural Product Chemistry, Leibniz Institute of Plant Biochemistry, Weinberg 3, D-06120 Halle (Saale), Germany; wbrandt@ipb-halle.de
[4] Laboratory of Growth Regulators, Centre of the Region Haná for Biotechnological and Agricultural Research, Institute of Experimental Botany AS CR & Palacký University, Šlechtitelů 11, CZ-78371 Olomouc, Czech Republic
* Correspondence: bhause@ipb-halle.de

**Abstract:** Jasmonates (JAs) are lipid-derived signals in plant stress responses and development. A crucial step in JA biosynthesis is catalyzed by allene oxide cyclase (AOC). Four genes encoding functional AOCs (AOC1, AOC2, AOC3 and AOC4) have been characterized for *Arabidopsis thaliana* in terms of organ- and tissue-specific expression, mutant phenotypes, promoter activities and initial *in vivo* protein interaction studies suggesting functional redundancy and diversification, including first hints at enzyme activity control by protein-protein interaction. Here, these analyses were extended by detailed analysis of recombinant proteins produced in *Escherichia coli*. Treatment of purified AOC2 with SDS at different temperatures, chemical cross-linking experiments and protein structure analysis by molecular modelling approaches were performed. Several salt bridges between monomers and a hydrophobic core within the AOC2 trimer were identified and functionally proven by site-directed mutagenesis. The data obtained showed that AOC2 acts as a trimer. Finally, AOC activity was determined in heteromers formed by pairwise combinations of the four AOC isoforms. The highest activities were found for heteromers containing AOC4 + AOC1 and AOC4 + AOC2, respectively. All data are in line with an enzyme activity control of all four AOCs by heteromerization, thereby supporting a putative fine-tuning in JA formation by various regulatory principles.

**Keywords:** Arabidopsis allene oxide cyclase isoforms; heteromerization; protein structure analysis; site-directed mutagenesis; activity regulation

## 1. Introduction

The lipid-derived signaling compounds of the jasmonate family, among them jasmonic acid (JA) and octadecanoids, such as *cis*-(+)-12-oxophytodienoic acid (OPDA), are most active in plant stress responses and development. All enzymes involved in the biosynthesis of JA have been cloned and characterized from numerous plant species. Important components in JA perception and signaling have been elucidated (reviewed by [1]). Among them are repressors, such as the so-called JAZ (jasmonate ZIM domain) proteins and the F-box protein COI1 (coronatine insensitive1). Both of them are constituents of the SCF$^{COI1}$ JAZ-co-receptor complex, which function as an E3 ubiquitin ligase (reviewed by [1]). This receptor has a regulatory role in JA biosynthesis, since genes encoding enzymes in JA biosynthesis are JA-inducible. Any proteasomal degradation of JAZ proteins will switch on expression of JA biosynthesis genes, thus being equivalent to a positive feedback loop in gene expression [2].

Since all enzymes of JA biosynthesis are constitutively present in most of the tissues, another level of regulation takes place by substrate availability [1]. α-Linolenic acid (α-LeA) is the initial substrate in the biosynthesis of JA, which is only generated upon the release of α-LeA from chloroplast membranes. Such release takes place by environmental stimuli, as wounding, or during development. The requirement for substrate generation was shown by transgenic lines overexpressing enzymes in JA biosynthesis [3,4]. Such lines contain high levels of biosynthetic proteins, but do not show enhanced basal levels of JA. However, upon stimulus leading to release of α-LeA, JA is synthesized to higher levels than in wild-type plants. The tissue specificity of the occurrence of JA biosynthesis enzymes is an additional regulatory level. The tissue-specific and common occurrence of enzymes is attributed to the capacity of different cells and tissues for JA formation and JA signaling [4–7]. Finally, post-translational regulation is suggested to be attributed to the regulation of JA biosynthesis since the first hints at protein-protein interaction among the biosynthetic enzymes were obtained [8,9]. Moreover, employing labelled JA or the synthetic JA-Ile mimic coronalon to distinguish exogenous and endogenous jasmonates, no jasmonate biosynthesis/accumulation could be found supporting a post-translational regulation [10,11].

JA biosynthesis takes place by a sequential action of a 13-lipoxygenase (13-LOX), 13-hydroperoxide-converting 13-allene oxide synthase (13-AOS), cyclopentenone-forming allene oxide cyclase (AOC), the cyclopentanone-forming OPDA reductase3 (OPR3), as well as the fatty acid β-oxidation machinery, including an acyl-CoA-oxidase (ACX1), in the shortening of the carboxylic acid side chain. In Arabidopsis, these enzymes are encoded by single copy genes (AOS, OPR3) or by small gene families (13-LOX, AOC, ACX1). In case the enzymes are active as multimers, such gene families open the possibility for regulation by multimerization, if all or several isoforms are present in the same tissue or compartment. The first indication for such a regulation has been described for OPR3 and AOC, which were shown to be, upon crystallization, a dimer [8] and a trimer [12], respectively. These studies analyzed predominantly the binding pocket of OPR3 and AOC2, respectively. Less is known, however, for the structural requirement of trimerization of AOC2 and putative heteromerization of AOCs. The *AOC* gene family in Arabidopsis comprises four members, which encode functional AOC1, AOC2, AOC3 and AOC4 in a COI1- and JA-dependent manner, including wound-induced expression [13]. For the four *AOC* promoters, redundant, as well as organ- and isoform-specific activities have been detected [9]: (i) in roots, only *AOC3* and *AOC4* showed promoter activity; (ii) in fully-developed leaves, *AOC1*, *AOC2* and *AOC3* promoters were active, whereas the *AOC4* promoter activity was confined to the vascular bundles; (iii) *AOC1* and *AOC4* promoter activities were found to be partially specific in flower development. Single and double *AOC* loss of function mutants were without any altered phenotype, further supporting the assumption of redundancies among the AOCs of Arabidopsis. Analyses by bimolecular fluorescence complementation (BiFC) indicated *in planta* interaction of all four AOCs to each other [9]. Whether this interaction of different isoforms results in altered enzymatic activities could, however, not be clarified.

Here, we complement and extend this work by analysis of the trimer formation of AOC2 and addressed the question of whether heteromers between different AOC isoforms do show altered enzymatic activities. The formation of AOC2 trimers has been analyzed by using selective treatments with sodium dodecyl sulfate (SDS) and chemical cross-linking, as well as by molecular modelling studies, leading to the identification of salt bridges and a hydrophobic core, which have been proven by site-directed mutageneses. Finally, the activities of recombinantly-expressed AOC heteromers were determined in comparison to the activity of homomers. The data are in line with an enzyme activity control of AOCs by heteromerization, thereby supporting a fine-tuning in the formation of JAs by an additional regulatory principle.

## 2. Results

### 2.1. Trimer Formation of Recombinant AOC2

The AOC2 has been determined as a trimer in crystals, and SDS-stable trimer formation was suggested for recombinantly-expressed AOC exhibiting an N-terminal His-tag, whereas the C-terminally tagged AOC2 appeared in monomers only [12]. To clarify the role of AOC2 trimer formation, we extended here these analyses. *AOC2* was fused N- and C-terminally to His-tag and expressed in *E. coli*. After purification of AOC2 using Ni-nitrilotriacetic acid (Ni-NTA), proteins were treated with SDS at room temperature, 42 °C and 96 °C for 5 min and separated by SDS-polyacrylamide gel electrophoresis (PAGE) (Figure 1A,B). After treatment at room temperature and 42 °C, most of the recombinant AOC2 was found at around 65 kDa, pointing to the appearance as a trimer independently on the site of the His-tag. After treatment at 96 °C, however, AOC2 was mostly in the monomeric form, exhibiting the expected size of 26 kDa. Lastly, solutions containing AOC2 were incubated with the chemical cross-linking agent ethylene glycol-bis(succinic acid *N*-hydroxysuccinimide ester) (EGS) and analyzed after treatment with SDS at 96 °C. In these experiments, EGS, as a zero-length spacer, forms covalent intermolecular bonds [14] between subunits in AOC2 multimers, allowing definitive identification of the subunit structure(s) of AOC2 in the presence or absence of detergent (Figure 1C). Cross-linking with EGS led to an increase in the amount of AOC2 in the trimeric state, which was not dissolved by treatment with SDS at 96 °C. This was also supported by size exclusion chromatography (Figure S1). These data confirm that the predominant form of recombinant, His-tagged AOC2 is a trimer.

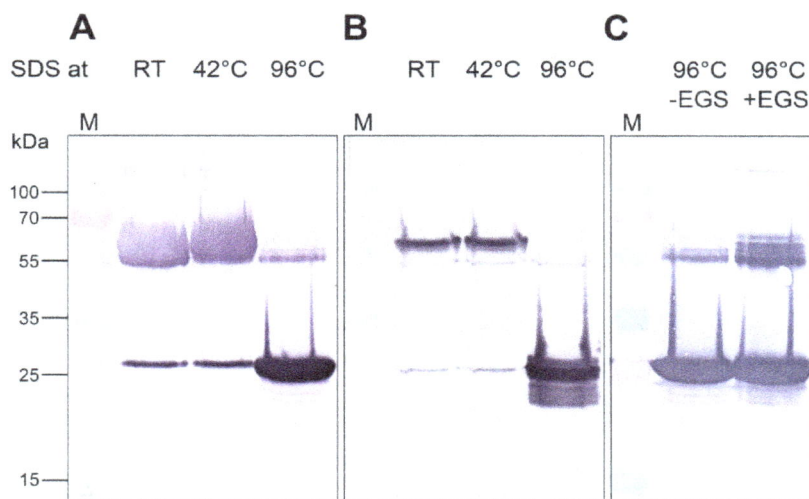

**Figure 1.** Recombinant Arabidopsis allene oxide cyclase 2 (AOC2) is a trimer. His-tagged proteins were purified on Ni-NTA, treated with SDS at given temperatures for 5 min and separated by SDS-PAGE. Detection of proteins was performed by immunolabeling using an anti-His-tag antibody. (**A**) Recombinant AOC2 protein with N-terminal His-tag; (**B**) recombinant AOC2 protein with C-terminal His-tag; (**C**) recombinant AOC2 protein with N-terminal His-tag cross-linked or not with 200 µM glycol-bis(succinic acid *N*-hydroxysuccinimide ester) (EGS) for 40 min. Note that only treatment at 96 °C resulted in the predominant occurrence of monomers (~26 kDa), whereas trimers resisted treatments at lower temperatures. Accordingly, cross-linking with EGS prevented separation of trimers by SDS treatment at 96 °C (C). M = size marker.

### 2.2. Analysis of the Quaternary Structure of AOC2

To clarify whether the formation of AOC2 trimer is essential for enzymatic activity, the structure of the trimer was analyzed by molecular modelling methods, including determination of activity after site-directed mutagenesis. For the modelling approach, the crystal structure of AOC2 available from the Protein Data Bank and deposited as 2Q4I was used (Figure 2A). This structure resulted from an improved refinement protocol using the original data of the Center for Eukaryotic Structural Genomics (CESG) and showed the most consistent R-free values at high resolution. The calculation of the interaction energies of all single monomers (chains A–C) from the remaining dimers (*i.e.*, the interactions of chain A with the dimer consisting of chains B and C and all other combinations) yielded in average −250 kcal/mol (±5 kcal/mol). This rather high interaction energy is a strong indication of the stability of the trimer complex, such as crystallized.

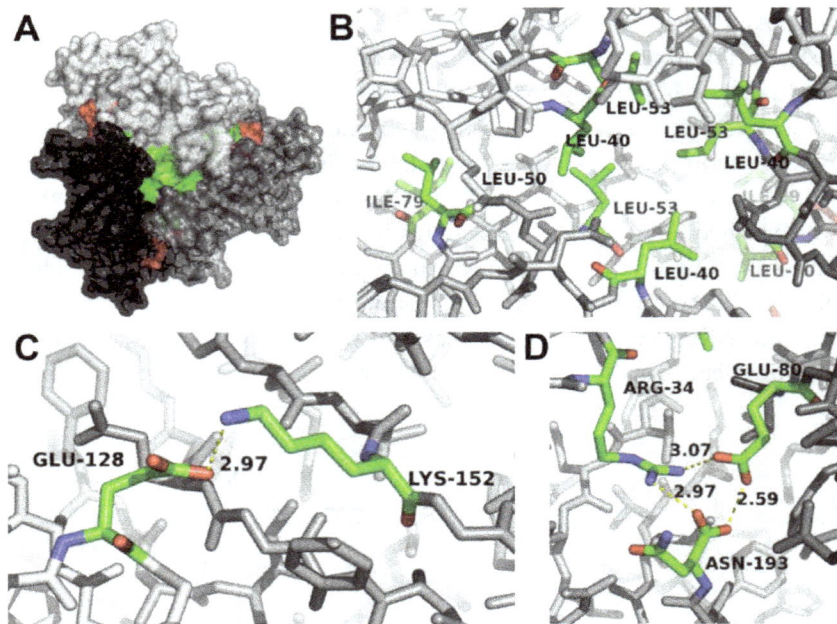

**Figure 2.** Identification of amino acids at the interaction sites between AOC2 monomers within the trimer. (**A**) Surface representation of the AOC2 trimer (PDB Code 2Q4I). The view is along the trimer axis. The three monomers are given in different gray scales. The hydrophobic core (detailed in (**B**)) and the salt bridges (detailed in (**C**)) are shown in green and red, respectively. Salt bridges on the back are detailed in (**D**); see Figure S1. (B) Location of Leu40, Leu50, Leu53 and Ile79 of all three monomers building the hydrophobic core of the trimer. (C) Salt bridge (yellow dashed line) between Lys152 of one monomer and Glu128 of the neighboring monomer. The distance between atoms building the salt bridge is given in Å. (D) Hydrogen bonds between Arg34 and Asn193 of one monomer and salt bridges (yellow dashed lines) to Glu80 of the neighboring monomer. The distances between atoms building a salt bridge are given in Å. Interacting faces are shown in stick representation (oxygen, red; nitrogen, blue; carbon, green; atoms not involved in the interaction, gray).

Inspection of the contact surfaces between the different chains showed the formation of hydrophobic interaction sites building a hydrophobic core (Figure 2A,B). The hydrophobic core is formed by the amino acids Leu40, Leu50, Leu53 and Ile79 of each monomer. In addition, different salt bridges between the monomers were found and are located on both sites of the trimer (Figure 2A and Figure S2). On the one hand, strong hydrogen bonds were detected between the side chains of Arg34 and Asn193 of one monomer and a salt bridge with Glu80 of the neighboring monomer (Figure 2C). These interactions have been identified previously by Hofmann *et al.* [12], but due to the other crystal structure used (PDB Code 2BRJ), the involved amino acids are numbered there as Arg29, Asn188 and Glu75. On the other hand, additional salt bridges were identified on the "back-side" of the trimer (Figure S1) and are formed between Glu128 of one monomer and Lys152 of the neighboring monomer (Figure 2D).

To test whether monomeric AOC2 exhibits enzymatic activity, the identified interaction sites were mutated to prevent the formation of a trimeric quaternary structure (Table 1). All three mutants and the AOC2 wild-type form were expressed as His-tagged versions in *E. coli* and purified (Figure 3A), whereby the yield of purified proteins was very low in the case of all mutant versions. The question could not be clarified, however, why there was a diminished production of recombinant proteins. Nevertheless, the purified mutant proteins did not appear different from wild-type AOC2 in regard to stability. Therefore, the quaternary state was compared upon SDS treatment of AOC2 proteins at 42 °C for 5 min (Figure 3A). Whereas the wild-type protein is clearly visible as a trimer with an apparent mass of about 65 kDa, the mutated AOC2 variants are monomeric. All purified proteins were used in equal amounts (10 μg purified protein each) to determine enzymatic activity by using the coupled enzymatic test by incubation with recombinant HvAOS [15] and 13(*S*)-hydroperoxyoctadeca-9,11,15-trienoic acid (HPOT) as the substrate. The production of *cis*-OPDA was recorded as enzymatically-formed OPDA and was measured in high amounts for wild-type AOC2 (Figure 3B). In contrast, proteins mutated in amino acid residues involved in the formation of salt bridges, in the hydrophobic core, or in both, reduced the AOC2 activity dramatically. Here, the detectable amount of *cis*-OPDA did not exceed that of the control performed by omitting any AOC protein. All of these data are in line with an essential role of trimer formation for AOC2 activity.

**Table 1.** Mutants of Arabidopsis AOC2 and expected alterations in the interactions of monomers.

| Mutant | Exchange of Amino Acids | Expected Effects |
|---|---|---|
| AB | K152A, E80A | Disruption of salt bridges |
| C | L53S | Disruption of hydrophobic core |
| ABC | K152A, E80A; L53S | Disruption of salt bridges and hydrophobic core |

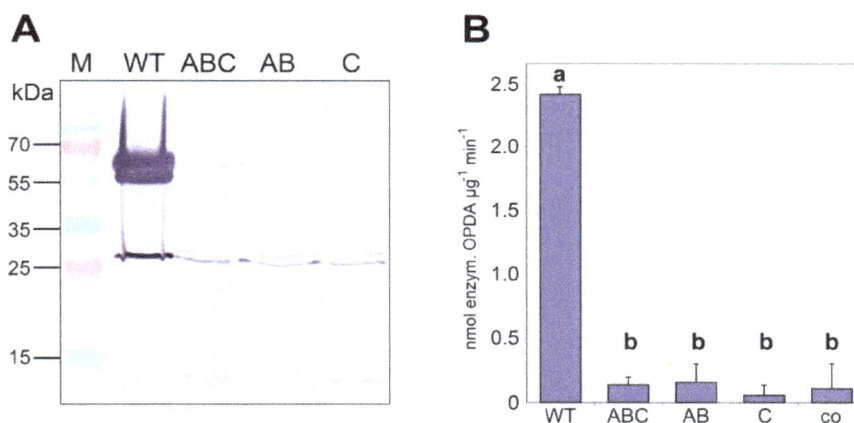

**Figure 3.** Monomers of AOC2 do not exhibit enzymatic activity. AOC2 was mutated to prevent trimer formation by the exchange of amino acids involved in salt bridge formation (K152A/E80A = "AB"), the formation of the hydrophobic core (L53S = "C") or both (K152A/E80A/L53S = "ABC") (see Table 1). (**A**) His-tagged recombinant proteins were treated with SDS at 42 °C for 5 min and separated on SDS-PAGE. Note that all mutant proteins are predominantly detectable as monomers. M = size marker. (**B**) AOC enzyme activity of recombinant wild-type AOC2 and mutant proteins. Each value is given as nmol of enzymatically-formed *cis*-(+)-12-oxophytodienoic acid (OPDA) per μg protein and min and is represented by the mean of three independent replicates ± SD. Different letters designate statistically-different values (one-way ANOVA with Tukey's HSD test, $p < 0.01$). co = control done without the addition of protein to the reaction mixture.

*2.3. Heteromerization of AOC Gene Family Members and Its Effect on AOC Activity*

The trimer formation of AOC2 raises the question of whether the activity of AOCs might be affected by putative heteromerization of different AOCs. The formation of heteromers between AOC family members has been detected upon *in planta* protein interaction studies using BiFC [9]. Even if interactions among all AOCs were found, the effect on activity by an *in planta* proof was, however, masked by overexpression conditions [9].

To check the influence of heteromerization on AOC activity, we purified the four recombinant AOCs as homomers using N-terminally His-tagged versions expressed in *E. coli*. To get heteromers, different pairwise combinations of all AOCs were expressed using one vector, thereby one AOC marked by a His-tag and the other AOC with a Strep-tag (Figure 4A). Heteromers were purified first using Ni-NTA for the His-tag followed by StrepTactin for the Strep-tag. After treatment with SDS at 96 °C and separation by SDS-PAGE, the homomers and heteromers of the AOCs were detected by immunoblot analyses (Figure 4A). The His-tag-labeled homomers of the four AOCs (Figure 4A, top left) were clearly purified, exhibiting slightly different molecular masses according to previously-published results [13]. Purification of the heteromers carrying His-tagged and the Strep-tagged monomers is indicated by their appearance in immunoblot analysis with both tag-specific antibodies (Figure 4A, right, top and bottom). All of these purified protein variants were used for the determination of activity (Figure 4B). The AOC activities of homomeric and heteromeric AOCs showed significant differences. Among the homomers, AOC4 exhibited the highest activity, whereas AOC1 and AOC3 showed similar, but low activity. Most importantly, the activities of several heteromeric AOCs exceeded that of homomeric AOCs. The highest activity was found for heteromers of AOC4 and AOC2, as well as AOC4 and AOC1 (Figure 4B). These data strongly suggest that AOC activity is altered by heteromerization of the AOC family members.

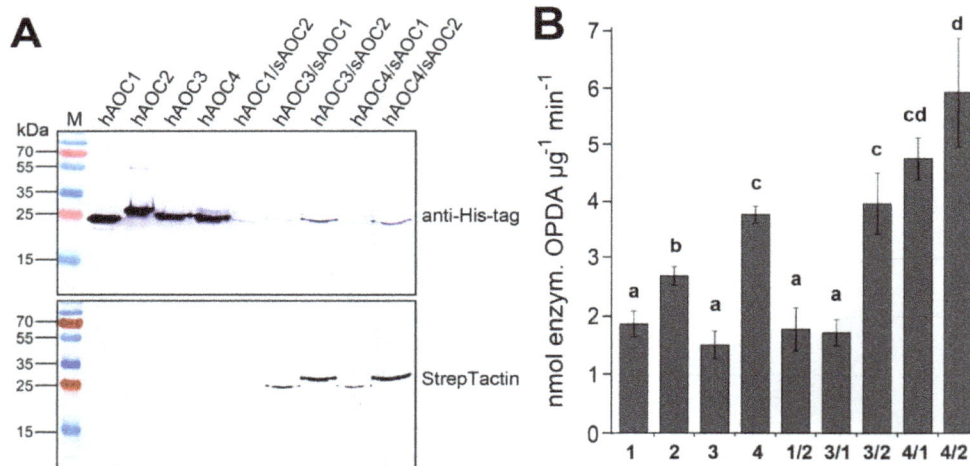

**Figure 4.** Heteromerization of AOC results in altered activities. (**A**) Purification of His-tagged homomers was done using Ni-NTA (left). Heteromers were purified using Ni-NTA (for His-tag) and StrepTactin (for Strep-tag) subsequently. Immunoblots of purified recombinant proteins treated with SDS at 96 °C show homomers with His-tag (hAOC) and heteromers exhibiting one isoform with His-Tag (hAOC) and the other with Strep-tag (sAOC). Note that the heteromers were detectable by both immuno-decorations. (**B**) Activity of recombinant homomeric and heteromeric AOCs. Each value is given as nmol of enzymatically-formed OPDA per μg protein and min and is represented by the mean of three independent replicates (±SD) obtained from independent protein preparations. Different letters designate statistically-different values (one-way ANOVA with Tukey's HSD test, $p < 0.05$).

## 3. Discussion

In the biosynthesis of jasmonates, AOC is of crucial importance, since it establishes the enantiomeric structure of the cyclopentenone ring. Crystal structures for AOCs from Arabidopsis [12,16,17] and *Physcomitrella patens* [18] have been obtained and revealed mechanistic information on catalysis, but additionally showed that these AOCs form a trimeric quaternary structure. Among the four isoforms occurring in Arabidopsis, AOC2 has been described as the most active form and has been intensively studied in terms of structure-function analysis regarding the active center [19]. The main structural feature of the AOC2 monomer is the central eight-stranded antiparallel β-barrel. The active site is located inside the barrel cavity and is mostly lined by hydrophobic and aromatic amino acids that help to coordinate the positioning of the substrate. The stereospecific catalysis was found to be a direct result of isomerization of the substrate *cis*-12,13S-epoxy-9Z,11Z,15Z-octadecatrienoic acid (12,13-EOT) due to this protein environment [19]. Within the crystallized trimer, the barrel axes of the monomers are tilted ~30° with respect to the trimer axis and pack closely with their barrel walls [12]. However, the question of whether the formation of trimeric structures is essential for AOC activity and how heteromers formed by different isoforms could affect the activity was not addressed yet.

### 3.1. Recombinant AOC2 of Arabidopsis Is Active as a Trimer

To better understand the AOC2 multimerization, we inspected the trimer formation of recombinant AOC2 of *A. thaliana* with respect to its activity. The oligomeric state of AOC2 was examined in the presence of detergent at various temperatures followed by gel-electrophoresis. We could not find any influence of the location of His-tag on the trimeric state of the protein, and the treatments with detergent and analyses by cross-linking using EGS supported the occurrence of the trimer as the predominant form of AOC2. This is in line with the crystallization data, but also with the detection of AOC2 from plant extracts in SDS PAGE or immunoblots, although these statements were given without data presentation [12].

Calculation of the interaction energy between the monomers strongly supported the stability of the trimer. Two sites at each monomer that form salt bridges and a hydrophobic core built by Leu40, Leu50, Leu53 and Ile79 of each monomer were found (Figure 2). All amino acids involved in this interaction are located apart from the hydrophobic binding pocket for the substrate. The trimer formation seems to be essential for activity, since site-directed mutagenesis of amino acid residues involved in trimer formation led to nearly complete loss of activity (Figure 3). This is in contrast to Hofmann *et al.* [12], where trimer formation was not assumed to be required for activity, but suggested for the improvement of overall protein stability. Interestingly, residues involved in the interaction between the monomers are predominantly conserved in all known AOC sequences [19], and AOC1 from Arabidopsis (PDB Code 1ZVC) and AOC1 and AOC2 from *P. patens* [18] crystallize as a trimer, as well. This supports the hypothesis that the trimer represents a relevant quaternary structure of enzymatically-active AOCs.

### 3.2. Heteromers between Different AOC Isoforms Exhibit Altered Activity

The four members of the AOC gene family of *A. thaliana* exhibit a redundant expression pattern, as also shown by promoter activity analyses [9]. Such a redundant expression pattern raises the question of the putative heteromerization of the various AOC isoforms. Indeed, the BiFC analysis showed an *in vivo* interaction of all AOCs with each other [9]. The *in planta* proof for the effects on activity done by monitoring the wound-induced transcript accumulation of a JA-responsive gene, however, did not deliver unequivocal results. *In planta*, homomers and heteromers could be formed simultaneously, leading to an overall high level of AOC protein and, thereby, overriding the putative effect of activity control. Consequently, putative heteromerization was analyzed here by pairwise combination of the four recombinantly-produced isoforms. This system allows the determination of the activity of the single AOC monomers and combinations of heteromers without effects by endogenous plant proteins.

Heteromer formation could be clearly documented in immunoblot analysis via detection of His-tag and Strep-tag, respectively (Figure 4A). The significant differences among the AOC activities of heteromeric combination strongly suggest the influence of heteromerization on AOC activity (Figure 4B). The highest activity was detected for heteromers of recombinant AOC4 together with AOC1 or with AOC2. This finding is further underscored by *in vivo* interaction studies of Arabidopsis AOCs, where preferentially, these pairs showed the strongest signals in the BiFC analyses [9]. Moreover, promoters of *AtAOC4* and *AtAOC2* are both active in the vasculature of mature leaves [9], the tissue that is most important for the production of jasmonates upon wounding [20]. Furthermore, promoters of *AtAOC4* and *AtAOC1* are commonly active in filaments of stamen [9]. In filaments, biosynthesis of JA is of special importance, since JA-deficient and -insensitive mutants are male sterile, showing a defect in filament elongation [21].

The mechanism on which the activity shift is based, however, remains unclear. The amino acid sequences of all four AOCs are highly similar [13], with all amino acids postulated to be involved in the cyclization reaction being conserved [12]. Therefore, at least the three-dimensional structure of their monomers, including the substrate-binding pocket, appears to be nearly identical [17]. Substrate binding within the active site does not involve an induced fit mechanism [17], since it is facilitated by the hydrophobic protein environment, as described above. Therefore, it is questionable whether the rigid barrel might be changed in its structure upon the interaction between different monomers. Nevertheless, despite their similarity, already, all four AOCs of Arabidopsis convert their substrate with different efficiencies, with the highest activity for AOC4 and the lowest activities for AOC1 and AOC3 (Figure 4).

## 4. Experimental Section

### 4.1. Cloning, Mutagenesis, Recombinant Expression and Purification of AOCs

cDNAs of *AtAOC1* (At3g25760), *AtAOC2* (At3g25770), *AtAOC3* (At3g25780) and *AtAOC4* (At1g13280) were used without the plastid signal sequence [13]. Mutated versions of AOC2 were created *in silico*, and DNA was synthesized by GENEART (Thermo Fisher Scientific, Dreieich, Germany, www.thermofisher.com). All constructs were introduced into pQE30 (N-terminal His-tag) or pQE60 (C-terminal His-tag). To obtain heteromeric versions, different combinations of two wild-type AOCs were cloned together into pQE30, whereby one AOC was tagged with His-tag, and the other AOC was N-terminally tagged with Strep-tag-II. All plasmids were transformed into the host strain *E. coli* M15. The total protein of isopropyl-β-thiogalactopyranoside (IPTG)-induced cultures was isolated, and AOC proteins were purified using Ni-NTA-agarose (GE Healthcare, Munich, Germany, www.gehealthcare.com) for His-tagged versions, as described [22]. To purify the heteromeric combinations of AOCs, the eluate from Ni-NTA-agarose was directly used for the second purification step. Here, StrepTrap HP (GE Healthcare) was applied to purify Strep-tagged proteins according to the manufacturer's instructions. The amount of all purified proteins was determined by the bicinchoninic acid (BCA) assay (Sigma-Aldrich, Taufkirchen, Germany, www.sigmaaldrich.com). Equal amounts of proteins were used directly for SDS-PAGE and the determination of activity, as described below.

### 4.2. Structure Analysis and Modelling of AOC Trimers

The X-ray structure of the AOC2 from *Arabidopsis thaliana* (At3g25770, PDB Code 2Q4I, [16]) was used for the modelling calculations. This structure was crystallized as a homo-trimer. To gain insight into the stability of the homo-trimeric structure, interaction energies between a monomer with the remaining dimer have been calculated. For this purpose hydrogen atoms were added to the X-ray structure with the help of MOE (Molecular Operating Environment Version 2014.09, chemical computing Group, Cologne, Germany, https://www.chemcomp.com) using the protonate

3D module. Subsequently, the homo-trimer structures were energy optimized using the CHARMM27 force field [23] with Born solvation [24]. The interaction energies of each monomer with the remaining dimer were calculated by subtracting the force field energies of the mono and dimer structures from the energy of the entire trimer. Additionally, the structure of AOC2 was analyzed using Pymol (www.pymol.org). To determine non-covalent interactions between the subunits of the homo-trimeric structure, electrostatic interactions were addressed between oppositely-charged amino acids Asp or Glu with Arg, Lys or His (salt bridges) and increased abundance of hydrophobic amino acids (Leu, Ile, Val) on the surface of the interacting monomers. To validate the interaction between promising candidates, measurements were performed using the software Pymol. A salt bridge was defined as an ion pair, if the centroids of the side-chain charged-group atoms in the residues lie within 4.0 Å of each other and at least one pair of Asp or Glu side-chain carbonyl oxygen and side-chain nitrogen atoms of Arg, Lys or His are also within this distance [25].

### 4.3. Cross-Linking, Size Exclusion Chromatography, SDS-PAGE and Immunoblot Analysis

The cross-linking reagent, ethylene glycol-bis(succinic acid N-hydroxysuccinimide ester) (EGS), was purchased from Sigma–Aldrich. Purified AOC2 protein was treated with 200 μM EGS for 40 min at room temperature. The reaction was stopped by adding 0.1 volumes of 1 M Tris-Cl/1 M glycine (pH 7.5). Cross-linked proteins, as well as AOC2 proteins directly after purification were separated on a HiLoad 16/60 Superdex 200 prep grade column (17-1069-01, GE Healthcare Life Science, Freiburg, Germany, www.gelifesciences.com) using an Äkta Explorer System (GE Healthcare Life Science). The molecular weight of fractions was calculated using a calibration curve (see Figure S1), whereby dextran blue (MW 2000 kDa) was used to obtain $v_o$. For SDS-PAGE, proteins were treated with 1% sodium dodecyl sulfate (SDS) at different temperatures for 5 min and separated on a 12% polyacrylamide gel [26]. Immunoblot analyses using an anti-His-tag antibody from mouse (C 0409, Novagen, Darmstadt, Germany, www.merckmillipore.com) or StrepTactin coupled to horseradish peroxidase (IBA Biotagnology, Goettingen, Germany, www.iba-lifesciences.com) were performed as described [13].

### 4.4. Determination of AOC Activity

AOC activity was determined according to [27] with modifications described in [28]. Briefly, 10 μg of purified recombinant proteins were incubated with recombinant HvAOS [15] and 13(S)-hydroperoxyoctadeca-9,11,15-trienoic acid (HPOT) at 4 °C for 10 min. The reaction was stopped by acidification, and Me-OPDA was added as the internal standard. Extraction with diethyl ether and evaporation of extract was performed followed by treatment with 0.2 M NaOH (in methanol) to activate trans-isomerization of cis-(+)-OPDA. After incubation at 4 °C for 60 min, the reaction was stopped by neutralization with 2 N HCl. The reaction mixtures were extracted with 2 mL of diethyl ether, evaporated and subjected to chiral phase HPLC, as described. The absolute content of OPDA was calculated using the internal standard. The percentage of enzymatically-formed cis-OPDA was calculated according to [29].

## 5. Conclusions

The regulation of enzyme activity by heteromerization is a repeatedly observed property of enzymes. A prominent example in plant hormone biosynthesis is ACC synthase, where different combinations of nine isoforms are attribute to altered ethylene formation [30,31]. The altered AOC activity by the heteromerization of members of the AOC family can be regarded as an additional regulatory principle in jasmonate biosynthesis, where substrate generation, posttranslational modifications and tissue specificity were discussed so far to have a regulatory role [1,11,32,33].

**Acknowledgments:** This work was supported by the Excellence Cluster Initiative of the Federal State of Sachsen-Anhalt (Project 13) and by the Ministry of Education, Youth and Sports of the Czech Republic, Grant Nos. MSM6198959216 and LO1204 (National Program for Sustainability I).

**Author Contributions:** M.O., C.W. and B.H. conceived of and designed the experiments. M.O., C.N. and W.B. performed the experiments and analyzed the data. C.W. and B.H. wrote the paper. All authors discussed the results and substantially contributed to this work.

**Conflicts of Interest:** The authors declare no conflict of interest.

## References

1.    Wasternack, C.; Hause, B. Jasmonates: Biosynthesis, perception, signal transduction and action in plant stress response, growth and development. An update to the 2007 review in Annals of Botany. *Ann. Bot.* **2013**, *111*, 1021–1058. [CrossRef] [PubMed]

2.    Chung, H.S.; Koo, A.J.K.; Gao, X.; Jayanty, S.; Thines, B.; Jones, A.D.; Howe, G.A. Regulation and function of Arabidopsis *JASMONATE ZIM*-domain genes in response to wounding and herbivory. *Plant Physiol.* **2008**, *146*, 952–964. [CrossRef] [PubMed]

3.    Laudert, D.; Schaller, F.; Weiler, E. Transgenic *Nicotiana tabacum* and *Arabidopsis thaliana* plants overexpressing allene oxide synthase. *Planta* **2000**, *211*, 163–165. [CrossRef] [PubMed]

4.    Stenzel, I.; Hause, B.; Maucher, H.; Pitzschke, A.; Miersch, O.; Ziegler, J.; Ryan, C.; Wasternack, C. Allene oxide cyclase dependence of the wound response and vascular bundle-specific generation of jasmonates in tomato - amplification in wound signaling. *Plant J.* **2003**, *33*, 577–589. [CrossRef] [PubMed]

5.    Hause, B.; Stenzel, I.; Miersch, O.; Maucher, H.; Kramell, R.; Ziegler, J.; Wasternack, C. Tissue-specific oxylipin signature of tomato flowers: Allene oxide cyclase is highly expressed in distinct flower organs and vascular bundles. *Plant J.* **2000**, *24*, 113–126. [CrossRef] [PubMed]

6.    Chauvin, A.; Caldelari, D.; Wolfender, J.-L.; Farmer, E.E. Four 13-lipoxygenases contribute to rapid jasmonate synthesis in wounded *Arabidopsis thaliana* leaves: a role for lipoxygenase 6 in responses to long-distance wound signals. *New Phytol.* **2013**, *197*, 566–575. [CrossRef] [PubMed]

7.    Gasperini, D.; Chételat, A.; Acosta, I.F.; Goossens, J.; Pauwels, L.; Goossens, A.; Dreos, R.; Alfonso, E.; Farmer, E.E. Multilayered organization of jasmonate signalling in the regulation of root growth. *PLoS Genetics* **2015**, *11*, e1005300. [CrossRef] [PubMed]

8.    Breithaupt, C.; Kurzbauer, R.; Lilie, H.; Schaller, A.; Strassner, J.; Huber, R.; Macheroux, P.; Clausen, T. Crystal structure of 12-oxophytodienoate reductase 3 from tomato: Self-inhibition by dimerization. *Proc. Natl. Acad. Sci. USA* **2006**, *103*, 14337–14342. [CrossRef] [PubMed]

9.    Stenzel, I.; Otto, M.; Delker, C.; Kirmse, N.; Schmidt, D.; Miersch, O.; Hause, B.; Wasternack, C. *ALLENE OXIDE CYCLASE (AOC)* gene family members of *Arabidopsis thaliana*: tissue- and organ-specific promoter activities and *in vivo* heteromerization. *J. Exper. Bot.* **2012**, *63*, 6125–6138. [CrossRef] [PubMed]

10.   Miersch, O.; Wasternack, C. Octadecanoid and jasmonate signaling in tomato (*Lycopersicon esculentum* Mill.) leaves: Endogenous jasmonates do not induce jasmonate biosynthesis. *Biol. Chem.* **2000**, *381*, 715–722. [CrossRef] [PubMed]

11.   Scholz, S.S.; Reichelt, M.; Boland, W.; Mithöfer, A. Additional evidence against jasmonate-induced jasmonate induction hypothesis. *Plant Sci.* **2015**, *239*, 9–14. [CrossRef] [PubMed]

12.   Hofmann, E.; Zerbe, P.; Schaller, F. The crystal structure of *Arabidopsis thaliana* allene oxide cyclase: Insights into the oxylipin cyclization reaction. *Plant Cell* **2006**, *18*, 3201–3217. [CrossRef] [PubMed]

13.   Stenzel, I.; Hause, B.; Miersch, O.; Kurz, T.; Maucher, H.; Weichert, H.; Ziegler, J.; Feussner, I.; Wasternack, C. Jasmonate biosynthesis and the allene oxide cyclase family of *Arabidopsis thaliana*. *Plant Mol. Biol.* **2003**, *51*, 895–911. [CrossRef] [PubMed]

14.   Sinz, A. Chemical cross-linking and FTICR mass spectrometry for protein structure characterization. *Anal. Bional. Chem.* **2005**, *381*, 44–47. [CrossRef] [PubMed]

15.   Ziegler, J.; Stenzel, I.; Hause, B.; Maucher, H.; Hamberg, M.; Grimm, R.; Ganal, M.; Wasternack, C. Molecular cloning of allene oxide cyclase: The enzyme establishing the stereochemistry of octadecanoids and jasmonates. *J. Biol. Chem.* **2000**, *275*, 19132–19138. [CrossRef] [PubMed]

16.   Levin, E.J.; Kondrashov, D.A.; Wesenberg, G.E.; Phillips, G.N. Ensemble refinement of protein crystal structures: validation and application. *Structure* **2007**, *15*, 1040–1052. [CrossRef] [PubMed]

17.  Hofmann, E.; Pollmann, S. Molecular mechanism of enzymatic allene oxide cyclization in plants. *Plant Physiol. Biochem.* **2008**, *46*, 302–308. [CrossRef] [PubMed]

18.  Neumann, P.; Brodhun, F.; Sauer, K.; Herrfurth, C.; Hamberg, M.; Brinkmann, J.; Scholz, J.; Dickmanns, A.; Feussner, I.; Ficner, R. Crystal structures of *Physcomitrella patens* AOC1 and AOC2: Insights into the enzyme mechanism and differences in substrate specificity. *Plant Physiol.* **2012**, *160*, 1251–1266. [CrossRef] [PubMed]

19.  Schaller, F.; Zerbe, P.; Reinbothe, S.; Reinbothe, C.; Hofmann, E.; Pollmann, S. The allene oxide cyclase family of *Arabidopsis thaliana*—localization and cyclization. *FEBS J.* **2008**, *275*, 2428–2441. [CrossRef] [PubMed]

20.  Gasperini, D.; Chauvin, A.; Acosta, I.F.; Kurenda, A.; Stolz, S.; Chételat, A.; Wolfender, J.-L.; Farmer, E.E. Axial and radial oxylipin transport. *Plant Physiol.* **2015**, *169*, 2244–2254. [CrossRef] [PubMed]

21.  Browse, J. The power of mutants for investigating jasmonate biosynthesis and signaling. *Phytochemistry* **2009**, *70*, 1539–1546. [CrossRef] [PubMed]

22.  Maucher, H.; Hause, B.; Feussner, I.; Ziegler, J.; Wasternack, C. Allene oxide synthases of barley (*Hordeum vulgare* cv. Salome): Tissue specific regulation in seedling development. *Plant J.* **2000**, *21*, 199–213. [CrossRef] [PubMed]

23.  MacKerell, J.A.D.; Banavali, N.; Foloppe, N. Development and current status of the CHARMM force field for nucleic acids. *Biopolymers* **2001**, *56*, 257–265. [CrossRef]

24.  Wojciechowski, M.; Lesyng, B. Generalized born model: Analysis, refinement, and applications to proteins. *J. Phys. Chem. B* **2004**, *108*, 18368–18376. [CrossRef]

25.  Kumar, S.; Nussinov, R. Close-range electrostatic interactions in proteins. *ChemBioChem* **2002**, *3*, 604–617. [CrossRef]

26.  Laemmli, U.K. Cleavage of structural proteins during the assembly of the head of bacteriophage T4. *Nature* **1970**, *227*, 680–685. [CrossRef] [PubMed]

27.  Ziegler, J.; Hamberg, M.; Miersch, O.; Parthier, B. Purification and characterization of allene oxide cyclase from dry corn seeds. *Plant Physiol.* **1997**, *114*, 565–573. [PubMed]

28.  Lischweski, S.; Muchow, A.; Guthörl, D.; Hause, B. Jasmonates act positively in adventitious root formation in petunia cuttings. *BMC Plant Biol.* **2015**, *15*, 229. [CrossRef] [PubMed]

29.  Ziegler, J.; Wasternack, C.; Hamberg, M. On the specificity of allene oxide cyclase. *Lipids* **1999**, *34*, 1005–1015. [CrossRef] [PubMed]

30.  Tsuchisaka, A.; Theologis, A. Unique and overlapping expression patterns among the Arabidopsis 1-amino-cyclopropane-1-carboxylate synthase gene family members. *Plant Physiol.* **2004**, *136*, 2982–3000. [CrossRef] [PubMed]

31.  Tsuchisaka, A.; Theologis, A. Heterodimeric interactions among the 1-amino-cyclopropane-1-carboxylate synthase polypeptides encoded by the Arabidopsis gene family. *Proc. Natl. Acad. Sci. USA* **2004**, *101*, 2275–2280. [CrossRef] [PubMed]

32.  Wasternack, C. Jasmonates: An update on biosynthesis, signal transduction and action in plant stress response, growth and development. *Ann. Bot.* **2007**, *100*, 681–697. [CrossRef] [PubMed]

33.  Schaller, A.; Stintzi, A. Enzymes in jasmonate biosynthesis—Structure, function, regulation. *Phytochemistry* **2009**, *70*, 1532–1538. [CrossRef] [PubMed]

# High Constitutive Overexpression of Glycosyl Hydrolase Family 17 Delays Floral Transition by Enhancing FLC Expression in Transgenic *Arabidopsis*

**Shinichi Enoki, Nozomi Fujimori, Chiho Yamaguchi, Tomoki Hattori and Shunji Suzuki *** (iD)

Laboratory of Fruit Genetic Engineering, The Institute of Enology and Viticulture, University of Yamanashi, Yamanashi 400-0005, Japan; senoki@yamanashi.ac.jp (S.E.); noro.noro.n@live.jp (N.F.); g16lf017@yamanashi.ac.jp (C.Y.); g15de002@yamanashi.ac.jp (T.H.)

* Correspondence: suzukis@yamanashi.ac.jp

**Abstract:** *Vitis vinifera* glycosyl hydrolase family 17 (VvGHF17) is a grape apoplasmic β-1,3-glucanase, which belongs to glycosyl hydrolase family 17 in grapevines. β-1,3-glucanase is not only involved in plant defense response but also has various physiological functions in plants. Although *VvGHF17* expression is negatively related to the length of inflorescence in grapevines, the physiological functions of *VvGHF17* are still uncertain. To clarify the physiological functions of *VvGHF17*, we conducted a phenotypic analysis of VvGHF17-overexpressing *Arabidopsis* plants. VvGHF17-overexpressing *Arabidopsis* plants showed short inflorescence, similar to grapevines. These results suggested that *VvGHF17* might negatively regulate the length of inflorescence in plants. *VvGHF17* expression induced a delay of floral transition in *Arabidopsis* plants. The expression level of *FLOWERING LOCUS C* (*FLC*), known as a floral repressor gene, in inflorescence meristem of transgenic plants were increased by approximately 10-fold as compared with wild plants. These results suggest that *VvGHF17* induces a delay of floral transition by enhancing *FLC* expression and concomitantly decreases the length of plant inflorescence.

**Keywords:** grapevine; β-1,3-glucanase; inflorescence; *VvGHF17*; floral transition; *FLC*; *Arabidopsis*

## 1. Introduction

Higher plants such as vascular plants have certain defense mechanisms against plant pathogens such as fungi, bacteria, and viruses. Pathogenesis-related proteins (PRs) are induced against pathogen invasion in plants and play an important role in plant defense [1]. PRs are classified into 17 families (PR-1 to PR-17) according their characteristics [2]. β-1,3-Glucanases (glucan endo-1,3-β-D-glucosidase, EC 3.2.1.39) belong to the second (PR-2) of the 17 families and are also included in the glycosyl hydrolase family 17 (GHF17) due to their degradation style. Since they hydrolyze 1,3-β-D-glycosidic bonds in β-1,3-glucan as the main component of the cell wall of many fungi [3,4], β-1,3-glucanases are thought to play an important role in the plant defense response against pathogen infection. In our previous study [5], we isolated the apoplasmic β-1,3-glucanase secreted from grape cells and demonstrated that VvGHF17-overexpressing *Arabidopsis thaliana* acquired multiple resistance to phytopathogenic fungi.

β-1,3-glucan (referred to as callose in plants) is widespread in plant bodies. Thereby, β-1,3-glucanase has various physiological functions in addition to plant defense, such as cell division and elongation [6,7], flower formation [8,9], pollen germination and tube growth [10], fertilization [11], and fruit ripening [12,13]. In particular, understanding the influence of β-1,3-glucanase on traits such as flower formation and subsequent fruit ripening is very important, since their traits are directly linked to the fruit quality which determines the value of fruit trees. However, there are few reports that β-1,3-glucanase has physiological functions affecting plant growth in grapevines, and its physiological functions are still uncertain.

To understand the physiological functions of *VvGHF17* in the vegetative and/or reproductive growth of grapevines, we conducted a phenotypic analysis of VvGHF17-overexpressing *Arabidopsis* plants. The present study demonstrates that *VvGHF17* delays floral transition in *Arabidopsis* plant through enhancing *FLOWERING LOCUS C* (*FLC*) expression, which is the transcription factor functioning as a repressor of floral transition.

## 2. Results

### 2.1. VvGHF17 Expression Is Negatively Related to the Length of Grape Inflorescence

A simple linear regression analysis was performed to investigate the relationship between the inflorescence length and the gene expression level of endogenous *VvGHF17* in grape cultivars. Endogenous *VvGHF17* expression in young grape inflorescence showed a strong negative correlation with the length of mature grape inflorescence ($p = 0.0091$) (Figure 1a). This result suggested that *VvGHF17* might function in the vegetative and/or reproductive growth of grapevines.

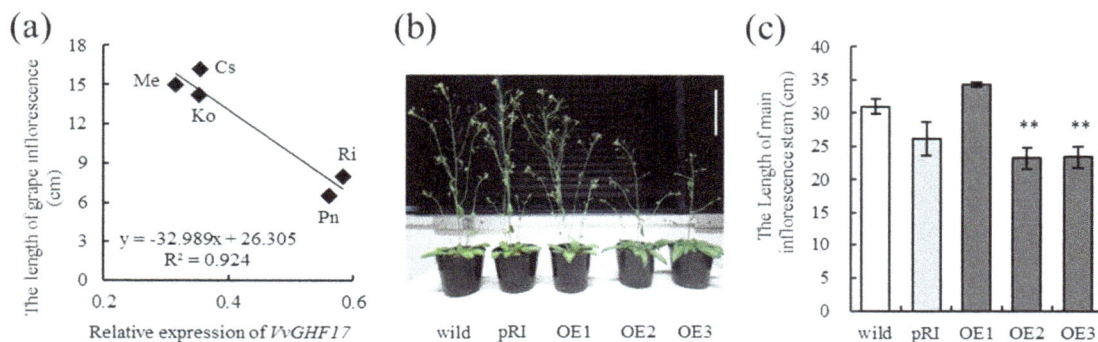

**Figure 1.** *VvGHF17* expression suppresses inflorescence growth in plants. (**a**) Regression line between the length of mature grape inflorescence and endogenous *VvGHF17* expression in young grape inflorescence. Averages ($n = 10$) are plotted in the graph. Cs, Cabernet Sauvignon; Ko, Koshu; Me, Merlot; Pn, Pinot Noir; Ri, Riesling grapevine cultivars. (**b**) VvGHF17-overexpressing *Arabidopsis* plants. Photograph was obtained at 31 days after sowing. Scale bar = 7.5 cm. (**c**) Length of main inflorescence of VvGHF17-overexpressing *Arabidopsis* plants. Bars indicate means ± standard errors ($n = 5$). ** $p < 0.01$ as compared with wild plants.

### 2.2. VvGHF17 Induce Delays of Floral Transition

The phenotypic analysis of VvGHF17-overexpressing transgenic *Arabidopsis* obtained by our previous study [5] was performed to clarify the detailed physiological function of *VvGHF17*. The growth of the main inflorescence stem of VvGHF17-overexpressing *Arabidopsis* plants (OE2 and OE3) tended to be poor (Figure 1b). The length of these stems was significantly lower than those of wild plants at 31 days after sowing (Figure 1c). In addition, the number of rosette leaves of transgenic plants (OE2 and OE3) was significantly higher compared with wild plants (Figure 2a,b). *VvGHF17* induced a delay of floral transition in VvGHF17-overexpressing *Arabidopsis* plants (Figure 2c). These results indicate a delay of floral transition in VvGHF17-overexpressing *Arabidopsis* plants and a lower the elongation of their inflorescence.

### 2.3. VvGHF17 Upregulates FLOWERING LOCUS C Expression

To determine the molecular mechanism on the delay of floral transition in the VvGHF17-overexpressing *Arabidopsis* plants, we analyzed the expression level of the *FLC* gene, which is a floral repressor gene [14], in each plant. *FLC* expression levels in OE2 and OE3 were increased by

10.12- and 8.76-fold compared to those of wild plants, respectively (Figure 2d). This result suggests that *VvGHF17* delays floral transition through the alternation of floral repression.

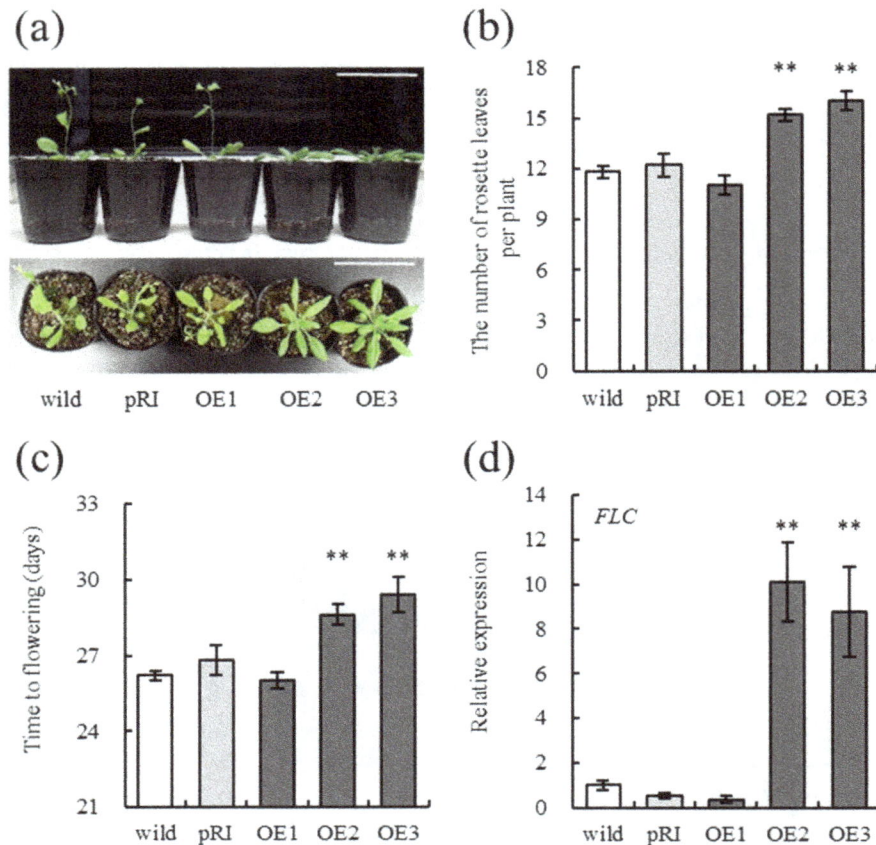

**Figure 2.** *VvGHF17* delays floral transition by enhancing *FLC* expression in VvGHF17-overexpressing *Arabidopsis* plants. (**a**) Photograph of VvGHF17-overexpressing *Arabidopsis* plants at 26 days after sowing. Scale bar = 7.5 cm. (**b**) Number of rosette leaves formed before the appearance of the inflorescence meristem. (**c**) Time to flowering. Bars indicate means $\pm$ standard errors ($n = 5$). (**d**) *FLC* expression. Total RNA was isolated from inflorescence meristems of 20-day-old *Arabidopsis* plants and subjected to real-time RT-PCR analysis. Bars indicate means $\pm$ standard errors ($n = 12$). ** $p < 0.01$ as compared with wild plants.

## 3. Discussion

We demonstrated that *VvGHF17* induces floral transition in *Arabidopsis* plants. The number of rosette leaves formed before the appearance of the inflorescence meristem was measured as an indicator of floral transition, because there is a positive correlation between the number of rosette leaves and floral transition [15]. The number of rosette leaves in VvGHF17-overxpressing plants (OE2 and OE3) was higher compared with that of wild plants (Figure 2a,b). Flowering day was delayed as well (Figure 2c). These results indicate that the vegetative growth period of VvGHF17-overexpressing *Arabidopsis* plants became longer than those of wild plants. However, there was no significant difference between OE1 and wild plants in this study. This may be due to the fact that the expression level of *VvGHF17* in OE1 is much lower than those of OE2 and OE3 in our previous report [5].

Figure 3 shows the relationship between the VvGHF17 and floral transition suggested from this study. Floral transition is controlled through four pathways: autonomous, vernalization, photoperiod, and gibberellin pathways. *FLC* encodes a MADS domain protein, which conserved sequence motif with many transcription factors, and integrates signals through autonomous and vernalization pathways.

FLC acts as a repressor of flowering [15]. VvGHF17 enhances the expression level of FLC directly or indirectly by an unknown mechanism. SUPPRESSOR OF CO OVEREXPRESSION 1 (SOC1), FLOWERING LOCUS T (FT), and LEAFY (LFY) are known as floral integrators and are upregulated by photoperiod and gibberellin pathways, respectively [16,17]. The expression of FLC downregulates the expression levels of SOC1, FT, and LFY. Therefore, FLC upregulation in VvGHF17-overexpressing plants induces a delay of floral transition. This is the first report, to our knowledge, that β-1,3-glucanase affects the timing of floral transition. However, since FLC is controlled by many genes in the autonomous and vernalization pathways [18], it is still unclear, according to this study, whether VvGHF17 directly or indirectly controls FLC through any of the genes of these two pathways.

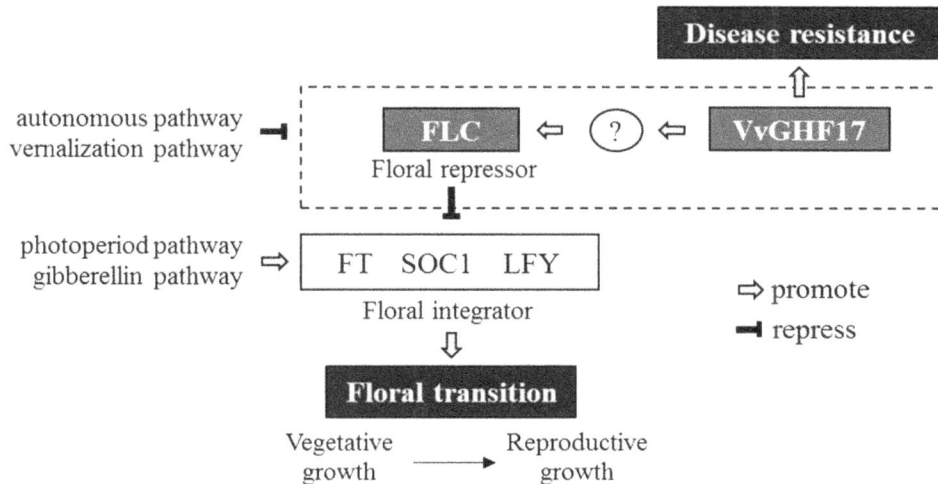

**Figure 3.** Theoretical model of VvGHF17-mediated delay of floral transition. Floral transition is controlled through four pathways: autonomous, vernalization, photoperiod, and gibberellin pathways. FLC encodes a MADS domain protein and integrates signal through autonomous and vernalization pathways. FLC acts as a repressor of flowering. VvGHF17 enhances the expression level of FLC directly or indirectly by an unknown mechanism. FLC represses the expression levels of SOC1, FT, and LFY, which are floral integrators. Therefore, VvGHF17 delays floral transition. FLC, FLOWERING LOCUS C; SOC1, SUPPRESSOR OF CO OVEREXPRESSION 1; FT, FLOWERING LOCUS T; LFY, LEAFY.

Endogenous GHF17 is highly expressed at the stage of flower formation [8] and fruit ripening [12]. Thus, multifunction of β-1,3-glucanase in vegetative and/or reproductive growth and in floral transition remains still unclear. So far, we could not demonstrate any mechanisms from *VvGHF17* expression to *FLC* expression. On the other hand, β-1,3-glucanase hydrolyses β-1,3-glucans is not only from fungal cell walls, but also from callose in plants. Flax overproducing β-1,3-glucanase changes cell wall composition and shows a decrease in callose (endogenic β-1,3-glucan) content as well as an increase in particular polysaccharides contents [19]. Therefore, the delay of floral transition in VvGHF17-overexpressing *Arabidopsis* plants might be due to the change of plant cell wall composition by *VvGHF17*, resulting in a change in cell growth. Further studies employing transcriptional analyses of genes located upstream of *FLC* in addition to surveys of cell wall composition in inflorescence would reveal the molecular mechanisms of the relationship between *VvGHF17* and floral transition.

*VvGHF17* has the function of multiple disease resistance against phytopathogenic fungi [5]. Generally, the longer the period of vegetative growth in grapevines, the higher the quality of the grape berries. Therefore, the production of VvGHF17-overexpressing grapevines could lead to the breeding of grapevines with disease resistance and good fruit quality. Although we revealed that *VvGHF17* influences floral transition, the influence of *VvGHF17* on fruit traits after flowering such as yield and fruit quality are still unknown. In the future, field research focusing disease resistance and quality and quantity of grape berries on VvGHF17-overexpressing grapevines would be required for an evaluation of genetically engineered grapevines.

## 4. Materials and Methods

### 4.1. Plant Materials

Grape cultivars in the test field of The Institute of Enology and Viticulture, University of Yamanashi (Japan), were used as plant materials. Grape bunches of each grapevine cultivar (*Vitis vinifera* cvs. Cabernet Sauvignon, Koshu, Merlot, Pinot Noir, and Riesling) were collected at young and mature stages in 2015. The length of mature grape inflorescences was measured.

*Arabidopsis thaliana* wild type (Col-0), pRI101-AN vector-transformed *Arabidopsis* plants (pRI) and VvGHF17-overexpressing *Arabidopsis* plants (OE1, OE2, and OE3), which were obtained in our previous study [5], were used as plant materials. T3 (third generation transgenic plant) homozygote seeds were sown in rockwool (2.5 cm × 2.5 cm × 3.8 cm) and grown in an incubator (11.8 W$^{-2}$/16 h/day, 22 °C). One week after sowing, the seedlings were moved to the soil together with the rock wool and grown in the incubator under same conditions.

### 4.2. Phenotypic Analysis in Arabidopsis Plants

After sowing each *Arabidopsis* plant, their phenotypes were observed daily. The length of the main inflorescence stem at 4 weeks after sowing, as well as the number of rosette leaves per plant at the appearance of an inflorescence, and the time to flowering was measured.

### 4.3. Isolation of Total RNA

We isolated the total RNA from the inflorescence meristem of grape and *Arabidopsis* plants. Young grape inflorescences of 3–8 mm in the longitudinal direction of each cultivar were used. The inflorescence meristems of 20-day-old *Arabidopsis* plants before the appearance of an inflorescence were used. After freezing these samples with liquid nitrogen, the samples were homogenized with an SK mill (SK-200) (Tokken, Kashiwa, Japan). According to the manufacturer's instructions, total RNA was isolated from these homogenized samples using Nucleospin RNA plant (Takara, Otsu, Japan).

### 4.4. Real-Time RT-PCR

First-strand cDNA were synthesized from the total RNA using a PrimeScript RT Reagent Kit with gDNA Eraser (Takara) and subsequently used for real-time (RT)-PCR analysis. RT-PCR analysis was performed using an SYBR Premix Ex Taq II (Takara) by Thermal Cycler Dice Real-Time System Single Software ver. 3.00 (Takara) and the standard curve method. Reaction conditions were as follows: 37 °C for 15 min, 85 °C for 5 s, 40 cycles at 95 °C for 5 s, and 60 °C for 30 s. Primer sequences were as follows: *Arabidopsis* FLC primers (5′-GAGCCAAGAAGACCGAACTCA-3′ and 5′-TCTCAGCTTCTGCTCCCACA-3′, GenBank accession no. NM_121052) and *Arabidopsis* actin primers (5′-GCCGACAGAATGAGCAAAGAG-3′ and 5′-AGGTACTGAGGGAGGCCAAGA-3′, GenBank accession no. NM_179953). *VvGHF17* primers were used as described previously [5]. *FLC* and *VvGHF17* expression levels were normalized to each *actin*, and relative expression of *FLC* in *Arabidopsis* plants were represented as values relative to the controls (wild plants).

### 4.5. Simple Linear Regression Analysis

Simple linear regression analysis between the length of mature grape inflorescence and the relative expression of *VvGHF17* in young grape inflorescence was conducted using Excel statistics software 2012 (Social Survey Research Information, Tokyo, Japan). The dependent and explanatory variables were the length of grape inflorescence and the relative expression of *VvGHF17*, respectively.

*4.6. Statistical Analysis*

The data are shown as means ± standard errors in the tests with *Arabidopsis* plants. These data were statistically analyzed by Dunnett's multiple comparison test using Excel statistics software 2012 (Social Survey Research Information, Tokyo, Japan).

**Acknowledgments:** Special thanks to Shiho Ishiai of University of Yamanashi for technical advice and assistance.

**Author Contributions:** Shinichi Enoki and Nozomi Fujimori conceived and designed the experiments; Shinichi Enoki and Nozomi Fujimori performed the experiments; Chiho Yamaguchi and Tomoki Hattori analyzed the data; Shinichi Enoki and Shunji Suzuki wrote the paper.

**Conflicts of Interest:** The authors declare no conflict of interest.

## References

1. Stintzi, A.; Heitz, T.; Prasad, V.; Wiedemann-Merdinoglu, S.; Kauffmann, S.; Geoffroy, P.; Legrand, M.; Fritig, B. Plant 'pathogenesis-related' proteins and their role in defense against pathogens. *Biochimie* **1993**, *75*, 687–706. [CrossRef]

2. Van Loon, L.C.; Rep, M.; Pieterse, C.M.J. Significance of inducible defense-related proteins in infected plants. *Annu. Rev. Phytopathol.* **2006**, *44*, 135–162. [CrossRef] [PubMed]

3. Adams, D.J. Fungal cell wall chitinases and glucanases. *Microbiology* **2004**, *150*, 2029–2035. [CrossRef] [PubMed]

4. Simmons, C.R. The physiology and molecular biology of plant 1,3-β-D-glucanases and 1,3;1,4-β-D-glucanases. *Crit. Rev. Plant Sci.* **1994**, *13*, 325–387. [CrossRef]

5. Fujimori, N.; Enoki, S.; Suzuki, A.; Naznin, H.A.; Shimizu, M.; Suzuki, S. Grape apoplasmic β-1,3-glucanase confers fungal disease resistance in *Arabidopsis*. *Sci. Hortic.* **2016**, *200*, 105–110. [CrossRef]

6. Fulcher, R.G.; McCully, M.E.; Setterfield, G.; Sutherland, J. β-1,3-glucans may be associated with cell plate formation during cytokinesis. *Can. J. Bot.* **1976**, *54*, 539–542. [CrossRef]

7. Masuda, Y.; Wada, S. Effect of beta-3-glucanase on elongation growth of oat coleoptile. *Bot. Mag.* **1967**, *80*, 100–102. [CrossRef]

8. Akiyama, T.; Pillai, M.A.; Sentoku, N. Cloning, characterization and expression of OsGLN2, a rice endo-1,3-β-glucanase gene regulated developmentally in flowers and hormonally in germinating seeds. *Planta* **2004**, *220*, 129–139. [CrossRef] [PubMed]

9. Kauffmann, S.; Legrand, M.; Geoffroy, P.; Fritig, B. Biological function of 'pathogenesis-related' proteins: Four PR proteins of tobacco have 1,3-β-glucanase activity. *EMBO J.* **1987**, *6*, 3209–3212. [PubMed]

10. Meikle, P.J.; Bonig, I.; Hoogenraad, N.J.; Clarke, A.E.; Stone, B.A. The location of (1–3)-β-glucans in the walls of pollen tubes of *Nicotiana alata* using a (1–3)-β-glucan-specific monoclonal antibody. *Planta* **1991**, *185*, 1–8. [CrossRef] [PubMed]

11. Ori, N.; Sessa, G.; Lotan, T.; Himmelhoch, S.; Fluhr, R. A major stylar matrix polypeptide (sp41) is a member of the pathogenesis-related proteins superclass. *EMBO J.* **1990**, *9*, 3429–3436. [PubMed]

12. Deytieux, C.; Geny, L.; Lapailerie, D.; Claverol, S.; Bonneu, M.; Donèche, B. Proteome analysis of grape skins during ripening. *J. Exp. Bot.* **2007**, *58*, 1851–1862. [CrossRef] [PubMed]

13. Hinton, D.M.; Pressey, R. Glucanases in fruits and vegetables. *J. Am. Soc. Hortic. Sci.* **1980**, *9*, 499–502.

14. Michaels, S.D.; Amasino, R.M. FLOWERING LOCUS C encodes a novel MADS domain protein that acts as a repressor of flowering. *Plant Cell* **1999**, *11*, 949–956. [CrossRef] [PubMed]

15. Koornneef, M.; Hanhart, C.J.; van der Veen, J.H. A genetic and physiological analysis of late flowering mutants in *Arabidopsis thaliana*. *Mol. Gen. Genet.* **1991**, *229*, 57–66. [CrossRef] [PubMed]

16. Parcy, F. Flowering: A time for integration. *Int. J. Dev. Biol.* **2005**, *49*, 585–593. [CrossRef] [PubMed]

17. Simpson, G.G.; Dean, C. *Arabidopsis*, the Rosetta stone of flowering time? *Science* **2002**, *296*, 285–289. [CrossRef] [PubMed]

18.  Zhang, N.; Wen, J.; Zimmer, E.A. Expression patterns of AP1, FUL, FT and LEAFY orthologs in *Vitaceae* support the homology of tendrils and inflorescences throughout the grape family. *J. Syst. Evol.* **2015**, *53*, 469–476. [CrossRef]

19.  Wojtasik, W.; Kulma, A.; Dymińska, L.; Hanuza, J.; Żebrowski, J.; Szopa, J. Fibres from flax overproducing β-1,3-glucanase show increased accumulation of pectin and phenolics and thus higher antioxidant capacity. *BMC Biotechnol.* **2013**, *13*, 10. [CrossRef] [PubMed]

# The Occurrence of Flavonoids and Related Compounds in Flower Sections of *Papaver nudicaule*

Bettina Dudek, Anne-Christin Warskulat and Bernd Schneider *

Max Planck Institute for Chemical Ecology, Hans-Knöll-Straße 8, 07745 Jena, Germany;
bdudek@ice.mpg.de (B.D.); awarskulat@ice.mpg.de (A.-C.W.)
* Correspondence: schneider@ice.mpg.de

Academic Editor: Ulrike Mathesius

**Abstract:** Flavonoids play an important role in the pigmentation of flowers; in addition, they protect petals and other flower parts from UV irradiation and oxidative stress. Nudicaulins, flavonoid-derived indole alkaloids, along with pelargonidin, kaempferol, and gossypetin glycosides, are responsible for the color of white, red, orange, and yellow petals of different *Papaver nudicaule* cultivars. The color of the petals is essential to attract pollinators. We investigated the occurrence of flavonoids in basal and apical petal areas, stamens, and capsules of four differently colored *P. nudicaule* cultivars by means of chromatographic and spectroscopic methods. The results reveal the specific occurrence of gossypetin glycosides in the basal spot of all cultivars and demonstrate that kaempferol glycosides are the major secondary metabolites in the capsules. Unlike previous reports, the yellow-colored stamens of all four *P. nudicaule* cultivars are shown to contain not nudicaulins but carotenoids. In addition, the presence of nudicaulins, pelargonidin, and kaempferol glycosides in the apical petal area was confirmed. The flavonoids and related compounds in the investigated flower parts and cultivars of *P. nudicaule* are profiled, and their potential ecological role is discussed.

**Keywords:** flower pigmentation; flavonoids; kaempferol; pelargonidin; nudicaulins; *Papaver nudicaule*

## 1. Introduction

Ubiquitous in angiosperms, flavonoids are extremely diverse in their chemical structure, color, and biological function. Flavonols, flavones, flavanones, flavanols, and anthocyanidins are just a few examples of the wide range of subclasses [1]. Anthocyanidins, in particular, and their corresponding glycosides (anthocyanins) are flower and fruit pigments, which enhance pollination and seed dispersal [2]. These red to blue pigments are often accompanied by pale yellow or colorless flavonols, which serve as co-pigments and may play a role in UV protection, disease resistance, or hormone signaling [3]. In flowers, the occurrence and distribution of flavonoids is likely connected to their specific function.

In 1931, the first flower pigments of *Papaver nudicaule*, a poppy species originating from Siberia [4] but commonly known as the Iceland poppy, were identified as pelargonidin glycosides in red and orange petals [5,6]. The specific substitution patterns of these anthocyanins were investigated later on by means of mass spectrometry (MS) and nuclear magnetic resonance (NMR) spectroscopy [7,8].

Yellow and orange flowers of *P. nudicaule* contain the yellow nudicaulins, an unusual group of indole alkaloids with three glucose substituents (Figure 1), that are derived from the indole and flavonoid biosynthetic pathways [8,9]. According to Schliemann *et al.* (2006), in yellow petals, nudicaulins are accompanied by gossypetin 7-*O*-glucoside (gossypitrin) and seven kaempferol glycosides whose substitution patterns correspond to those of the indole alkaloids [10]. In the absence of other pigments, the colorless kaempferol glycosides and the pale yellow gossypitrin contribute to the white or ivory appearance of petals [11]. Additionally, the presence of kaempferol in pollen was linked to the production of functional pollen tubes and to a successful germination process in maize and petunia, but its absence does not automatically imply sterility [12].

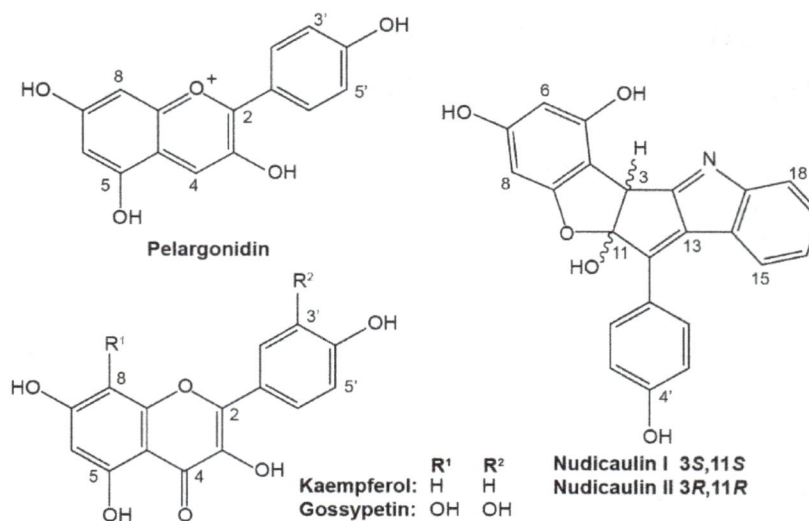

**Figure 1.** Aglycone structures of flavonoids and nudicaulins from *P. nudicaule* flowers.

In 1962, it was shown that yellow compounds are present in both the petals and stamens of wild-type and cultivars of *P. nudicaule*; pelargonidin glycosides were limited to the petals of garden varieties [13]. Furthermore, a yellow compound that was named nudicaulin was reported in the filament of stamens in various *Papaver* species, but there was no accompanying detailed chemical analysis [14].

In the present study, we report on the occurrence of flavonoids and the biosynthetically related nudicaulins in two petal areas, capsules, and stamens (Figure 2) of white, yellow, orange, and red flowers of *P. nudicaule* cultivars. The potential biological and ecological function of the constituents of *P. nudicaule* is discussed in the context of current knowledge and hypothetical considerations.

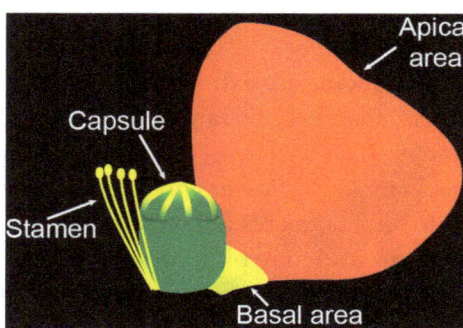

**Figure 2.** Scheme of a *P. nudicaule* flower and the four investigated flower parts.

## 2. Results

### 2.1. Apical and Basal Petal Areas

High-performance liquid chromatography–photodiode array detection (HPLC-PDA) analysis of fresh extracts obtained from the apical petal areas of the white, yellow, orange, and red cultivars confirmed the presence of nudicaulins, kaempferol, and pelargonidin glycosides. In previous studies, the aglycone structures and the substitution patterns of the apical pigments were already elucidated by LC-MS (Figure S1) and NMR [7,10,15]. Taking this knowledge into account and considering the analytical value of characteristic UV/Vis absorption spectra, we suppose that the major flavonoid substance classes are those shown in Figure 3. These UV/Vis absorption spectra were also used to characterize the compounds in extracts of the other flower parts. Based on these data, kaempferol glycosides are assumed to be the only flavonols occurring in the apical petal area of all *P.nudicaule* cultivars. Two of these glycosides are present in all flower samples (indicated by blue boxes in Figure 3). Likewise, pelargonidin glycosides are assumed to be the only group of anthocyanidins in the flowers of this plant.

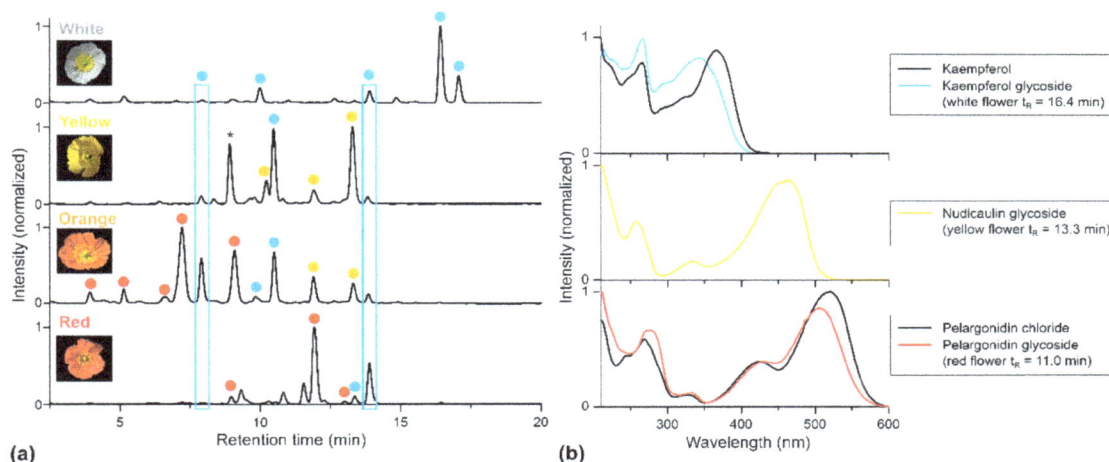

**Figure 3.** HPLC-PDA analysis of extracts of apical petal parts of four *P. nudicaule* cultivars. (**a**) Chromatograms recorded at 254 nm. Peaks representing the same aglycone (previously identified by LC-MS and NMR [7,10,15] and here classified by the corresponding UV/Vis absorption spectra) are marked with the same color: • Kaempferol glycoside, • nudicaulin, • pelargonidin glycoside. The peak marked with an asterisk (yellow flower, $t_R$ = 8 min) shows no flavonoid absorption spectrum and may be a degradation product. (**b**) UV/Vis absorption spectra of representative glycosides and authentic aglycones of kaempferol and pelargonidin. Deviations between UV/Vis absorption spectra of the references (kaempferol, pelargonidin chloride), and the glycosides are an effect of the substitution. Nudicaulin aglycone is not available due to instability. The obtained nudicaulin UV/Vis absorption spectrum matches the one reported by Tatsis *et al.* [15].

White flowers possess the most kaempferol glycosides but lack pelargonidin glycosides and nudicaulins. The two major kaempferols elute after a retention time ($t_R$) of 16 min, indicating a lower polarity and a reduced degree of glycosylation compared with kaempferol glycosides occurring in petals of other cultivars.

Furthermore, the nudicaulins are present exclusively in yellow and orange flowers, where they have identical $t_R$ and, consequently, identical substitution patterns. This confirms previous studies [8]. In contrast, pelargonidin glycosides occur solely in orange and red petals. Due to varying substituents, the pelargonidin glycosides found in red flowers are less polar than those found in orange ones, which is consistent with reported data [15].

Although the basal petal part appears yellowish, the presence of nudicaulins in this tissue was not confirmed. The only flavonoids present in this area, gossypetin glycosides, are absent in the apical part of the flowers. Overall, the HPLC-PDA profiles of the basal petal areas of all four *P. nudicaule* cultivars are very similar (Figure 4).

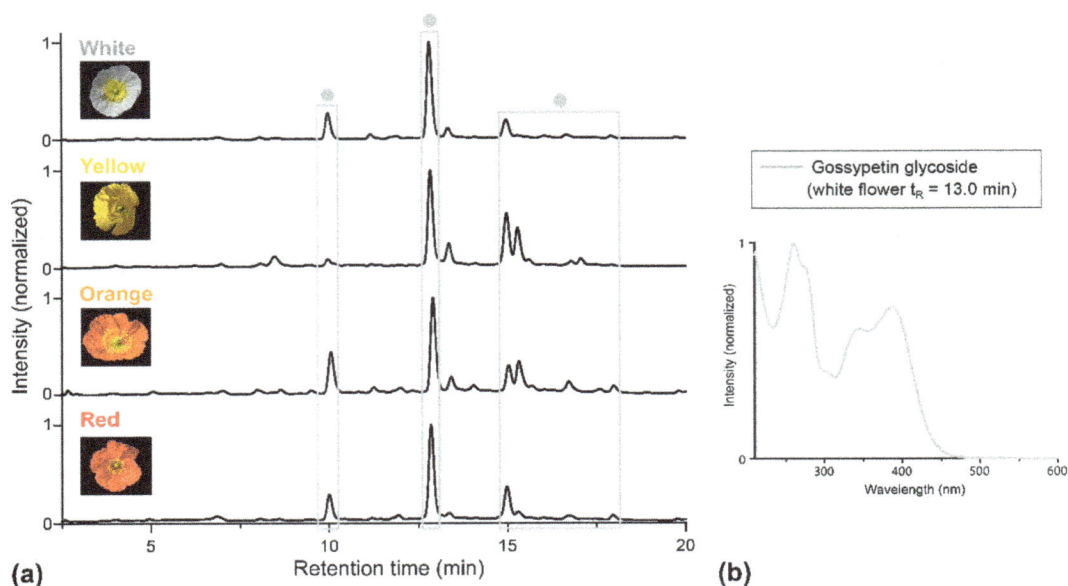

**Figure 4.** HPLC-PDA analysis of extracts of basal petal parts of four *P. nudicaule* cultivars. (**a**) Chromatograms recorded at 351 nm. Gossypetin glycosides (identified by UV/Vis absorption spectra) are marked with dots •. (**b**) UV/Vis absorption spectrum of one representative gossypetin glycoside. The spectrum matches the one reported by Suzuki *et al.* [16].

## 2.2. Stamens and Capsules

The stamen extracts contain a huge diversity of flavonoids or, more precisely, flavonols (Figure 5). In this flower section, kaempferol and gossypetin glycosides occur side by side, and all four *P. nudicaule* cultivars possess the same substances. Pelargonidin glycosides and nudicaulins are missing in the stamens.

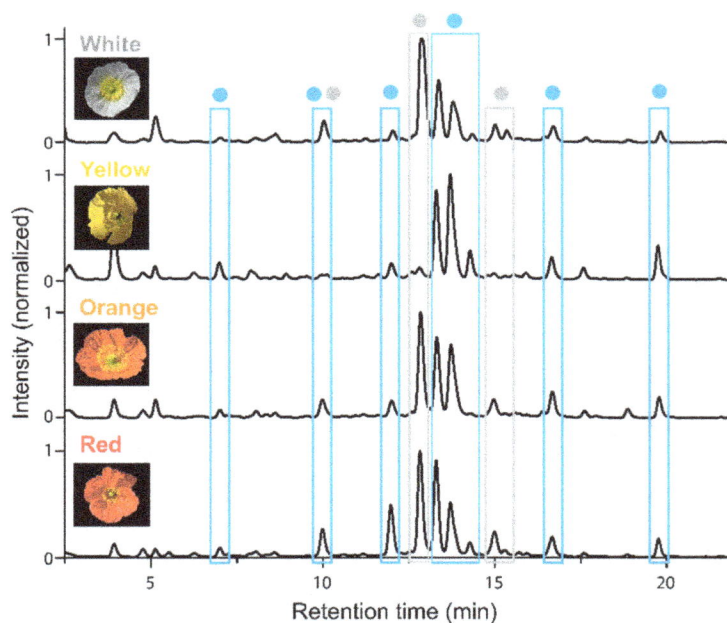

**Figure 5.** HPLC-PDA chromatograms of stamen extracts of four *P. nudicaule* cultivars recorded at 254 nm. Peaks with the same aglycone (identified by UV/Vis absorption spectra) are marked with the same color: • Kaempferol glycoside, • gossypetin glycoside.

Extracts of the capsules of all four cultivars contain only traces of kaempferol glycosides, indicating that flavonoids likely do not serve a function in this area or are only present at an earlier stage of capsule development. Moreover, because the yellow pigments of stamens and the upper area of the capsule could not be extracted from the tissue by a water-methanol mixture (Method 1; see Section 4.2), it was unlikely that nudicaulins are responsible for the yellow color of these tissues.

When hexane was used as an extractant (Method 2; see Section 4.3), it was possible to retrieve the yellow color from stamens and the upper capsule. As expected, the UV/Vis absorption spectra did not match with those of the flavonoids and nudicaulins obtained from the other flower parts. In contrast, the UV/Vis spectra corresponded to the known absorption characteristics of carotenoids (Figure 6) [17]. We conclude that carotenoids serve as yellow pigments in the stamens and upper capsules of all four *P. nudicaule* cultivars.

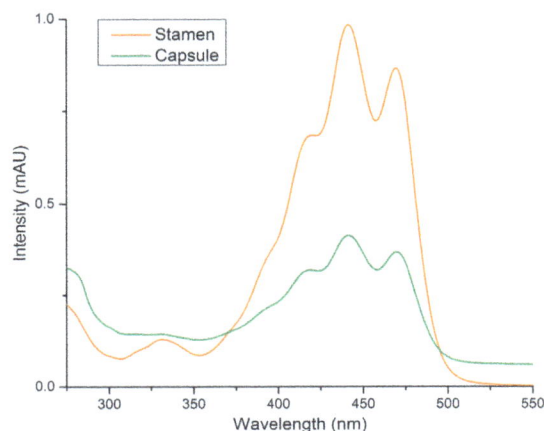

**Figure 6.** UV/Vis absorption spectra of pigments obtained by extraction with hexane from stamens and capsules (Method 2). The spectra of extracts of yellow flower parts are shown as representative examples for all cultivars.

## 3. Discussion

This study investigated the occurrence of UV/Vis absorbing pigments in the apical and basal petal areas, capsules, and stamens of *P. nudicaule* flowers. Glycosides of two flavonols (*i.e.*, kaempferol, gossypetin), one anthocyanidin (*i.e.*, pelargonidin), and flavonoid-derived indole alkaloids (nudicaulins), as well as carotenoids were detected. Their distribution in different flower parts of the examined white, yellow, orange, and red cultivars is summarized in Table 1.

**Table 1.** Pigment distribution in the four studied flower parts.

| *P. nudicaule* Cultivar | Apical Area | Basal Area | Capsule | Stamen |
|---|---|---|---|---|
| White | ● | ● | ○ ● | ● ● ● |
| Yellow | ● ● | ● | ○ ● | ● ● ● |
| Orange | ● ● ● | ● | ○ ● | ● ● ● |
| Red | ● ● | ● | ○ ● | ● ● ● |

Colored dots indicate main compounds detected and colored circles (○) represent traces. ● Kaempferol glycosides, ● gossypetin glycosides, ● nudicaulins, ● pelargonidin glycosides, ● carotenoids.

Pelargonidin glycosides and nudicaulins are present in the apical parts of red, orange, and yellow *P. nudicaule* flowers. This is in agreement with results from previous studies [7,8,10].

Since pelargonidin glycosides represent biosynthetic precursors of the nudicaulins [9], the fact that orange and red cultivars contain different pelargonidin glycosides raises questions: Does the substitution pattern affect the conversion to nudicaulins or is the formation of nudicaulins limited by the availability of indole, the second ultimate biosynthetic precursor?

The identified kaempferol glycosides are the main flavonoids in the apical area of white petals and may function as UV pigments, as reported for other plants [11]. Additionally, kaempferol glycosides could serve as UV screens in the petal tissue, although the lack of a second hydroxyl group in ring B renders them less optimal for such a function [18]. Since in all four cultivars these compounds are produced in the developing petals before the buds open for flowering, it is more likely that the kaempferol glycosides play a role in growth processes than in the protection of the plants against UV light. In addition, in yellow-, orange- and red-colored petals, kaempferol glycosides might act as co-pigments [19].

Pigments coloring the apical area serve an important ecological function by attracting pollinators such as bees and bumblebees. Differently appearing petals might attract different pollinators and therefore improve reproductive success [2]. Furthermore, because the flowers of the *Papaver* plants are open during both day and night, a broad pollinator audience with varying color preferences may be attracted. Consequently, the various flower constituents in *P. nudicaule* cultivars may affect different processes, each of which contributes to the survival of the species.

The present study describes for the first time the occurrence of several gossypetin glycosides exclusively in the basal area of the petals of all four cultivars. Until now, only gossypitrin, the gossypetin 7-O-glucoside, had been reported in yellow petals [10]. The fact that we detected more gossypetin glycosides than Schliemann *et al.* (2006) is considered to be a consequence of our modified experimental protocol, which includes separate extractions of basal and apical petal areas [10]. Gossypitrin is responsible for the yellow color of *Chrysanthemum segetum* flowers and, along with herbacitrin (missing OH group in position 3'), of *P. radicatum* [20,21]. It is unlikely, however, that gossypetin glycosides serve primarily as pigments in the basal area of *P. nudicaule*, because the basal flower part is visually covered by stamens and the capsule. Similarly, the functions as a UV protector (for which gossypetins would have the catechol moiety as a favorable structural precondition) or as a visual nectar guide for insects, as described for example in *Rudbeckia hirta* [22], are not likely.

However, it is noticeable that gossypetin glycosides also occur, together with kaempferol glycosides, in the stamens of all four *P. nudicaule* cultivars. It may be that in *P. nudicaule* the found flavonols play a role in pollen germination or pollen tube development, as it was previously reported in maize and petunia [23].

Furthermore, contrary to preceding reports [14], we show that carotenoids, not nudicaulins, are coloring the yellow stamen and the upper part of the capsule. Carotenoids are widespread pollen and stamen pigments and are also known from some poppy species, for example, *Eschscholzia californica* Cham. [24]. The yellow pigments of stamens of various *Papaver* species—tentatively reported by Tétényi to be nudicaulins—may have been confused with these indole alkaloids because of their similar color. As of now, nudicaulins are known only from petals of *P. nudicaule*, *Papaver alpinum*, and *Meconopsis cambrica* [8].

Our results show a defined distribution of weakly colored kaempferol and gossypetin glycosides in all four *P. nudicaule* cultivars, while strongly colored pelargonidin glycosides are restricted to apical petal areas of the orange and red cultivars and nudicaulins to apical petal areas of the yellow and orange flowers. This distribution pattern suggests that each group of compounds serves a different function. While the role in pigmentation is obvious, it may also be that some of the flavonoids have specific functions in growth and reproduction processes. Certain compounds may represent important metabolites that are present during petal development rather than biosynthesis products assigned to functions in the flowering stage. For example, it was already shown in yellow *P. nudicaule* flowers that pelargonidin glycosides decrease in favor of nudicaulin accumulation during petal development [9]. Future efforts should test these hypotheses and identify all substitution patterns of the flavonoids and carotenoids.

## 4. Materials and Methods

### 4.1. Plant Material

Seeds of *Papaver nudicaule* cultivars were purchased from Jelitto Staudensamen GmbH ("Summer Breeze", yellow and orange) and Syringia ("Wonderland White" and "Matador Red"). The plants were reared in soil in the greenhouse facility of the Max Planck Institute for Chemical Ecology at temperatures between 21 °C and 23 °C during the day and between 19 °C and 21 °C during the night, with an average humidity of 55%. Phillips Sun-T Agro 400 Na lights were used to ensure a daily light period of 14 h. Plants were watered daily for 10 min. After sowing, a period of six to eight weeks passed before multiple buds developed on individual plants.

### 4.2. HPLC-PDA Analysis of Flavonoids and Nudicaulins

After freshly opened flowers were harvested, petals were separated, washed with deionized water and wiped dry, and the apical and basal areas were dissected. Every flower compartment was weighed and coarsely ground under liquid nitrogen. For extraction (Method 1), an amount of solvent in a 1:1 (*v:v*) ratio of water to methanol was added to achieve a uniform concentration of 1 mg of material per 10 µL of solvent (or, for capsules, per 5 µL). After 30 min of extraction in the ultrasonic bath, samples were centrifuged (4 °C, 13,200 rpm), and the supernatant was used for HPLC-PDA analysis. The entire procedure was carried out with three biological replicates.

The analytical HPLC-PDA system consisted of an Agilent series HP1100 (binary pump G1312A, auto sampler G1313A, and photodiode array detector G1315B, 200–700 nm) equipped with an EC250/4 Nucleodur C18 HTec column from Macherey-Nagel (5 µm; injection volume 20 µL). The method included a 21-min gradient from 20% to 67% of methanol in acidified water (0.1% trifluoroacetic acid) with a subsequent washing step (100% methanol) and equilibration to starting conditions. The flow rate was 1 mL· $min^{-1}$, and the detection wavelengths were 211, 254, 281, 351, and 460 nm. The aglycone standards of kaempferol and pelargonidin chloride were purchased from Sigma-Aldrich.

### 4.3. UV/Vis Analysis of Carotenoids

For recording UV/Vis absorption spectra of carotenoids, the stamen and the upper capsule were separated, hexane was added, and the samples were homogenized using Bertin Minilys (60 s, full speed) (Method 2). After extraction in the ultrasonic bath for 3 min and centrifugation (10 °C, 13,200 rpm), spectra of the supernatant were measured against a hexane reference. Samples were applied inside a quartz cuvette with an internal width of 1 cm to a Jasco V-550 UV/Vis spectrophotometer.

**Acknowledgments:** The authors wish to thank Emily Wheeler for editorial assistance in the preparation of this manuscript.

**Author Contributions:** Bettina Dudek and Anne-Christin Warskulat conceived and designed the experiments; Bettina Dudek performed the experiments; Bettina Dudek and Anne-Christin Warskulat analyzed the data; Bettina Dudek, Anne-Christin Warskulat and Bernd Schneider wrote the paper.

**Conflicts of Interest:** The authors declare no conflict of interest.

## References

1. Parihar, A.; Grotewold, E.; Doseff, A.I. Flavonoid Dietetics: Mechanisms and Emerging Roles of Plant Nutraceuticals. In *Pigments in Fruits and Vegetables*; Chen, C., Ed.; Springer New York: New York, NY, USA, 2015; pp. 93–126.
2. Winkel-Shirley, B. Biosynthesis of flavonoids and effects of stress. *Curr. Opin. Plant Biol.* **2002**, *5*, 218–223. [CrossRef]
3. Taylor, L.P.; Grotewold, E. Flavonoids as developmental regulators. *Curr. Opin. Plant Biol.* **2005**, *8*, 317–323. [CrossRef] [PubMed]

4.  Hanelt, P. Die Typisierung von *Papaver nudicaule* L. und die Einordnung von *P. nudicaule* hort. non L. *Die Kulturpflanze* **1970**, *18*, 73–88. [CrossRef]

5.  Robinson, G.M.; Robinson, R. A survey of anthocyanins. I. *Biochem. J.* **1931**, *25*, 1687–1705. [CrossRef] [PubMed]

6.  Robinson, G.M.; Robinson, R. A survey of anthocyanins. II. *Biochem. J.* **1932**, *26*, 1647–1664. [CrossRef] [PubMed]

7.  Cornuz, G.; Wyler, H.; Lauterwein, J. Pelargonidin 3-malonylsophoroside from the red Iceland poppy, *Papaver nudicaule*. *Phytochemistry* **1981**, *20*, 1461–1462. [CrossRef]

8.  Tatsis, E.C.; Böhm, H.; Schneider, B. Occurrence of nudicaulin structural variants in flowers of papaveraceous species. *Phytochemistry* **2013**, *92*, 105–112. [CrossRef] [PubMed]

9.  Warskulat, A.-C.; Tatsis, E.C.; Dudek, B.; Kai, M.; Lorenz, S.; Schneider, B. Unprecedented utilization of pelargonidin and indole for the biosynthesis of plant indole alkaloids. *ChemBioChem* **2016**, *17*, 318–327. [CrossRef] [PubMed]

10. Schliemann, W.; Schneider, B.; Wray, V.; Schmidt, J.; Nimtz, M.; Porzel, A.; Böhm, H. Flavonols and an indole alkaloid skeleton bearing identical acylated glycosidic groups from yellow petals of *Papaver nudicaule*. *Phytochemistry* **2006**, *67*, 191–201. [CrossRef] [PubMed]

11. Mol, J.; Grotewold, E.; Koes, R. How genes paint flowers and seeds. *Trends Plant Sci.* **1998**, *3*, 212–217. [CrossRef]

12. Mo, Y.; Nagel, C.; Taylor, L.P. Biochemical complementation of chalcone synthase mutants defines a role for flavonols in functional pollen. *Proc. Natl. Acad. Sci. USA* **1992**, *89*, 7213–7217. [CrossRef] [PubMed]

13. Acheson, R.M.; Jenkins, C.L.; Harper, J.L.; McNaughton, I.H. Floral pigments in *Papaver* and their significance in the systematics of the genus. *New Phytol.* **1962**, *61*, 256–260. [CrossRef]

14. Tétényi, P. Chemodifferentiation of Papavereae from coasts of the Black-Sea to the Atlantic. In *Natural Products in the New Millennium: Prospects and Industrial Application*; Rauter, A.P., Palma, F.B., Justino, J., Araújo, M.E., Santos, S.P., Eds.; Springer Netherlands: Dordrecht, The Netherlands, 2002; pp. 173–181.

15. Tatsis, E.C.; Schaumlöffel, A.; Warskulat, A.C.; Massiot, G.; Schneider, B.; Bringmann, G. Nudicaulins, yellow flower pigments of *Papaver nudicaule*: Revised constitution and assignment of absolute configuration. *Org. Lett.* **2013**, *15*, 156–159. [CrossRef] [PubMed]

16. Suzuki, H.; Sasaki, R.; Ogata, Y.; Nakamura, Y.; Sakurai, N.; Kitajima, M.; Takayama, H.; Kanaya, S.; Aoki, K.; Shibata, D.; *et al.* Metabolic profiling of flavonoids in *Lotus japonicus* using liquid chromatography Fourier transform ion cyclotron resonance mass spectrometry. *Phytochemistry* **2008**, *69*, 99–111. [CrossRef] [PubMed]

17. Rodriguez-Amaya, D.B.; Kimura, M. *HarvestPlus Handbook for Carotenoid Analysi*; HarvestPlus Technical Monograph 2; HarvestPlus: Washington, DC, USA; Cali, CO, USA, 2004.

18. Di Ferdinando, M.; Brunetti, C.; Fini, A.; Tattini, M. Flavonoids as antioxidants in plants under abiotic stresses. In *Abiotic Stress Responses in Plants*; Ahmad, P., Prasad, M., Eds.; Springer New York: New York, NY, USA, 2012; pp. 159–179.

19. Cooper-Driver, G.A. Contributions of Jeffrey Harborne and co-workers to the study of anthocyanins. *Phytochemistry* **2001**, *56*, 229–236. [CrossRef]

20. Stich, K.; Halbwirth, H.; Wurst, F.; Forkmann, G. UDP-glucose: Flavonol 7-*O*-glucosyltransferase activity in flower extracts of *Chrysanthemum segetum*. *Z. Naturforsch. C* **1997**, *52*, 153–158. [PubMed]

21. Wind, O.; Christensen, S.B.; Mølgaard, P. Colouring agents in yellow and white flowered *Papaver radicatum* in Northern Greenland. *Biochem. Syst. Ecol.* **1998**, *26*, 771–779. [CrossRef]

22. Thompson, W.R.; Meinwald, J.; Aneshansley, D.; Eisner, T. Flavonols: Pigments responsible for ultraviolet absorption in nectar guide of flower. *Science* **1972**, *177*, 528–530. [CrossRef] [PubMed]

23. Pollak, P.E.; Vogt, T.; Mo, Y.; Taylor, L.P. Chalcone synthase and flavonol accumulation in stigmas and anthers of *Petunia hybrida*. *Plant. Physiol.* **1993**, *102*, 925–932. [PubMed]

24. Barrell, P.J.; Wakelin, A.M.; Gatehouse, M.L.; Lister, C.E.; Conner, A.J. Inheritance and epistasis of loci influencing carotenoid content in petal and pollen color variants of California poppy (*Eschscholzia californica* Cham.). *J. Hered.* **2010**, *101*, 750–756. [CrossRef] [PubMed]

# 4

# Caffeoylquinic Acids from the Aerial Parts of *Chrysanthemum coronarium* L.

**Chunpeng Wan [1,2,†], Shanshan Li [1,†], Lin Liu [1], Chuying Chen [1,2] and Shuying Fan [1,\*]**

[1]    College of Agronomy, Jiangxi Agricultural University, Nanchang 330045, China;
       lemonwan@126.com (C.W.); liss0824@126.com (S.L.); linliu0960@126.com (L.L.); ccy0728@126.com (C.C.)
[2]    Collaborative Innovation Center of Post-Harvest Key Technology and Quality Safety of Fruits and
       Vegetables in Jiangxi Province, Jiangxi Agricultural University, Nanchang 330045, China
\*    Correspondence: chunpengwan@jxau.edu.cn or fansy12@126.com
†    These authors contributed equally to this work.

Academic Editor: Milan S. Stankovic

**Abstract:** To elucidate the chemical compositions of the aerial parts of *Chrysanthemum coronarium* L., the ethanol extracts of *Ch. coronarium* L. were firstly isolated by the MCI-gel resin column. The caffeoylquinic acid-rich fractions were further purified by various chromatographic columns including silica gel, Sephadex LH-20, and semi-preparative HPLC to yield the compounds. The purified compounds were characterized by [1]H-Nuclear Magnetic Resonance ([1]H-NMR), [13]C-NMR, and high resolution electrospray ionisation mass spectral (HR-ESI-MS) spectroscopy. Seven caffeoylquinic acid (CQA) compounds were isolated from this plant. Their structures were clarified by spectrometric methods and identified as 3-*O*-caffeoylquinic acid (**1**), 5-*O*-caffeoylquinic acid (**2**), 4-*O*-caffeoylquinic acid (**3**), 3,4-di-*O*-caffeoylquinic acid (**4**), 1,5-di-*O*-caffeoylquinic acid (**5**), 3,5-di-*O*-caffeoylquinic acid (**6**), and 4,5-di-*O*-caffeoylquinic acid (**7**). Caffeoylquinic acids were the major constituents present in the aerial parts of *Ch. coronarium* L. All of the isolates except for compounds **2** and **6** were reported for the first time from this species. Moreover, compounds **3–5**, and **7** were identified from the *Chrysanthemum* genus for the first time.

**Keywords:** *Chrysanthemum coronarium* L.; aerial parts; caffeoylquinic acids

---

## 1. Introduction

*Chrysanthemum coronarium* L., commonly known as "Tonghao", is used as an edible vegetable and medicinal plant, and belongs to the genus of *Chrysanthemum* (Compositae) [1]. Many previous studies have reported the isolation, identification, and the biological activities of the plant. The results have shown that flavonoids [2], phenolic acids [3], sesquiterpene lactone [4], monoterpene [5], diterpene [6], glycosyldiglycerides [7], alkaloid [8], phytosterol [9], heterocyclic compounds [8], polyacetylenes [9], and essential oils [10,11] were the major chemical constituents present in the plant. *Ch. coronarium* L. has shown a variety of biological activities including the elimination of phlegm, plant allelopathy, nematicidal activity, cytotoxic activity, antioxidant and free radical scavenging, insect antifeedant, hepatic protection, and antimicrobial properties [1]. Flavonoids and phenolic acids are responsible for the plant allelopathy and antioxidant and free radical scavenging activities [3]; polyacetylenes and essential oil are responsible for the insect antifeedant [12] and antimicrobial activities [10]; and terpene, in particular sesquiterpene lactone, is responsible for the cytotoxic [13] and antimicrobial activities [6].

Flavonoids and polyphenols are the characteristic constituents of *Ch. coronarium* L., which are also the main bioactive constituents of the Compositae family [2,3]. Caffeoylquinic acids (CQAs) are cinnamate conjugates, which are biosynthesized through the phenylpropanoid pathway. These phenolic compounds are generally involved in plant disease-resistance responses to biotic or abiotic stress [14]. Preliminary studies on the chemical constituents of the aerial parts of *Ch. coronarium* L. indicated that the caffeoylquinic acids were the major components in the plant, however, only three CQA compounds were elucidated until now; 3,5-di-*O*-caffeoyl-4-succinylquinic acid, chlorogenic acid, and 3,5-di-*O*-caffeoylquinic acid [3].

Herein, the isolation and structure elucidation of caffeoylquinic acids (CQAs) from the aerial parts of *Ch. coronarium* was achieved.

## 2. Results

Caffeoylquinic acid derivatives showed typical UV spectra peaks at 327, 298 (sh), and 246 nm [15]. The HPLC profile of the ethanolic extract (TH) and of its five purified fractions (THA to THE) indicated that THA−THC were caffeoylquinic acid derivative-rich fractions (Figure 1). Many components in the THB fraction overlapped with the THA and THC fractions based on the HPLC profile. Thus, the THA and THC fractions were further isolated to yield the pure compounds.

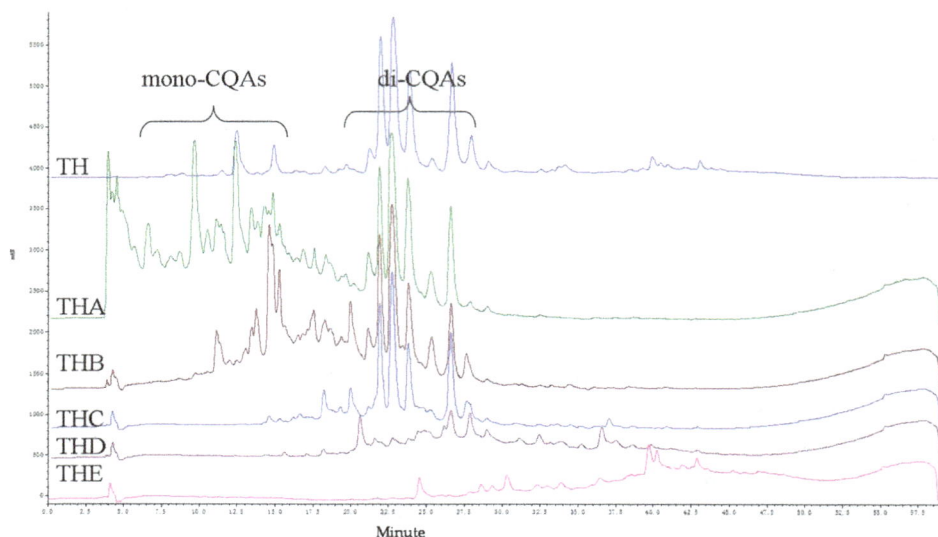

**Figure 1.** HPLC chromatogram of ethanol extract (TH) and MCI fractions (THA–THE) of *Ch. coronarium* L.

The structures of the isolated caffeoylquinic acid derivatives were elucidated by HR-ESI-MS analysis, 1D-NMR data (Tables 1 and 2), and comparison of these data with the literature. Compounds 1–7 (Figure 2) were identified as 3-*O*-caffeoylquinic acid (**1**) [16], 5-*O*-caffeoylquinic acid (**2**) [17], 4-*O*-caffeoylquinic acid (**3**) [16], 3,4-di-*O*-caffeoylquinic acid (**4**) [18,19], 1,5-di-*O*-caffeoylquinic acid (**5**) [20], 3,5-di-*O*-caffeoylquinic acid (**6**) [21], and 4,5-di-*O*-caffeoylquinic acid (**7**) [18,19]. Among them, compounds **1**, **3–5**, and **7** were isolated from this species for the first time. Moreover, compounds **3–5**, and **7** were reported from the genus *Chrysanthemum* for the first time.

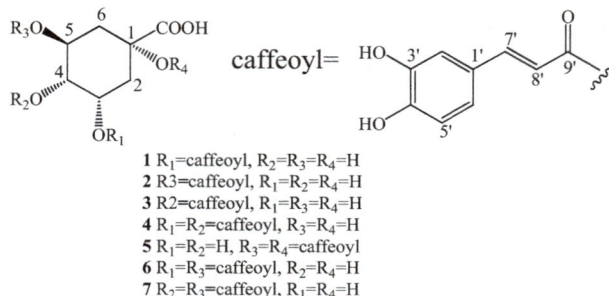

**Figure 2.** The chemical structures of the compounds **1–7** isolated from *Ch. coronarium* L.

Caffeoylquinic Acids from the Aerial Parts of Chrysanthemum coronarium L.

29

**Table 1.** $^1$H-NMR ($^1$H-Nuclear Magnetic Resonance, 400 MHz, MeOH-$d_4$) characteristics of the caffeoylquinic acid derivatives **1–7** isolated from aerial parts of *Ch. coronarium* L.

| No. | 1 δH (J Hz) | 2 δH (J Hz) | 3 δH (J Hz) | 4 δH (J Hz) | 5 δH (J Hz) | 6 δH (J Hz) | 7 δH (J Hz) |
|---|---|---|---|---|---|---|---|
| 2 | 1.93–2.22 (2H, m) | 2.04–2.25 (2H, m) | 1.98–2.23 (2H, m) | 2.07–2.35 (2H, m) | 2.54 (1H, dd, 3.5, 10.2) 2.43 (1H, m) | 2.12–2.32 (2H, m) | 2.14–2.30 (2H, m) |
| 3 | 5.36 (1H, brd, 2.9) | 4.18 (1H, brd, 2.9) | 4.29 (1H, brs) | 5.61 (1H, brd, 3.4) | 4.28 (1H, brd, 3.5) | 5.36 (1H, brd, 5.6) | 4.40 (1H, brs) |
| 4 | 3.65 (1H, dd, 2.8, 8.5) | 3.75 (1H, dd, 2.3, 8.0) | 4.80 (1H, dd, 2.3, 9.0) | 4.96 (1H, dd, 3.1, 9.4) | 3.76 (1H, dd, 3.0, 8.1) | 3.96 (1H, dd, 3.2, 7.3) | 5.10 (1H, d, 8.5) |
| 5 | 4.14 (1H, ddd, 3.6, 8.5, 8.5) | 5.35 (1H, ddd, 3.5, 8.0, 8.0) | 4.27 (1H, ddd, 9.0, 9.0, 4.5) | 4.35 (1H, ddd, 4.2, 9.4, 9.4) | 5.37 (1H, ddd, 3.7, 8.1, 8.1) | 5.42 (1H, ddd, 3.2, 7.3, 7.3) | 5.63 (1H, ddd, 4.2, 8.5, 8.5) |
| 6 | 1.93–2.22 (2H, m) | 2.04–2.25 (2H, m) | 1.98–2.23 (2H, m) | 2.07–2.35 (2H, m) | 2.05 (1H, dd, 11.1, 13.8) 2.43 (1H, m) | 2.12–2.32 (2H, m) | 2.14–2.30 (2H, m) |
| 2'/2'' | 7.04 (1H, d, 1.4) | 7.05 (1H, brs) | 7.06 (1H, d, 1.4) | 7.02/7.00 (each 1H, d, 1.8) | 7.04 (each 1H, brs) | 7.05 (each 1H, s) | 7.00/6.97 (each 1H, s) |
| 5'/5'' | 6.78 (1H, d, 8.0) | 6.78 (1H, d, 8.0) | 6.78 (1H, d, 8.0) | 6.76/6.72 (each 1H, d, 7.9) | 6.78/6.76 (each 1H, d, 8.1) | 6.77/6.75 (each 1H, d, 8.1) | 6.73/6.71 (each 1H, d, 8.1) |
| 6'/6'' | 6.94 (1H, dd, 1.4, 8.0) | 6.95 (1H, d, 8.0) | 6.97 (1H, dd, 1.4, 8.0) | 6.90/6.85 (each 1H, dd, 1.8, 7.9) | 6.96/6.94 (each 1H, d, 8.1) | 6.96/6.94 (each 1H, m) | 6.89/6.87 (each 1H, d, 8.1) |
| 7'/7'' | 7.58 (1H, d, 15.9) | 7.56 (1H, d, 15.9) | 7.63 (1H, d, 15.9) | 7.57/7.52 (each 1H, d, 15.9) | 7.58/7.55 (each 1H, d, 15.9) | 7.60/7.56 (each 1H, d, 15.9) | 7.57/7.49 (each 1H, d, 15.9) |
| 8'/8'' | 6.30 (1H, d, 15.9) | 6.27 (1H, d, 15.9) | 6.37 (1H, d, 15.9) | 6.27/6.23 (each 1H, d, 15.9) | 6.27/6.24 (each 1H, d, 15.9) | 6.33/6.24 (each 1H, d, 15.9) | 6.26/6.17 (each 1H, d, 15.9) |

**Table 2.** $^{13}$C-NMR Data for Compounds **2**, **5–6** (100 MHz, MeOH-$d_4$).

| No. | Compounds | | |
|---|---|---|---|
| | **2** | **5** | **6** |
| 1 | 74.7 | 80.9 | 73.3 |
| 2 | 36.8 | 35.7 | 34.6 |
| 3 | 69.9 | 69.4 | 71.1 |
| 4 | 72.1 | 72.8 | 69.2 |
| 5 | 70.5 | 71.6 | 70.6 |
| 6 | 37.4 | 36.9 | 36.2 |
| 7 | 175.6 | 174.8 | 175.9 |
| 1′ | 126.4 | 127.8/127.8 | 126.5/126.4 |
| 2′ | 113.8 | 115.3/115.3 | 114.2/113.9 |
| 3′ | 145.3 | 147.6/147.4 | 145.3/145.3 |
| 4′ | 148.1 | 149.7/149.7 | 148.1/148.0 |
| 5′ | 115.1 | 116.5/116.5 | 115.1/115.1 |
| 6′ | 121.6 | 123.1/123.1 | 121.6/121.6 |
| 7′ | 145.7 | 147.4/146.9 | 145.9/145.6 |
| 8′ | 113.8 | 115.2/115.1 | 113.7/113.7 |
| 9′ | 167.3 | 168.7/168.0 | 167.4/167.0 |

## 3. Discussion

Compounds **1–3** were obtained as white power. The HR-ESI-MS yielded a quasi-molecular ion peak [M-H]$^-$ at $m/z$ 353.08. The UV spectrum showed $\lambda_{max}$ at 328, 298 (shoulder), and 246 nm, which suggested that compounds **1–3** were single caffeoyl substituted quinic acid derivatives. The $^1$H-NMR spectrum (400 MHz, MeOH-$d_4$) of compounds **1–3** showed caffeoyl signals at $\delta$ 7.56, 7.58, 7.63 (1H, d, $J$ = 15.9 Hz, H-7′), 6.27, 6.30, 6.37 (1H, d, $J$ = 15.9, H-8′), 7.04, 7.05, 7.06 (1H, H-2′), 6.94, 6.95, 6.97 (1H, H-6′), 6.78, 6.78, 6.78 (1H, H-5′), and quinic acid signals at [**1**: $\delta$ 5.36 (1H, brd, $J$ = 2.9, H-3), 3.65 (1H, dd, $J$ = 8.5, 2.8, H-4), 4.14 (1H, ddd, $J$ = 8.5, 8.5, 3.6, H-5), 1.93–2.22 (4H, m, H-2, H-6); **2**: $\delta$ 4.18 (1H, brd, $J$ = 2.9, H-3), 3.75 (1H, dd, $J$ = 8.0, 2.3, H-4), 5.35 (1H, ddd, $J$ = 8.0, 8.0, 3.5, H-5), 2.04–2.25 (4H, m, H-2, H-6); **3**: $\delta$ 4.29 (1H, brs, H-3), 4.80 (1H, dd, $J$ = 9.0, 2.3, H-4), 4.27 (1H, ddd, $J$ = 9.0, 9.0, 4.5, H-5), 1.98–2.23 (4H, m, H-2, H-6)]. The substituted position of caffeoyl can be determined by the analysis of the chemical shift and coupling constants of the oxygenated methine protons of the quinic acid core. Once the oxygenated methine of quinic acid was acylated by caffeoyl, the proton signal will shift downfield. The coupling constant of the downfield shifted proton was then applied to the acylation position. Generally, the H-3 signal has a small coupling constant and shows a brd or brs type peak, the H-4 signal showed a dd type peak with coupling constants at 8.0–9.0 Hz and 2.0–3.0 Hz, while the H-5 signal showed a ddd type peak with coupling constants at 8.0–9.0 Hz, 8.0–9.0 Hz, and 3.0–5.0 Hz. Based on these rules, the structures of compounds **1–3** were determined as depicted.

Compounds **4–7** were obtained as white powder. The ESI-MS yielded a quasi-molecular ion peak [M-H]$^-$ at $m/z$ 515.11, and the UV spectrum showed $\lambda_{max}$ at 327, 298 (shoulder), and 245 nm, suggesting that compounds **4–7** were double caffeoyl substituted quinic acid derivatives. The $^1$H-NMR spectrum showed similar signal patterns to compounds **1–3**, but one more caffeoyl signal was observed. The $^1$H-NMR spectrum of compounds **4–7** showed two sets of caffeoyl signals [7.49–7.60 (each 1H, d, $J$ = 15.9 Hz, H-7′ and H-7″), 6.97–7.05 (each 1H, H-2′ and H-2″), 6.85–6.96 (each 1H, H-6′ and H-6″), 6.71–6.78 (each 1H, H-5′ and H-5″), 6.17–6.33 (each 1H, d, $J$ = 15.9 Hz, H-8′ and H-8″)], and quinic acid signals (see Table 1). Similar to compounds **1–3**, the acylation positions were determined by the chemical shift and coupling constants of the oxygenated methine protons of quinic acid. As only the H-5 signal of compound **5** was observed with a downfield shift, another substituted position was tentatively assigned to C-1 of quinic acid.

The chemical compositions of the plant are characterized by flavonoids and phenolic acids [22], which showed typical UV spectra based on the HPLC-DAD. In the current study, seven CQAs including three mono-CQAs and four di-CQAs were isolated and identified. Previously, two phenolic acids as plant growth inhibitors were isolated from *Ch. coronarium* L. and identified as isoferulic acid and methyl parahydroxybenzoats [12]. Ferulic acid methyl ester was also detected in *Ch. coronarium* L., which showed low-density lipoprotein (LDL) oxidation inhibited activity [8]. Only three quinic acid derivatives, namely, chlorogenic acid (5-*O*-caffeoylquinic acid, **2**), 3,5-di-*O*-caffeoylquinic acid (**6**), and 3,5-di-*O*-caffeoyl-4-succinylquinic acid were detected by the HPLC method [3]. This is the first report of the isolation of CQAs except for compounds **2** and **6**, while compounds **3–5**, and **7** were identified from the Chrysanthemum genus for the first time. Additionally, the di-CQAs are the major phenolic acid constituents present in this plant.

## 4. Materials and Methods

### 4.1. Plant Material

The aerial parts of *Ch. coronarium* L. were purchased from a local market, Nanchang City, Jiangxi Province, China, and identified by Prof. Shuying Fan (College of Agronomy, Jiangxi Agricultural University, Nanchang, China). A voucher specimen (TH-2015041) has been deposited in the Department of Horticulture, College of Agronomy, Jiangxi Agricultural University (Nanchang, Jiangxi, China).

### 4.2. Equipment and Reagents

$^1$H and $^{13}$C-NMR were detected on a Varian 400 MHz spectrometer in CD$_3$OD with Tetramethylsilane (TMS) as the internal standard. HR-ESI-MS data were obtained on a 6538 Ultra High Definition (UHD) Accurate-Mass Q-TOF LC/MS system (Agilent, Santa Clara, CA, USA). High performance liquid chromatography (HPLC) was performed on a Hitachi Elite Chromaster system including a 5110 pump, 5210 autosampler, 5310 column oven, a 5430 diode array detector, and operated by EZChrom Elite software. Luna C18 (2) column (5 μm, 4.6 × 250 mm) for analysis and Luna C18 (2) column (10 μm, 10 × 250 mm) for HPLC preparation were purchased from Phenomenex Inc (Torrance, CA, USA). The HPLC grade solvents were purchased from Sigma (Sigma, St. Louis, MO, USA). All analytical solvents were purchased from Tansoole (Shanghai, China). Silica gel (250 mesh; Qingdao Haiyang Chemical Co., LTD, Qingdao, China) was used as normal phase, whereas YMC Pack ODS-A (50 μm; YMC) was used as reversed phase column material. MCI gel CHP20P (75–150 μm; Mitsubishi Chemical Corp, Tsukuba, Japan) and Sephadex LH-20 (GE Healthcare, Uppsala, Sweden) were also used for column chromatography.

### 4.3. Extraction and Chromatography

The fresh aerial parts of *Ch. coronarium* L. (20 kg) were dried in air, yielding a crude dry material which amounted to about 2.2 kg. The dried material (2.0 kg) was ground and extracted using an ultrasonic-assisted method with 95% ethanol (3 × 50 L) at 45 °C for 2 h. The dried ethanol extract (TH, 118.5 g) was subjected to MCI gel column chromatography (4.0 × 25 cm), eluted with water, 10% methanol, 30% methanol, 50% methanol, 70% methanol, and 90% methanol, respectively (each, 2.0 L). Lastly, the MCI gel column was washed with acetone. Six fractions were yielded (THA–THF).

### 4.4. Purification of the Caffeoylquinic Acid Derivatives

The THA fraction (13.2 g) was subjected to ODS C18 column chromatography (3.0 × 25 cm) eluting with 5% methanol, 15% methanol, 25% methanol, 35% methanol, and 50% methanol (each 1.0 L), respectively. Five fractions (THA-**1**–THA-**5**) were obtained after being pooled according to their HPLC profiles.

Fraction THA-3 (3.3 g) was further subjected to Sephadex LH-20 (2.0 × 100 cm) elution with MeOH to furnish fractions THA-3A–3D. Fraction THA-3C was purified by semi-preparative HPLC (10 μm, 10 × 250 mm), eluting with MeOH-H$_2$O (0–23 min: 18:82 to 45:55; *v/v*, 3 mL/min) and yielding compounds **1** (12.5 mg), **2** (26.8 mg), and **3** (14.5 mg).

The THC fraction (6.57 g) was subjected to silica gel chromatography (4.0 × 26 cm) using CHCl$_3$-MeOH (100:1 to 2:1, $v/v$) for elution to yield six fractions THC-**1–6** according to their TLC profiles. THC-**6** (1.2 g) was further subjected to Sephadex LH-20 (2.0 × 100 cm) elution with MeOH to furnish fractions THC-6A–6G. THC-6D was purified by semi-preparative HPLC (10 μm, 10 × 250 mm), eluting with an isocratic elution of MeOH-H$_2$O (34:66; $v/v$, 3 mL/min) yielding compounds **4** (8.5 mg) and **6** (9.8 mg). THC-6F was purified by semi-preparative HPLC (10 μm, 10 × 250 mm), eluting with an isocratic elution of MeOH-H$_2$O(35:65; $v/v$, 3 mL/min) yielding compounds **5** (13.5 mg) and **7** (14.6 mg) (Figure 3).

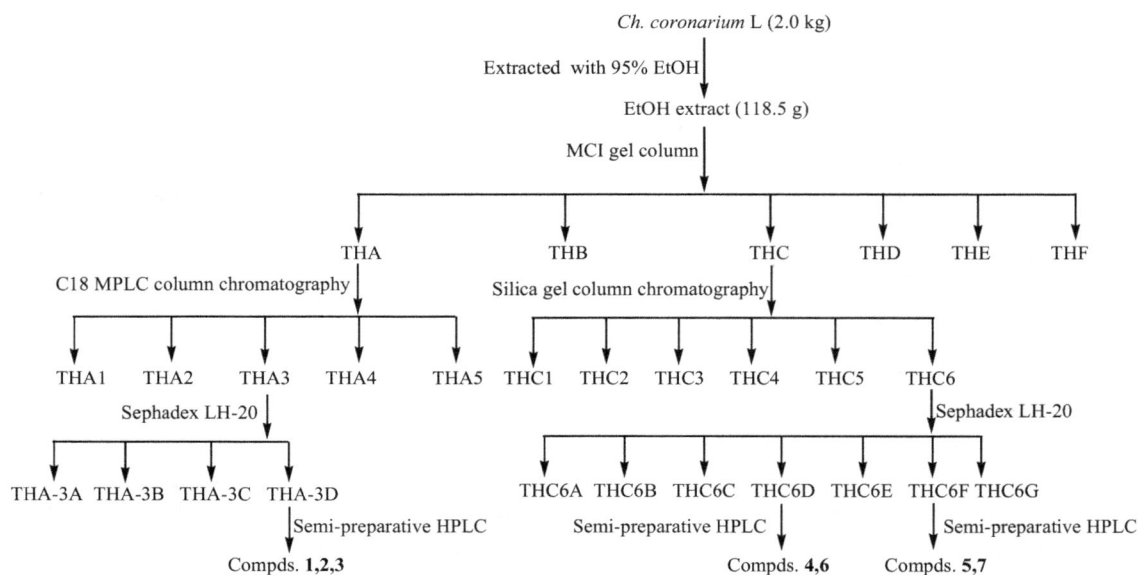

**Figure 3.** The procedure for the extraction and isolation of compounds from *Ch. coronarium* L.

## 5. Conclusions

Caffeoylquinic acids (CQAs) were the major phenolic constituents present in the aerial parts of *Ch. coronarium* L. Seven CQAs including three mono-CQAs and four di-CQAs were isolated from this plant. They were 3-*O*-caffeoylquinic acid (**1**), 5-*O*-caffeoylquinic acid (**2**), 4-*O*-caffeoylquinic acid (**3**), 3,4-di-*O*-caffeoylquinic acid (**4**), 1,5-di-*O*-caffeoylquinic acid (**5**), 3,5-di-*O*-caffeoylquinic acid (**6**), and 4,5-di-*O*-caffeoylquinic acid (**7**), respectively. All of the isolates except for **2** and **6** were isolated from this species for the first time. Moreover, compounds **3–5**, and **7** were identified from the Chrysanthemum genus for the first time.

**Acknowledgments:** This project was supported by the National Natural Science Foundation of China (31360487) and the Natural Science Foundation of Jiangxi Province (20132BAB204016).

**Author Contributions:** Chunpeng Wan and Shuying Fan conceived and designed the experiments; Chunpeng Wan, Shanshan Li, Lin Liu, and Chuying Chen performed the experiments; Chunpeng Wan and Shuying Fan analyzed the data; Chunpeng Wan and Shuying Fan wrote the paper.

**Conflicts of Interest:** The authors declare no conflict of interest.

## References

1.  Dokuparthi, S.K.; Manikanta, P. Phytochemical and pharmacological studies on *Chrysanthemum coronarium* L.: A review. *J. Drug Discov. Ther.* **2015**, *3*, 11–16.
2.  Ibrahim, L.F.; El-Senousy, W.M.; Hawas, U.W. NMR spectral analysis of flavonoids from *Chrysanthemum coronarium*. *Chem. Nat. Compd.* **2007**, *43*, 659–662. [CrossRef]
3.  Chuda, Y.; Suzuki, M.; Nagata, T.; Tsushida, T. Contents and cooking loss of three quinic acid derivatives from garland (*Chrysanthemum coronarium* L.). *J. Agric. Food Chem.* **1998**, *46*, 1437–1439. [CrossRef]

4. El-Masry, S.; Abou-Donia, A.H.; Darwish, F.A.; Abou-Karam, M.A.; Grenz, M.; Bohlmann, F. Sesquiterpene lactones from *Chrysanthemum coronarium*. *Phytochemistry* **1984**, *23*, 2953–2954. [CrossRef]

5. Song, M.C.; Kim, D.H.; Hong, Y.H.; Kim, D.K.; Chung, I.S.; Kim, S.H.; Baek, N.I. Terpenes from the aerial parts of *Chrysanthemum coronarium* L. *Agric. Chem. Biotechnol.* **2003**, *46*, 118–121.

6. Ragasa, C.Y.; Natividad, G.M. An Antimicrobial Diterpene from *Chrysanthemum coronarium*. *Kimica* **1998**, *14*, 17–20.

7. Song, M.C.; Yang, H.J.; Lee, D.G.; Kim, D.K.; Ahn, E.M.; Woo, Y.M.; Baek, N.I. Glycosyldiglycerides from the Aerial Parts of Garland (*Chrysanthemum coronarium*). *J. Korean Soc. Appl. Biol. Chem.* **2009**, *52*, 88–91. [CrossRef]

8. Song, M.C.; Yang, H.J.; Jeong, T.S.; Kim, K.T.; Baek, N.I. Heterocyclic compounds from *Chrysanthemum coronarium* L. and their inhibitory activity on hACAT-1, hACAT-2, and LDL-oxidation. *Arch. Pharm. Res.* **2008**, *31*, 573–578. [CrossRef] [PubMed]

9. Song, M.C.; Kim, D.H.; Hong, Y.H.; Yang, H.J.; Chung, I.S.; Kim, S.H.; Baek, N.I. Polyacetylenes and Sterols from the Aerial Parts of *Chrysanthemum coronarium* L. (Garland). *Front. Nat. Prod. Chem.* **2005**, *1*, 163–168. [CrossRef]

10. Alvarez-Castellanos, P.P.; Bishop, C.D.; Pascual-Villalobos, M.J. Antifungal activity of the essential oil of flowerheads of garland chrysanthemum (*Chrysanthemum coronarium*) against agricultural pathogens. *Phytochemistry* **2001**, *57*, 99–102. [CrossRef]

11. Basta, A.; Pavlović, M.; Couladis, M.; Tzakou, O. Essential oil composition of the flowerheads of *Chrysanthemum coronarium* L. from Greece. *Flavour Fragr. J.* **2007**, *22*, 197–200. [CrossRef]

12. Mahahiro, T.; Kazuhiro, C. Novel plant growth inhibitors and an insect antifeedant from *Chrysanthemum coronarium* L. *Agric. Biol. Chem.* **1984**, *48*, 1367–1369.

13. Lee, K.D.; Park, K.H.; Kim, H.; Kim, J.H.; Rim, Y.S.; Yang, M.S. Cytotoxic Activity and Structural Analogues of Guaianolide Derivatives from the Flower of *Chrysanthemum coronarium* L. *Agric. Chem. Biotechnol.* **2003**, *46*, 29–32.

14. Mondolot, L.; La, Fisca, P.; Buatois, B.; Talansier, E.; De Kochko, A.; Campa, C. Evolution in caffeoylquinic acid content and histolocalization during Coffea canephora leaf development. *Ann. Bot.* **2006**, *98*, 33–40. [CrossRef] [PubMed]

15. Wan, C.; Yu, Y.; Zhou, S.; Tian, S.; Cao, S. Isolation and identification of phenolic compounds from Gynura divaricata leaves. *Pharmacogn. Mag.* **2011**, *7*, 101–108. [PubMed]

16. Nakatani, N.; Kayano, S.; Kikuzaki, H.; Sumino, K.; Katagiri, K.; Mitani, T. Identification, Quantitative Determination, and Antioxidative Activities of Chlorogenic Acid Isomers in Prune (*Prunus domestica* L.). *J. Agri. Food Chem.* **2000**, *48*, 5512–5516. [CrossRef]

17. Wan, C.; Yuan, T.; Cirello, A.L.; Seeram, N.P. Antioxidant and α-glucosidase inhibitory phenolics isolated from highbush blueberry flowers. *Food Chem.* **2012**, *135*, 1929–1937. [CrossRef] [PubMed]

18. Shi, S.; Huang, K.; Zhang, Y.; Zhao, Y.; Du, Q. Purification and identification of antiviral components from Laggera pterodonta by high-speed counter-current chromatography. *J. Chromatogr. B* **2007**, *859*, 119–124. [CrossRef] [PubMed]

19. Chen, J.; Mangelinckx, S.; Ma, L.; Wang, Z.; Li, W.; De Kimpe, N. Caffeoylquinic acid derivatives isolated from the aerial parts of *Gynura divaricata* and their yeast α-glucosidase and PTP1B inhibitory activity. *Fitoterapia* **2014**, *99*, 1–6. [CrossRef] [PubMed]

20. Carnat, A.; Heitz, A.; Fraisse, D.; Carnat, A.P.; Lamaison, J.L. Major dicaffeoylquinic acids from Artemisia vulgaris. *Fitoterapia* **2000**, *71*, 587–589. [CrossRef]

21. Kodoma, M.; Wada, H.; Otani, H.; Kohmoto, K.; Kimura, Y. 3,5-Di-*O*-caffeoylquinic acid, an infection-inhibiting factor from *Pyrus pyrifolia* induced by infection with *Alternaria alternata*. *Phytochemistry* **1998**, *47*, 371–373. [CrossRef]

22. Wan, C.; Liu, Q.; Zhang, X.; Fan, S. A Review of the Chemical Composition and Biological Activities of the Edible and Medicinal Plant *Chrysanthemum coronarium* L. *Mod. Food Sci. Technol.* **2014**, *30*, 282–288.

# 5

# FLOWERING LOCUS T Triggers Early and Fertile Flowering in Glasshouse Cassava (*Manihot esculenta* Crantz)

Simon E. Bull [1],*, Adrian Alder [1], Cristina Barsan [2], Mathias Kohler [1], Lars Hennig [3], Wilhelm Gruissem [1] and Hervé Vanderschuren [1,2],*

[1]  Plant Biotechnology, Department of Biology, ETH Zürich, 8092 Zürich, Switzerland; adrianalder@gmx.ch (A.A.); math.kohler@gmail.com (M.K.); wgruissem@ethz.ch (W.G.)
[2]  Gembloux Agro-Bio Tech, University of Liège, 5030 Gembloux, Belgium; cibarsan@ulg.ac.be
[3]  Department of Plant Biology and Linnean Centre for Plant Biology, PO Box 7080, The Swedish University of Agricultural Sciences, SE-750 07 Uppsala, Sweden; Lars.Hennig@slu.se
*  Correspondence: sbull@ethz.ch (S.E.B.); hvanderschuren@ethz.ch or herve.vanderschuren@ulg.ac.be (H.V.)

Academic Editor: Milan S. Stankovic

**Abstract:** Accelerated breeding of plant species has the potential to help challenge environmental and biochemical cues to support global crop security. We demonstrate the over-expression of *Arabidopsis FLOWERING LOCUS T* in *Agrobacterium*-mediated transformed cassava (*Manihot esculenta* Crantz; cultivar 60444) to trigger early flowering in glasshouse-grown plants. An event seldom seen in a glasshouse environment, precocious flowering and mature inflorescence were obtained within 4–5 months from planting of stem cuttings. Manual pollination using pistillate and staminate flowers from clonal propagants gave rise to viable seeds that germinated into morphologically typical progeny. This strategy comes at a time when accelerated crop breeding is of increasing importance to complement progressive genome editing techniques.

**Keywords:** cassava; *Manihot esculenta* Crantz; flowering; *FLOWERING LOCUS T*; breeding; biotechnology; grafting; seed; recalcitrant crops

## 1. Introduction

Rapid improvement and commercialization of woody perennial plant species is often hampered by lengthy breeding cycles [1]. This obstacle has gained prominence in recent years with the widespread uptake of genome editing techniques, notably CRISPR-Cas9 [2] and the need to segregate out T-DNA after site-specific editing of the genome [3,4]. Genome editing is revolutionizing agricultural breeding, but for many crops including cassava (*Manihot esculenta* Crantz), lengthy life cycles and limited fertility are major bottlenecks in harnessing this technology. Cassava is a prime candidate for accelerated breeding and genomic selection [5] because the starch-rich storage roots are both a staple food, particularly in sub-Saharan Africa, and a multi-billion dollar commodity in countries such as Brazil, China, and Thailand [6,7].

In cassava, time to flowering remains highly dependent on genotype and environmental conditions. Many farmer-preferred cultivars are non-branching to facilitate farming practices and to maximize stem growth for subsequent vegetative propagation. However, erect architecture is also associated with poor flowering capacity, with many cassava cultivars taking more than nine months to establish flowers in the field [8,9] and almost never flower under glasshouse conditions. In vitro manipulation of cytokinins have induced flowering in the laboratory [10] but controlled induction of stable inflorescence and seed production has so far remained elusive. The inefficacy of seed production is exacerbated by asynchronous flowering time, whereby monoecious pistillate flowers open one to two weeks prior to staminate flowers [11,12]. This gives rise to a highly heterozygous gene pool and complicates breeding such that introgression of desirable traits can take up to 15 years [9,13]. Overcoming poor seed production has been the focus in the development of alternative technologies, including synthetic seeds, permitting rapid multiplication and dissemination of cassava cultivars [14].

Flowering is a highly complex developmental process but the identification of FLOWERING LOCUS T (FT) [15] has prompted manipulation for advanced breeding initiatives. FT is a small globular protein produced in phloem companion cells where it interacts with FT-INTERACTING PROTEIN1 for movement to the sieve elements. Once in the phloem, FT is translocated to the shoot apical meristem where interaction with the bZIP transcription factor FD and phospholipid phosphatidylcholine [16] results in nuclear localization and activation of *LEAFY* (*LFY*), *APETALA1* (*AP1*), and *SUPPRESSOR OF OVEREXPRESSION OF CONSTANS1* (*SOC1*) to trigger flower development [17–19]. With improved understanding of flowering mechanisms, over-expression of *FT* has been exploited to induce precocious flowering in various plant species [20–25] thus enabling a more rapid and refined approach to breeding.

The expeditious advancement of genome editing technology, e.g., CRISPR-Cas9, has elicited a revival of tissue culture studies and rapid breeding strategies to enable segregation of editing tools in crop development [3,26]. Here we demonstrate the capacity to induce early flowering in glasshouse-cultivated cassava via the over-expression of *Arabidopsis FT* and, importantly, to enable sexual reproduction, yielding viable seeds that germinate into healthy progeny.

## 2. Results

A binary expression cassette comprising the coding sequence of *AtFT* constitutively expressed by a *CaMV35S* promoter was assembled. Putative transgenic in vitro plantlets derived from *Agrobacterium*-mediated transformation of the model cultivar 60444 were screened for the presence of the *CaMV35S:AtFT* construct using a rooting assay, PCR amplification of the selection marker *hptII*, the *AtFT* transgene (Figure S1), and Southern blot analysis using a DIG-labelled probe to *hptII* (Figure S2). RT-PCR of selected plantlets revealed transgene expression in lines FT-11, FT-13, and FT-14 (Figure S3). These lines, a control (line FT-7, which contains T-DNA with *hptII* but lacking *AtFT*), and wild-type cultivar 60444 (60444 WT) were propagated in vitro and six or seven plants per line were transferred to soil in the glasshouse. After only six weeks, flower development was observed in two plants of line FT-14 (Figure 1a), although these flowers did not develop to maturity. Approximately 15 weeks under glasshouse cultivation, all six of the FT-13 plants and six of the seven FT-14 plants had floral buds or flowers. Flowering was predominantly at the apical growing region but a branched phenotype with terminal inflorescence and typical arrangement and number of pistillate and staminate flowers was observed (Figure 1b,c). No flowers developed in any of the seven plants of line FT-11 but this observation was not investigated further. As expected, no flowers were observed in any of the FT-7 and 60444 WT plants. Due to the asynchronous flowering of plants of lines FT-13 and FT-14, it was not possible to manually pollinate pistillate flowers.

Due to the asynchronous flowering of cassava, stems made from the mature plants were vegetatively propagated to increase plant number and improve the likelihood of concurrent fertile cyathia production to allow pollination studies. Flower buds were first observed in a plant of FT-14 at 16 weeks after planting (Figure 1d), yielding morphologically normal flowers (Figure 1e). Flowering peaked in weeks 18–20 post propagation when 13 plants from lines FT-13 and FT-14 flowered (Figure 1d). In-keeping with results from the previous experiment, no plants of lines FT-7 and FT-11 produced flowers. Manual pollination (Figure S4) of pistillate flowers of FT-13 was performed successfully during a six-month period. Dehiscence of the capsules released two seeds from each crossing after approximately 11 weeks of maturation (Figure 1f,g), with the exception of the final cross in which three seed were formed. In total, 12 from 13 seeds germinated to yield morphologically normal cassava; the exception was a seed from the three-seed capsule that failed to germinate and was presumed aborted.

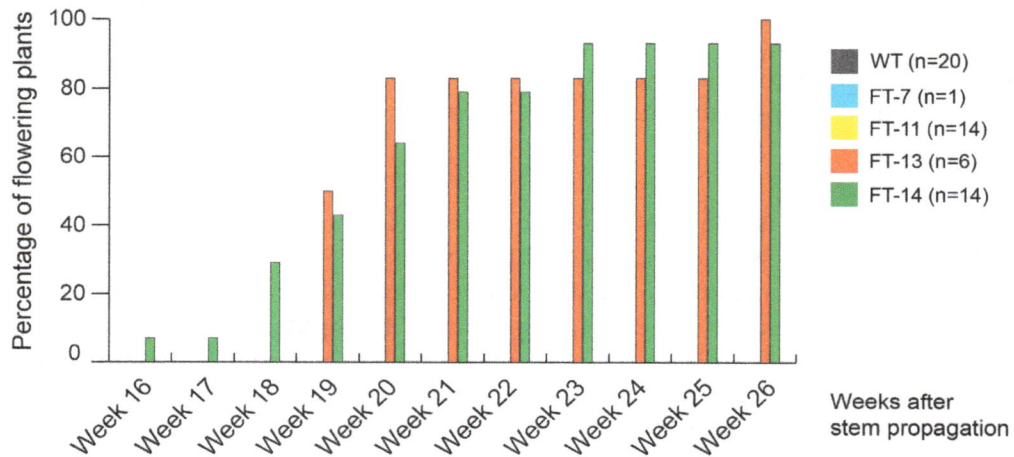

**Figure 1.** Early flowering in glasshouse-cultivated cassava. (**a**) FT-14 plant with flower buds evident after six weeks of growth in soil; (**b**) FT-13 plant at five months of growth; and (**c**) enlarged image showing pistillate and staminate buds on a branch. (**d**) Percentage of flowering plants per line following stem propagation; (**e**) Staminate flower from transgenic line FT-13; (**f**) Seed capsule developing after manual pollination of a pistillate flower with pollen of transgenic line FT-13 (selfing); (**g**) Harvested cassava seed resulting from manual pollination. Glasshouse conditions were 26 °C, 60% humidity and 14 h photoperiod with supplementary lighting.

To investigate whether further vegetative propagation of the *AtFT* expressing transgenic lines affected flowering capacity, multiplication of stems of FT-13 (14 plants), FT-14 (18 plants), and 60444 WT (47 plants) was performed. As observed previously, inflorescence initiation was noted after only six weeks of growth with flowers forming at the apical growing region. After approximately 12 weeks, 64% of FT-13 plants and 33% of FT-14 plants had developing flowers. None of the 47 wild-type plants produced flower bud primordia. This data suggests vegetative propagation of the transgenic material does not alter the capacity of the *AtFT* transgene to induce flowering.

With the successful accelerated flowering in the transgenic lines FT-13 and FT-14, we tested if grafting of a non-flowering 60444 WT scion on the rootstock of an FT line could induce flowering. 60444 WT scions were grafted to FT-13, FT-14, or 60444 WT control rootstocks (3, 7, and 10 plants per line, respectively) and maintained under glasshouse conditions. However, even after 10-months growth, none of the grafted plants flowered. It is not clear why flowering was halted, but it may be associated with an inability of rootstocks to generate sufficient amounts of mobile FT, which is normally loaded into the phloem in leaves. This notion is consistent with the observation that transgenic rootstocks retained a limited (approximately one to three) number of leaves under glasshouse conditions due to natural leaf shedding observed in cassava over a 10-month period. Additionally, rootstocks with a single stem were chosen to improve grafting success.

## 3. Discussion

Our results demonstrate the capacity of stably transformed cassava cultivar 60444 expressing *AtFT* to initiate flower development in glasshouse conditions. Moreover, propagation of plants to maturity allows fertile flowers to be crossed and viable seed to be harvested within one year. To our knowledge, this is the first report of accelerated flowering and viable seed production in cassava using *Arabidopsis FT*. This study complements other reports of induced flowering [20–25,27], further highlighting the importance to continually develop efficient methods to foster flowering in cassava and other important crops. Similar to our grafting experiment, a recent field study also revealed a lack of flowering in non- or late-flowering cultivars grafted to rootstocks of a profuse, early flowering (non-transgenic) cultivar [28]. Whilst progress has been made, it is apparent further studies are required to improve our understanding of floral signals [29] in this important crop.

To date, advances in conventional breeding and biotechnology have been slowed by the poor flowering capacity of many farmer-preferred cultivars and exacerbated by heterogeneity, leading to lengthy breeding programs to introgress desirable traits. Successful induction of flowering in cassava would help maximize breeding programs [30,31], not only in tropical regions but also for crossing of varieties in environments where flowering is seldom seen, namely, glasshouses in temperate countries. Cassava breeding programs have sourced improved traits from wild relatives, including resistance traits against viral pathogens [14,32]. We anticipate that the proof-of-concept *AtFT* expression in cassava will also become instrumental to either facilitate introgression of improved traits or accelerate domestication of wild *Manihot* species. For example, there is tremendous scope to utilize CRISPR-Cas9 for genome editing of domestication genes in parallel to accelerated flowering lines, the progeny of which can be screened for null mutants for the transgene but which have improved traits [33]. The development of crop varieties using these new techniques offers an opportunity to improve agricultural diversity and sustainability.

## 4. Materials and Methods

### 4.1. Expression Vector Assembly and Bacterial Transformation

*Arabidopsis FLOWERING LOCUS T* (*AtFT*; AT1G65480) coding sequence was cloned into the pENTR™/D Gateway™ vector (Invitrogen, Carlsbad, CA, United States) and then into the pMDC32 Cassette 1 (pMDC32-C1) destination vector [34] via an LR reaction (Invitrogen guidelines). *AtFT* expression was controlled by the proximal enhanced *Cauliflower mosaic virus* (*CaMV*) 35S promoter and nopaline synthase (*nos*) transcription terminator sequence. Plant selection utilized the bacterial *hptII* (hygromycin B) resistance gene. The vector was transformed into chemically competent *Escherichia coli* (One Shot® TOP10 competent cells; Invitrogen guidelines) and colonies were grown on LB agar plates containing 50 mg L$^{-1}$ kanamycin antibiotic. Plasmid DNA was purified from a liquid culture (GeneJET, ThermoFisher Scientific, Waltham, MA, United States) and used to electroporate *Agrobacterium tumefaciens* strain LBA4404. Colonies were recovered on LB agar plates containing rifampicin 25 mg L$^{-1}$, streptomycin 100 mg L$^{-1}$ and 50 mg L$^{-1}$ kanamycin. Intact plasmid (*pMDC32-AtFT*) uptake was verified by PCR amplification.

### 4.2. Agrobacterium-*Mediated Cassava Transformation*

Cassava cultivar 60444 (a Nigerian bred line) was transformed as described [35]. In brief, friable, embryogenic callus (FEC) derived from secondary somatic embryos was propagated on Gresshoff and Doy-based [36] medium (Duchefa Biochemie B.V., Haarlem, The Netherlands) supplemented with the synthetic auxin, picloram (12 mg L$^{-1}$). Two independent batches of FEC generated from cassava cultivar 60444 were used for transformation. *Agrobacterium* harboring the *pMDC32-AtFT* plasmid (see above) were cultured in LB until growth reached optimal density. Resuspended bacteria were co-cultivated with the FEC on solid medium prior to selection and regeneration via embryogenesis [35].

### 4.3. Cassava Regeneration and Glasshouse Cultivation

Developing cotyledonary embryos were transferred to Murashige and Skoog (MS)-based [37] medium (Duchefa Biochemie B.V., Haarlem, The Netherlands) supplemented with 6-Benzylaminopurine (a synthetic cytokinin) to promote plant growth and shoot development. Apical growing tips were isolated and planted in MS medium containing 10 mg L$^{-1}$ hygromycin B for selection; plantlets were screened after two weeks and only those that had developed roots were maintained for further analysis. In vitro plantlets were propagated via stem sections and cultured for four weeks (28 °C, 16 h photoperiod) prior to transfer to soil pots maintained in the glasshouse [35].

### 4.4. PCR Amplification to Screen Putative Transgenic Plantlets

Genomic DNA was extracted from in vitro leaf samples frozen in liquid nitrogen using a modified protocol [38] and used as template material in PCR. Reactions contained 1X DreamTaq buffer, 0.2 mM dNTPs, 1 μM SuperFT-F primer (5′-AGACCCTCTTATAGTAAGCAGAG-3′), 1 μM SuperFT-R primer (5′-TACACTGTTTGCCTGCCAAG-3′), sterile, distilled water, 200 ng genomic DNA, 2.5 U DreamTaq DNA Polymerase (ThermoFisher Scientific, Waltham, MA, United States). Reactions were cycled with an initial step 95 °C (1 min), then 35 cycles of 95 °C (1 min), 57 °C (1 min), 72 °C (1 min), then a final step of 72 °C (5 min). The products were resolved in a 1% TAE agarose electrophoresis gel containing ethidium bromide and visualized alongside a 1 Kb molecular marker (GeneRuler™).

### 4.5. Southern Blot Analysis

Genomic DNA was extracted from in vitro leaf samples frozen in liquid nitrogen using a modified protocol [38]. Quality and quantity of DNA samples were determined by NanoDrop™ (ThermoFisher Scientific, Waltham, MA, United States). 10 μg DNA was restriction enzyme digested (*Hind*III; New England Biolabs, Ipswich, MA, United States) for 16 h and subsequently ethanol precipitated and resuspended in 20 μL sterile, nuclease-free water prior to loading on a 1% TAE agarose gel, including a DIG-labeled marker (Roche, Basel, Switzerland). DNA was transferred to nylon membrane via Southern blotting [39] and hybridized with a DIG-labeled probe targeting *hptII*. T-DNA integration events were determined following exposure to autoradiograph film.

*4.6. RT-PCR for AtFT Transgene Expression*

Total RNA was extracted from in vitro leaf samples frozen in liquid nitrogen using a modified protocol [38]. Quality and quantity of RNA samples were determined by NanoDrop™ (ThermoFisher Scientific, Waltham, MA, United States) and subsequently validated in a 2% agarose electrophoresis gel. Samples were treated with DNase according to the manufacturer's guidelines (RQ1 RNase-Free DNase; Promega, Madison, WI, United States) and cDNA prepared using the RevertAid First Strand cDNA Synthesis Kit and random hexamer primer (ThermoFisher Scientific, Waltham, MA, United States). RT-PCR contained 1X DreamTaq buffer, 0.2 mM dNTPs, 1 μM SuperFT-F primer (5′-AGACCCTCTTATAGTAAGCAGAG-3′), 1 μM LH286-R primer (5′-CTAAAGTCTTCTTCCTCCGCA-3′), sterile, distilled water, 1 μL cDNA, 2.5 U DreamTaq DNA Polymerase (ThermoFisher Scientific, Waltham, MA, United States). Amplification of the *PP2A* reference sequence used oligonucleotides PP2A-F (5′-TGTGGAAATATGGCATCAATTTGG-3′) and PP2A-R (5′-GCAACAGAAAGCCGTGTCAC-3′). Reactions were cycled with an initial step 95 °C (1 min), then 35 cycles of 95 °C (30 s), 60 °C (*AtFT* primers) and 58 °C (*PP2A* primers) (30 s), 72 °C (15 s), then a final step of 72 °C (10 min). The products were resolved in a 1% (*AtFT* amplification products) or 2% (*PP2A* amplification products) TAE agarose electrophoresis gel containing ethidium bromide and visualized alongside a molecular marker (GeneRuler™).

*4.7. Flowering, Pollination, and Seed Germination*

In vitro plantlets and stem cuttings of each selected transgenic line and wild-type 60444, were grown in the glasshouse at 26 °C, 60% humidity and 14 h photoperiod with supplementary lighting. Upon flowering, staminate flowers were removed and used to pollinate the pistillate flowers (from a clonal propagant). After 10–12 weeks, seed was harvested and germinated on soil in the glasshouse. Plants were watered daily and fertilized (Wuxal®) fortnightly.

*4.8. Grafting of Cassava Stems*

Cassava scions were cleft grafted under glasshouse conditions following a modified procedure [40] and described previously for cassava [41]. Cassava scions with at least two internodes were grafted onto rootstocks with similar stem diameters. Grafts were joined using Parafilm® and the apical part of each scion was coated with wax.

**Acknowledgments:** The authors thank Irene Zurkirchen (ETH Zürich) for maintaining the glasshouse plants and with cassava pollination experiments. Thank you also to Kim Schlegel and Leen N. Abraham for technical assistance and Pawel Roszak for photographs (Figure 1).

**Author Contributions:** M.K., A.A., S.E.B., and H.V. performed the experiments. S.E.B., C.B., and H.V. analyzed the data and wrote the publication. L.H. advised and provided clones. L.H. and W.G. edited the manuscript. All authors have approved the publication. The project was funded by ETH Zürich. S.E.B. was supported by a PLANT FELLOWS post-doctoral fellowship (FP7-Marie Curie COFUND).

**Conflicts of Interest:** The authors declare no conflict of interest.

## References

1. Van Nocker, S.; Gardiner, S.E. Breeding better cultivars, faster: Applications of new technologies for the rapid deployment of superior horticultural tree crops. *Hortic. Res.* **2014**, *1*, 14022. [CrossRef] [PubMed]
2. Jinek, M.; Chylinski, K.; Fonfara, I.; Hauer, M.; Doudna, J.A.; Charpentier, E. A programmable dual-RNA-guided DNA endonuclease in adaptive bacterial immunity. *Science* **2012**, *337*, 816–821. [CrossRef] [PubMed]

3.   Altpeter, F.; Springer, N.M.; Bartley, L.E.; Blechl, A.E.; Brutnell, T.P.; Citovsky, V.; Conrad, L.J.; Gelvin, S.B.; Jackson, D.P.; Kausch, A.P.; et al. Advancing crop transformation in the era of genome editing. *Plant Cell* **2016**, *28*, 1510–1520. [CrossRef] [PubMed]

4.   Schaeffer, S.M.; Nakata, P.A. CRISPR/Cas9-mediated genome editing and gene replacement in plants: Transitioning from lab to field. *Plant Sci.* **2015**, *240*, 130–142. [CrossRef] [PubMed]

5.   De Oliveira, E.J.; Vilela de Resende, M.D.; da Silva Santos, V.; Fortes Ferreira, C.; Alvarenga Fachardo Oliveira, G.; Suzarte da Silva, M.; Alves de Oliveira, L.; Aguilar-Vildoso, C.I. Genome-wide selection in cassava. *Euphytica* **2012**, *187*, 263–276. [CrossRef]

6.   FAO. *Save and Grow: Cassava. A Guide to Sustainable Production Intensification*; FAO: Rome, Italy, 2013; pp. 1–142.

7.   Balat, M.; Balat, H. Recent trends in global production and utilization of bio-ethanol fuel. *Appl. Energy* **2009**, *86*, 2273–2282. [CrossRef]

8.   Byrne, D. Breeding cassava. *Plant Breed. Rev.* **1984**, *2*, 73–133.

9.   Ceballos, H.; Iglesias, C.A.; Pérez, J.C.; Dixon, A.G. Cassava breeding: Opportunities and challenges. *Plant Mol. Biol.* **2004**, *56*, 503–516. [CrossRef] [PubMed]

10.  Tang, A.F.; Cappadocia, M.; Byrne, D. In vitro flowering in cassava (*Manihot esculenta* Cranz). *Plant Cell Tissue Organ Cult.* **1983**, *2*, 199–206. [CrossRef]

11.  Halsey, M.E.; Olsen, K.M.; Taylor, N.J.; Chavarriaga-Aguirre, P. Reproductive biology of cassava (*Manihot esculenta* Crantz) and isolation of experimental field trials. *Crop Sci.* **2008**, *48*, 49–58. [CrossRef]

12.  Perera, P.I.; Quintero, M.; Dedicova, B.; Kularatne, J.D.; Ceballos, H. Comparative morphology, biology and histology of reproductive development in three lines of *Manihot esculenta* Crantz (Euphorbiaceae: Crotonoideae). *AoB Plant.* **2013**, *5*, pls046. [CrossRef] [PubMed]

13.  Rudi, N.; Norton, G.W.; Alwang, J.; Asumugha, G. Economic impact analysis of marker-assisted breeding for resistance to pests and post-harvest deterioration in cassava. *Afr. J. Agric. Resour. Econ.* **2010**, *4*, 110–122.

14.  Chavarriaga-Aguirre, P.; Brand, A.; Medina, A.; Prías, M.; Escobar, R.; Martinez, J.; Díaz, P.; López, C.; Roca, W.M.; Tohme, J. The potential of using biotechnology to improve cassava: A review. *In Vitro Cell. Dev. Biol. Plant* **2016**, *52*, 461–478. [CrossRef] [PubMed]

15.  Corbesier, L.; Vincent, C.; Jang, S.; Fornara, F.; Fan, Q.; Searle, I.; Giakountis, A.; Farrona, S.; Gissot, L.; Turnbull, C.; et al. FT protein movement contributes to long-distance signaling in floral induction of *Arabidopsis*. *Science* **2007**, *316*, 1030–1033. [CrossRef] [PubMed]

16.  Nakamura, Y.; Andrés, F.; Kanehara, K.; Liu, Y.C.; Dörmann, P.; Coupland, G. Arabidopsis florigen FT binds to diurnally oscillating phospholipids that accelerate flowering. *Nat. Commun.* **2014**, *5*, 3553. [CrossRef] [PubMed]

17.  Abe, M.; Kobayashi, Y.; Yamamoto, S.; Daimon, Y.; Yamaguchi, A.; Ikeda, Y.; Ichinoki, H.; Notaguchi, M.; Goto, K.; Araki, T. FD, a bZIP protein mediating signals from the floral pathway integrator FT at the shoot apex. *Science* **2005**, *309*, 1052–1056. [CrossRef] [PubMed]

18.  Wigge, P.A.; Kim, M.C.; Jaeger, K.E.; Busch, W.; Schmid, M.; Lohmann, J.U.; Weigel, D. Integration of spatial and temporal information during floral induction in *Arabidopsis*. *Science* **2005**, *309*, 1056–1059. [CrossRef] [PubMed]

19.  Andrés, F.; Coupland, G. The genetic basis of flowering responses to seasonal cues. *Nat. Rev. Genet.* **2012**, *13*, 627–639. [CrossRef] [PubMed]

20.  Klocko, A.L.; Ma, C.; Robertson, S.; Esfandiari, E.; Nilsson, O.; Strauss, S.H. FT overexpression induces precocious flowering and normal reproductive development in *Eucalyptus*. *Plant Biotechnol. J.* **2016**, *14*, 808–819. [CrossRef] [PubMed]

21.  McGarry, R.C.; Prewitt, S.; Ayre, B.G. Overexpression of *FT* in cotton affects architecture but not floral organogenesis. *Plant Signal. Behav.* **2013**, *8*, e23602. [CrossRef] [PubMed]

22.  Graham, T.; Scorza, R.; Wheeler, R.; Smith, B.; Dardick, C.; Dixit, A.; Raines, D.; Callahan, A.; Srinivasan, C.; Spencer, L.; et al. Over-expression of *FT1* in plum (*Prunus domestica*) results in phenotypes compatible with spaceflight: A potential new candidate crop for bioregenerative life support systems. *Gravit. Space Res.* **2015**, *3*, 39–50.

23.  Böhlenius, H.; Huang, T.; Charbonnel-Campaa, L.; Brunner, A.M.; Jansson, S.; Strauss, S.H.; Nilsson, O. *CO/FT* regulatory module controls timing of flowering and seasonal growth cessation in trees. *Science* **2006**, *312*, 1040–1043. [CrossRef] [PubMed]

24. Matsuda, N.; Ikeda, K.; Kurosaka, M.; Takashina, T.; Isuzugawa, K.; Endo, T.; Omura, M. Early flowering phenotype in transgenic pears (*Pyrus communis* L.) expressing the *CiFT* gene. *J. Jpn. Soc. Hortic. Sci.* **2009**, *78*, 410–416. [CrossRef]

25. Tränkner, C.; Lehmann, S.; Hoenicka, H.; Hanke, M.V.; Fladung, M.; Lenhardt, D.; Dunemann, F.; Gau, A.; Schlangen, K.; Malnoy, M.; et al. Over-expression of an *FT*-homologous gene of apple induces early flowering in annual and perennial plants. *Planta* **2010**, *232*, 1309–1324. [CrossRef] [PubMed]

26. Khatodia, S.; Bhatotia, K.; Passricha, N.; Khurana, S.M.P.; Tuteja, N. The CRISPR/Cas genome-editing tool: Application in improvement of crops. *Front. Plant Sci.* **2016**, *7*, 506. [CrossRef] [PubMed]

27. McGarry, R.C.; Klocko, A.L.; Pang, M.; Strauss, S.H.; Ayre, B.G. Virus-induced flowering: An application of reproductive biology to benefit plant research and breeding. *Plant Physiol.* **2017**, *173*, 47–55. [CrossRef] [PubMed]

28. Ceballos, H.; Jaramillo, J.J.; Salazar, S.; Pineda, L.M.; Calle, F.; Setter, T. Induction of flowering in cassava through grafting. *J. Plant Breed. Crop Sci.* **2017**, *9*, 19–29.

29. Blümel, M.; Dally, N.; Jung, C. Flowering time regulation in crops—What did we learn from Arabidopsis? *Curr. Opin. Biotechnol.* **2015**, *32*, 121–129. [CrossRef] [PubMed]

30. Ceballos, H.; Kawuki, R.; Gracen, V.; Yencho, G.C.; Hershey, C. Conventional breeding, marker-assisted selection, genomic selection and inbreeding in clonally propagated crops: A case study for cassava. *Theor. Appl. Genet.* **2015**, *128*, 1647–1667. [CrossRef] [PubMed]

31. Wolfe, M.D.; Pino del Carpio, D.; Alabi, O.; Egesi, C.; Ezenwaka, L.C.; Ikeogu, U.N.; Kawuki, R.S.; Kayondo, I.S.; Kulakow, P.; Lozano, R.; et al. Prospects for genomic selection in cassava breeding. *BioRxiv* **2017**. [CrossRef]

32. Rey, M.E.C.; Vanderschuren, H. Cassava mosaic and brown streak diseases: Current perspectives and beyond. *Ann. Rev. Virol.* **2017**, *4*, in press.

33. Østerberg, J.T.; Xiang, W.; Olsen, L.I.; Edenbrandt, A.K.; Vedel, S.E.; Christiansen, A.; Landes, X.; Andersen, M.M.; Pagh, P.; Sandøe, P.; et al. Accelerating the domestication of new crops: Feasibility and approaches. *Trends Plant Sci.* **2017**, *22*, 373–384. [CrossRef] [PubMed]

34. Curtis, M.; Grossniklaus, U. A Gateway™ cloning vector set for high-throughput functional analysis of genes in plants. *Plant Physiol.* **2003**, *133*, 462–469. [CrossRef] [PubMed]

35. Bull, S.E.; Owiti, J.A.; Niklaus, M.; Beeching, J.R.; Gruissem, W.; Vanderschuren, H. *Agrobacterium*-mediated transformation of friable embryogenic calli and regeneration of transgenic cassava. *Nat. Protoc.* **2009**, *4*, 1845–1854. [CrossRef] [PubMed]

36. Gresshoff, P.; Doy, C. Derivation of a haploid cell line from *Vitis vinifera* and the importance of the stage of meiotic development of the anthers for haploid culture of this and other genera. *Z. Pflanzenphysiol.* **1974**, *73*, 132–141. [CrossRef]

37. Murashige, T.; Skoog, F. A revised medium for rapid growth and bioassays with tobacco tissue cultures. *Physiol. Plant.* **1962**, *15*, 473–497. [CrossRef]

38. Soni, R.; Murray, J.A.H. Isolation of intact DNA and RNA from plant tissues. *Anal. Biochem.* **1994**, *218*, 474–476. [CrossRef] [PubMed]

39. Southern, E.M. Detection of specific sequences among DNA fragments separated by gel electrophoresis. *J. Mol. Biol.* **1975**, *98*, 503–517. [CrossRef]

40. Wojtusik, T.; Felker, P. Interspecific graft incompatibility in *Prosopis*. *For. Ecol. Manag.* **1993**, *59*, 329–340. [CrossRef]

41. Moreno, I.; Gruissem, W.; Vanderschuren, H. Reference genes for reliable potyvirus quantitation in cassava and analysis of *Cassava brown streak virus* load in host varieties. *J. Virol. Methods* **2011**, *177*, 49–54. [CrossRef] [PubMed]

# 6

# Paired Hierarchical Organization of 13-Lipoxygenases in *Arabidopsis*

**Adeline Chauvin** [1,2,†], **Aurore Lenglet** [2,†], **Jean-Luc Wolfender** [1] **and Edward E. Farmer** [2,*]

[1] School of Pharmaceutical Sciences, University of Lausanne, University of Geneva, quai Ernest-Ansermet 30, CH-1211 Geneva 4, Switzerland; chauvinunige@gmail.com (A.C.); jean-luc.wolfender@unige.ch (J.-L.W.)

[2] Department of Plant Molecular Biology, University of Lausanne, CH-1015 Lausanne, Switzerland; aurore.lenglet@unil.ch

[*] Correspondence: edward.farmer@unil.ch

[†] These authors contributed equally to this work.

Academic Editor: Eleftherios P. Eleftheriou

**Abstract:** Embryophyte genomes typically encode multiple 13-lipoxygenases (13-LOXs) that initiate the synthesis of wound-inducible mediators called jasmonates. Little is known about how the activities of these different LOX genes are coordinated. We found that the four *13-LOX* genes in *Arabidopsis thaliana* have different basal expression patterns. *LOX2* expression was strong in soft aerial tissues, but was excluded both within and proximal to maturing veins. *LOX3* was expressed most strongly in circumfasicular parenchyma. *LOX4* was expressed in phloem-associated cells, in contrast to *LOX6*, which is expressed in xylem contact cells. To investigate how the activities of these genes are coordinated after wounding, we carried out gene expression analyses in *13-lox* mutants. This revealed a two-tiered, paired hierarchy in which LOX6, and to a lesser extent LOX2, control most of the early-phase of jasmonate response gene expression. Jasmonates precursors produced by these two LOXs in wounded leaves are converted to active jasmonates that regulate *LOX3* and *LOX4* gene expression. Together with LOX2 and LOX6, and working downstream of them, LOX3 and LOX4 contribute to jasmonate synthesis that leads to the expression of the defense gene *VEGETATIVE STORAGE PROTEIN2* (*VSP2*). LOX3 and LOX4 were also found to contribute to defense against the generalist herbivore *Spodoptera littoralis*. Our results reveal that *13-LOX* genes are organised in a regulatory network, and the data herein raise the possibility that other genomes may encode LOXs that act as pairs.

**Keywords:** jasmonic acid; jasmonate; oxylipin; eicosanoid; wounding; defense; herbivore

## 1. Introduction

Lipoxygenases (LOXs) function to produce lipid mediators that operate in a broad range of processes, many of which are related to defense in animals [1] and in plants [2]. The *Arabidopsis thaliana* genome encodes six *LOX*s of which four are 13-LOXs, where "13" refers to the carbon atom in polyunsaturated 18-carbon fatty acids that are preferentially oxygenated by the LOX. 13-LOXs incorporate molecular oxygen into α-linolenic acid to produce its 13(*S*)-hydroperoxide [3], a molecule that is transformed into jasmonates which regulate wound-induced defense gene expression [4,5]. To complete jasmonate synthesis, fatty acid hydroperoxides formed through LOX action are dehydrated and cyclized to form the intermediate 12-oxo-phytodienoic acid (OPDA) in reactions catalysed by allene oxide synthase (AOS) and auxiliary proteins [6,7]. Further transformations of OPDA then result in the production of jasmonic acid (JA) and its biologically active derivatives, chief among which is jasmonyl-isoleucine (JA-Ile) [8].

Jasmonates (and/or their immediate precursors) produced in response to wounding do not stay where they are formed and can be transported efficiently within tissues. Following wounding, jasmonates/jasmonate precursors produced via LOX6 action in the vasculature of *Arabidopsis* leaves are highly mobile and move radially outwards from veins into the mesophyll [9]. In addition to jasmonate mobility, many 13-LOXs are themselves jasmonate-inducible. Therefore, in theory, jasmonates produced via the activity of any 13-LOX could be dispersed to different cell types capable of expressing other LOXs that also make jasmonate precursors. This raises an obvious question: how is the activity of the four *13-LOX* genes in *Arabidopsis* coordinated?

The roles of 13-LOXs in jasmonate-controlled defense responses have been studied in numerous plants, including, but not restricted to potato [10], wild tobacco [11], tomato [12], rice [13], and maize [14], as well as in *Arabidopsis*, a plant in which systematic *LOX* gene mutagenesis has been employed [15]. Intriguingly, while all four 13-LOXs encoded in the *A. thaliana* genome contribute to the synthesis of jasmonic acid [15], they each appear to have somewhat different functions in physically damaged leaves—the subject of the present work. For example, in addition to the initiation of JA synthesis in wounded leaves [16,17], LOX2 also plays a minor role in JA synthesis in undamaged leaves distal to wounds [18]. Furthermore, close to the site of damage, LOX2-derived hydroperoxides are also channelled into the synthesis of arabidopsides, galactolipids that carry one or more esterified OPDA or dinor-OPDA molecules [17,19–21]. Consistent with arabidopsides being defensive secondary metabolites, plants lacking LOX2 were more susceptible to the lepidopteran herbivore *Spodoptera littoralis* than is the wild type [17].

LOX6 also plays a role in leaf defense. The experiments that revealed this role involved the genetic removal of each of the three other 13-LOXs through producing a *lox2 lox3 lox4* triple mutant. In this plant, LOX6 functioning alone was capable of maintaining the defense of emerging leaves and shoot apical tissues in *Arabidopsis* rosettes [15]. Interestingly, the relative impact of LOX6 in early wound-stimulated jasmonate production in leaves increases with the distance from damage sites. That is, LOX6 was necessary for most of the rapid distal expression of the regulatory gene *JASMONATE ZIM-DOMAIN 10 (JAZ10)* when the rosette was wounded [15], making this LOX of particular relevance in studies of long distance wound signalling.

Finally, the LOX3/LOX4 pair contributes approximately 20% of the total JA pool that accumulates in leaves in the first three minutes after wounding [15], however, no roles for these two LOXs in leaves are known. Here, we investigated the relative contributions of *LOX2, LOX3, LOX4,* and *LOX6* to each other's expression, as well as to the expression of a defense gene. We then used herbivory assays to investigate LOX3 and LOX4 function in leaves. Our results revealed a lipoxygenase network that operates to coordinate jasmonate synthesis and defense responses in wounded leaves.

## 2. Results and Discussion

### 2.1. 13-LOX Expression Patterns in Unwounded Rosettes

The expression patterns of *Arabidopsis* LOXs have been examined at the seedling stage [22], but equivalent data for leaves were lacking. Is each *13-LOX* expressed in a different leaf cell type? To characterize basal *13-LOX* gene activity in unwounded leaves, each promoter was fused to a secretable β-glucuronidase (*GUS*) reporter gene. LOX6, principally expressed in xylem contact cells [9,15], served as a comparison with other *13-LOXs*, as shown in Figure 1. GUS staining in younger leaves was stronger than in older leaves for all *13-LOX* promoters, and sections of younger leaves were compared. *LOX2* had the only promoter among the four that was widely active in most tissues except in and near maturing veins. Because of this, transversal sections for visualizing *LOX2* reporter expression were cut nearer the leaf tip than for the other reporters (red bars in Figure 1). LOX2 protein is readily detectable in leaves [23] and *LOX2* expression was strong in mesophyll cells, bundle sheaths, and leaf-tip vasculature—but only at a distance from maturing veins. *LOX3* activity was perivascular and strongest in the xylem region, with weaker expression in the mesophyll. *LOX4* promoter activity was strongest in small cells in the phloem region. These might be companion cells, although their identity was not verified. Phloem is a known region of JA synthesis enzyme localisation [24]. We noted that *LOX2* expression was almost a "mirror image" of the *LOX3*, *LOX4*, and *LOX6*. The expression of these three *LOXs* (unlike for *LOX2*) was strong in the maturing vasculature of expanding leaves.

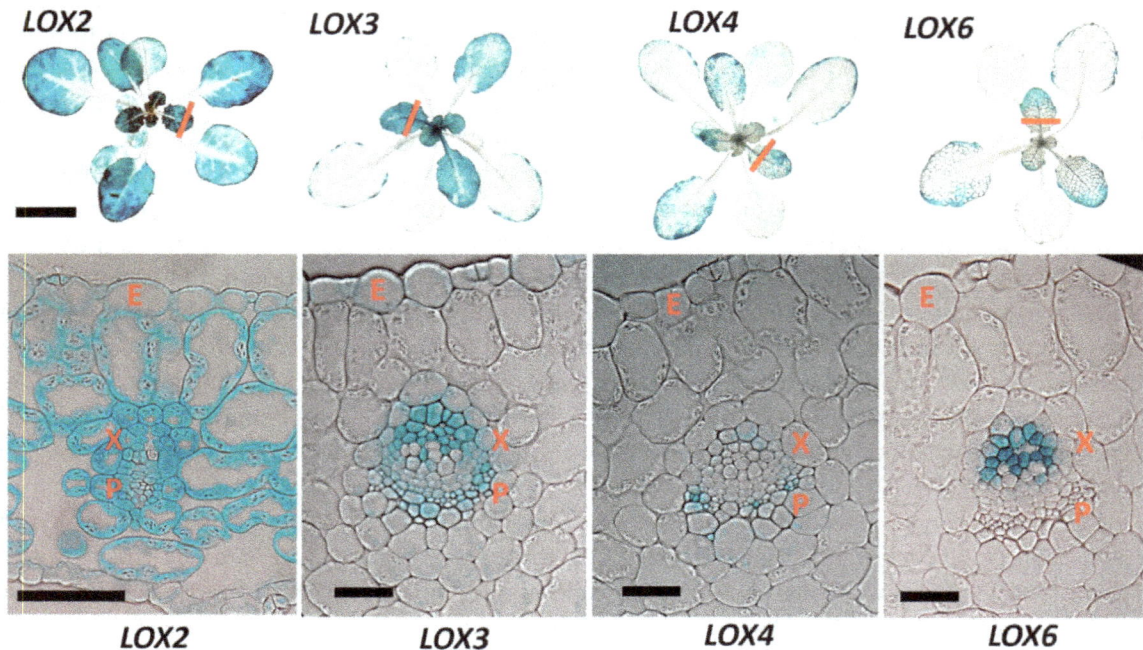

**Figure 1.** 13-lipoxygenases (*13-LOX*) promoter-driven β-glucuronidase (GUS) activity in 3.5 week-old plants. Upper images: rosettes. Scale bar = 1 cm. Red bars indicate approximate section locations shown in lower images. E = epidermal cell; X = xylem region; P = phloem region. Scale bars for sections = 30 μm. Note that the *LOX2* section was cut nearer the leaf tip than the other *LOX* sections.

Notably, all 13-LOXs are expressed in vascular tissues with only basal *LOX2* expression excluded from maturing veins. *LOX3* and *LOX6* expression is strongest near the xylem. *LOX4* is the only *13-LOX* to show basal expression almost exclusively in the phloem region. In terms of cellular space covered by promoter activities, *LOX4* and *LOX6* in unwounded leaves display a relatively small basal promoter activity domain, whereas the other two *13-LOXs* (*LOX2* and *LOX3*) have more extensive basal activity domains (Figure 1). GUS staining in the rosettes of wounded plants is shown in the Appendix (Figure A1).

## 2.2. LOX2 and LOX6 Regulate LOX3 and LOX4 Expression

*LOX6* transcripts are not wound-inducible in roots [25] or leaves [9,15]. *LOX2* expression was investigated in the *lox6* single mutant and the *lox3 lox4* double mutant. *LOX2* remained wound-inducible in each of the genetic backgrounds, however, in the absence of the functional *LOX6* gene, there was weakly reduced *LOX2* expression following wounding (Figure 2). LOX6 activity therefore contributes to *LOX2* gene expression in wounded leaves. The possibility that there are compensatory effects whereby above-WT activity of a particular *13-LOX* gene is stimulated by mutation of one or more of its homologues was tested. Figure A2 shows that basal *LOX3* or *LOX4* expression was not affected in unwounded leaves in the *lox2-* and *lox6*-containing backgrounds.

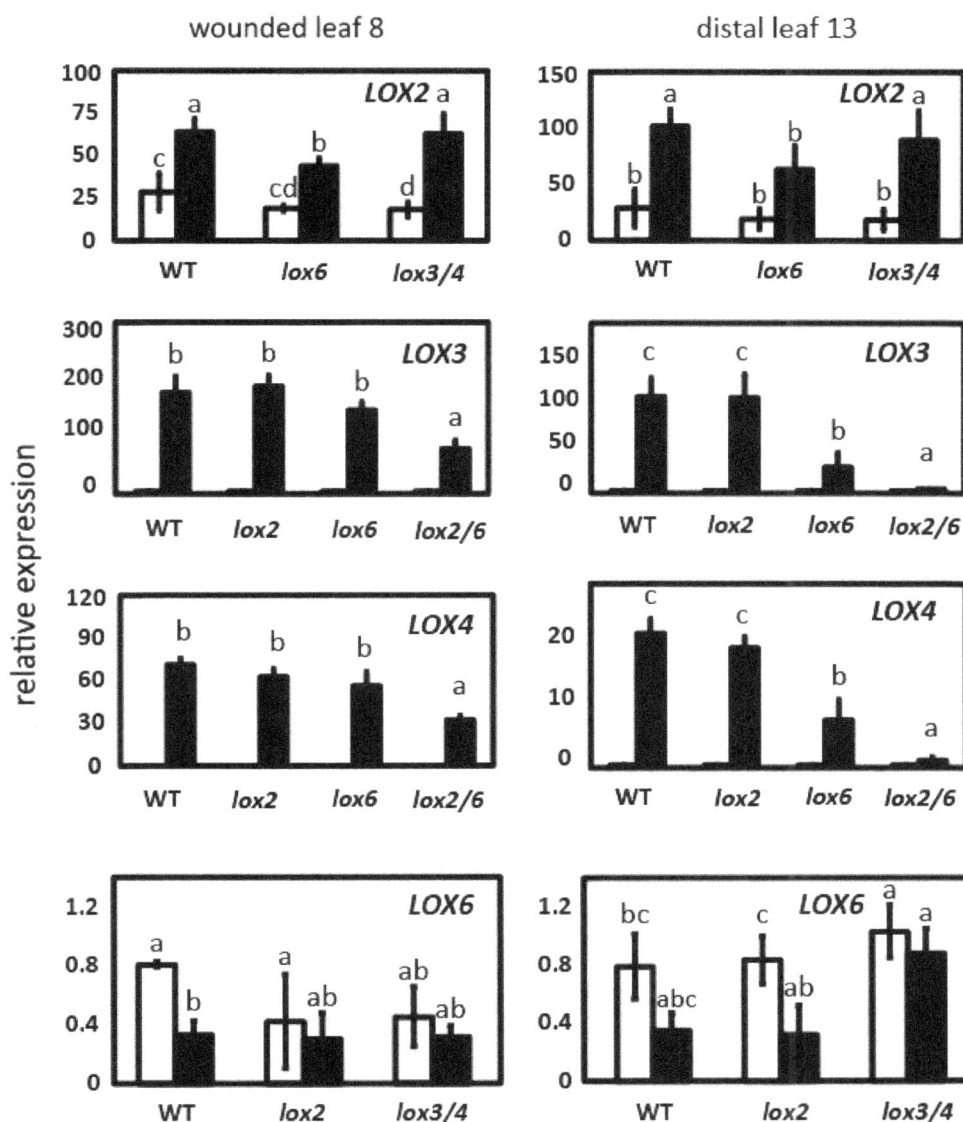

**Figure 2.** *13-LOX* gene expression in different *13-lox* mutant backgrounds. *LOX2* expression analysed in WT, in *lox6A* and in *lox3B lox4A* double mutants. *LOX3* and *LOX4* expression was analyzed in WT, *lox2-1*, and *lox6A* single mutants, and the *lox2-1 lox6A* double mutant. *LOX6* expression was analyzed in *lox2-1* and the *lox3B lox4A* double mutant. Leaves 8 (wounded) and 13 (distal) were snap-frozen before (unfilled bars) and 1 h after the wounding (filled bars). Data are from three (controls) and three to four (wounded) biological replicates (±SD). Letters (a, b, and c) refer to significant differences (ANOVA and *t*-test; *p* < 0.05).

*LOX3* and *LOX4* transcript levels in the WT and both *lox2* and *lox6* single mutants were similar in the wounded leaf, while *LOX6* was found to be required for full, wound-induced *LOX3* and *LOX4* transcript levels in the distal leaf (Figure 2). In the wounded leaf, the double mutant *lox2 lox6* displayed 2-fold lower inductions of *LOX3* and *LOX4* transcripts compared to the WT. In the distal leaf, *LOX3* and *LOX4* gene expression was reduced by 97.2% and 96.3%, respectively, in the *lox2 lox6* double mutant relative to the WT. In the distal leaf, the *lox6* single mutant displayed an approximately 20-fold induction of *LOX3* transcripts, but in *lox2* these transcripts were induced to a far higher level (approximately 100-fold)—that is to a level similar to that in the WT. Similarly, a strong effect on distal *LOX4* transcript accumulation was observed in *lox6* compared to *lox2*. *LOX6* was not wound-inducible in the WT or in *lox* mutant backgrounds (Figure 2). The low transcript levels and associated large error bars for this *LOX* limit the interpretation of this result.

### 2.3. 13-LOXs Contribute Differentially to Inducible VSP2 Defense Gene Expression

The expression of the jasmonate-regulated defense gene (*VEGETATIVE STORAGE PROTEIN2; VSP2*) was then investigated. *VSP2* is expressed at maximal levels several hours after wounding leaves [26] so a 4 h timepoint after wounding was chosen for initial experiments. *VSP2* transcript levels after wounding were reduced relative to the WT in the wounded leaf of the double mutant *lox3 lox4*, while in the distal leaf their levels were reduced in both the *lox3 lox4* double mutants and the *lox6* single mutant (Figure 3). Consistent with a role of all 13-LOXs in controlling *VSP2* expression through jasmonate production, the *lox2 lox6* double mutant, and the *lox2 lox3 lox4* and *lox3 lox4 lox6* triple mutants were unable to reach WT *VSP2* levels in either the wounded leaf or the distal undamaged leaf. Figure 3 shows that *VSP2* expression was significantly reduced in the *lox2 lox6* double mutant, whereas the *lox6* mutation alone had little effect on *VSP2* transcript levels.

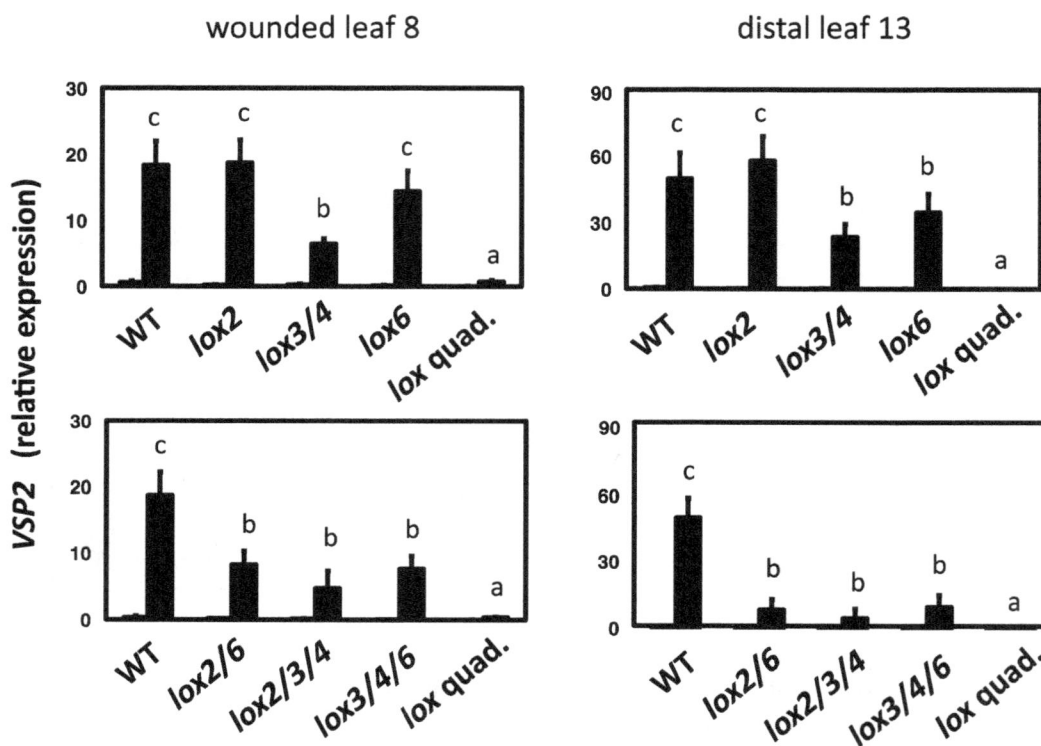

**Figure 3.** Wound-induced *VSP2* expression in *13-lox* mutants. Leaves 8 (wounded) and 13 (distal) were snap-frozen at 4 h after wounding (filled bars) or harvested at the same timepoint from unwounded plants (unfilled bars). Data from three (unwounded controls) to four (wounded samples) biological replicates (±SD). Letters (a, b, and c) refer to significant differences (ANOVA and t-test; $p < 0.05$). *VSP2* transcript levels in unwounded plants is low and statistics are shown for wounded treatments only. WT = wild type; *lox2* = *lox2-1*; *lox3/4* = the *lox3B lox4A* double mutant; *lox6* = *lox6A*; *lox2/6* = the *lox2-1 lox6A* double mutant; *lox2/3/4* = *lox2-1 lox3B lox4A* triple mutant; *lox3/4/6* = *lox3B lox4A lox6A* triple mutant; *lox* quad. = *lox2-1 lox3B lox4A lox6A* quadruple mutant.

### 2.4. Comparison of LOX3, LOX4 and VSP2 Expression in the WT and the lox2 lox6 Double Mutant

To gain insights into the temporal control of gene expression we compared the wound induction of *LOX3*, *LOX4*, and *VSP2* transcripts in the WT and the *lox2 lox6* double mutant. As shown in Figure 4, *LOX3* and *LOX4* transcript levels were upregulated rapidly (within 1 h) in the wounded leaf and in the distal leaf of the WT. However, *LOX3* and *LOX4* transcript accumulation in response to wounding was reduced in the *lox2 lox6* double mutant. The possibility that the *lox2 lox6* mutations might cause increased expression of *LOX3* and *LOX4* is, given the results in Figure 2, considered unlikely. Over the experimental period, *LOX3* and *LOX4* transcript levels in the distal leaf of *lox2 lox6* never reached WT levels. This is consistent with the LOX2/LOX6 pair acting upstream of the LOX3/LOX4 pair and contributing to their expression. Using an identical timeframe we observed that *VSP2* transcripts accumulated with different kinetics in the WT and the *lox2 lox6* double mutant. The wound-induced expression of *VSP2* was reduced in the *lox2 lox6* double mutant (*i.e.*, LOX2 and LOX6 are essential for full *VSP2* gene expression). This led us to test the roles of the *LOX3/LOX4* gene pair in herbivory assays.

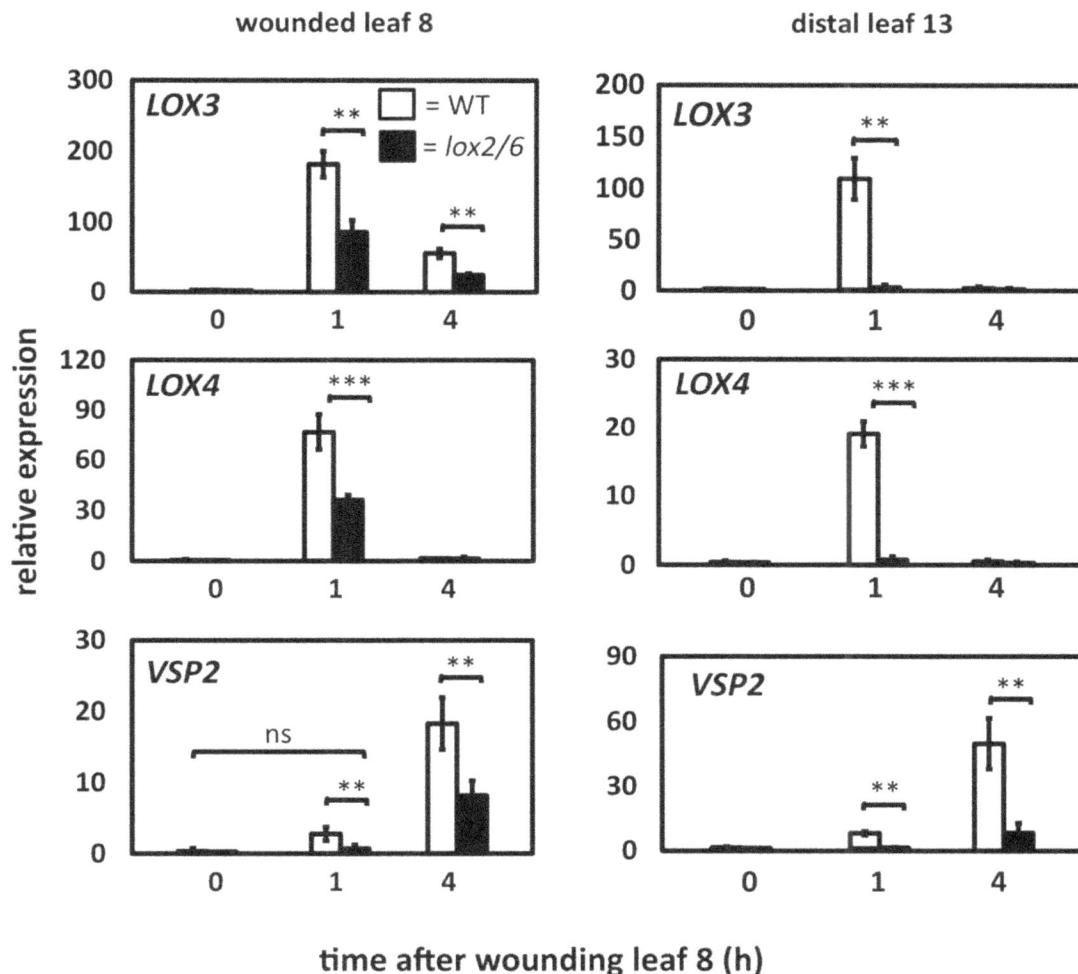

**Figure 4.** Wound-inducible *LOX3*, *LOX4*, and *VSP2* expression in WT (open bars) and the *lox2-1 lox6A* double mutant (black bars). Leaves 8 (wounded) and 13 (distal) were snap-frozen before and after wounding. Data are from three to four biological replicates (±SD). Data from a 2h timepoint also included in the original experiment are not shown. Asterisks refer to data significantly different from WT (ns: not significant; *, $p < 0.05$; **, $p < 0.01$, and ***, $p < 0.001$; *t*-test).

## 2.5. All 13-LOXs Act in Defense against a Lepidopteran Herbivore

The *lox2 lox3 lox4 lox6* quadruple mutant is known to display greatly reduced resistance to larvae of *Spodoptera littoralis* relative to the WT [15]. This genotype was used as a control to investigate the roles of *LOX2*, *LOX3*, and *LOX4* in defense. In all plants—except the quadruple mutant, where the center of the rosette was attacked—damage inflicted by *S. littoralis* was restricted to expanded leaves (Figure 5a).

Larvae also grew fast on the *lox2-1* single mutant (Figure 5b) as described in Glauser *et al.* [17]. To further investigate LOX2 function we used the LOX2 "mirror" mutant (*i.e.*, the triple mutant *lox3 lox4 lox6* that retains functional LOX2 as its only 13-LOX). As judged by measuring insect weight gain, *LOX2* on its own in the triple mutant mediated near-WT-levels of defense (Figure 5a,b). Since redundancy between *LOX3* and *LOX4* has been observed in early responses to leaf wounding [15], larval growth was examined on a *lox3 lox4* double mutant and on the "mirror" mutant *lox2 lox6* that retains functional LOX3 and LOX4 as its only 13-LOXs. Caterpillar growth was found to be similar to that on the WT for both the *lox3 lox4* double and *lox3 lox4 lox6* triple mutants. However, the insects gained 1.7 times more weight than they did on the WT when feeding on the *lox2 lox6* double mutant, but this weight gain was less than that seen on the quadruple mutant (3.1 times more weight gain than on the WT). That is, the presence of functional *LOX3* and *LOX4* in the *lox2 lox6* double mutant actively reduced insect weight gain.

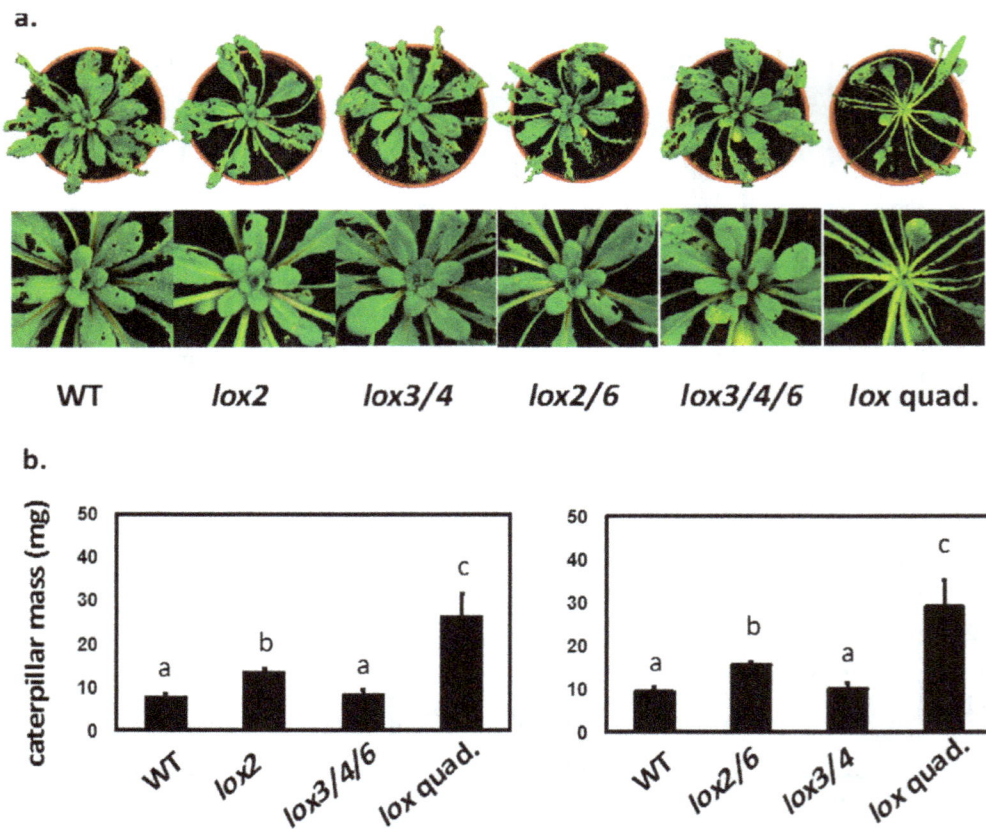

**Figure 5.** Contributions of 13-LOXs to defense against a chewing herbivore. (a) Damage to rosettes inflicted by *Spodoptera littoralis* larvae. WT = wild type; *lox3/4* = the *lox3B lox4A* double mutant; *lox2/6* = the *lox2-1 lox6A* double mutant; lox3/4/6 = *lox3B lox4A lox6A* triple mutant; lox quad. = *lox2-1 lox3B lox4A lox6A* quadruple mutant; (b) Larval mass after feeding. Insects were harvested at 11d. Letters (a, b, and c) refer to significant differences (ANOVA and t-test; $p < 0.05$).

In summary, while *LOX2* and *LOX6* are known to make different contributions to leaf defense [9,15], the new results show that *LOX3* and *LOX4* also help protect the rosette.

## 2.6. Functional Pairs of 13-LOXs in Leaf Defense

An outcome of the present work was the finding that 13-LOX activities appear to function in an organised regulatory network. Specifically, we provide evidence consistent with LOX2 and LOX6 functioning upstream of LOX3 and LOX4. This is shown in Figure 6.

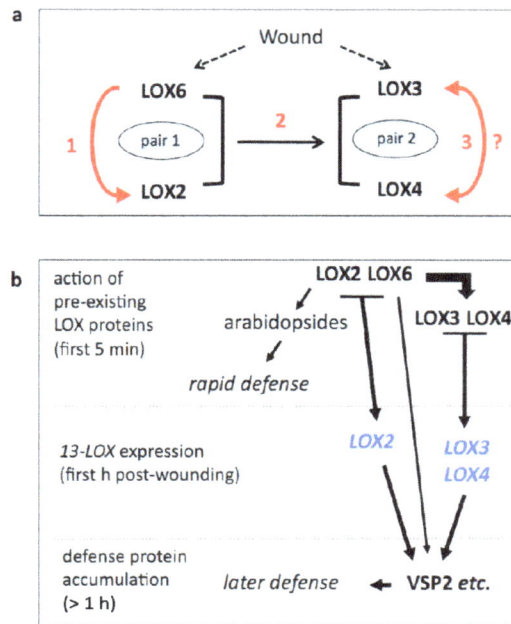

**Figure 6.** Figure 6. 13-LOX activities in wounded *Arabidopsis* leaves. (a) 13-LOXs interact through jasmonate production. 1. LOX6 participates in the production of jasmonate that upregulates *LOX2* expression (from Figure 2). 2. LOX6 and LOX2 together produce jasmonates that enhance *LOX3* and *LOX4* expression (from Figure 2). 3. Still hypothetical, LOX3 and LOX4 may enhance each other's expression through jasmonate production; (b) 13-LOX roles in the wound response. Preformed LOX2 rapidly produces arabidopsides, secondary compounds implicated in direct defense [17]. LOX6 and LOX2 produce jasmonates necessary for early gene expression (e.g., *JAZ10*; 9,15 and *LOX3/4* expression; Figures 2 and 4). LOX3 and LOX4 produce jasmonates that ensure correct late-phase *VSP2* gene expression (from Figures 3 and 4). Proteins, black; transcripts, blue.

While we suggest that LOX2/LOX6 and LOX3/LOX4 act in pairs in *Arabidopsis*, it is important to note that each enzyme has some distinct roles. LOX2 activity in the Columbia (Col) accession is dedicated in large part to the rapid production of arabidopsides in and near wounds [17]. LOX2 is closely related to a *Nicotiana attenuata* LOX that produces substrates for 2(*E*)-hexenal synthesis via HPL activity [27]. However, the *HPL* (*HYDROPEROXIDE LYASE*) gene, which uses 13-LOX products for the synthesis of volatile aldehydes, is mutated in Col accessions—and Col plants produce little or no 2(*E*)-hexenal upon wounding [28]—so it is possible that LOX2 would also produce this volatile in accessions with a functional *HPL* gene. Moreover, *LOX2* expression seems to be optimally placed for the defense of soft tissues: it would be difficult for a chewing herbivore to avoid damaging cells in its broad basal expression domain. As shown both previously [9] and herein, LOX2 also contributes to the production of biologically active jasmonates in leaves distal to wounds, and these jasmonates stimulate *LOX3*, *LOX4*, and *VSP2* expression.

LOX6 produces the precursors of biologically active jasmonates, and it is possible that this enzyme builds pre-formed OPDA pools that are mobilized on wounding [25]. The relative level of jasmonate originating from LOX6 activity increases at a distance from a wound [15], and LOX6 plays a more powerful role than LOX2 in this respect. LOX6, like LOX2, contributes to *VSP2* gene expression. Finally, *lox6* mutants are more drought-sensitive than the WT, hinting at other roles for this LOX [25]. LOX3 and LOX4 proteins also contribute to jasmonate synthesis that then leads to increased *LOX3*, *LOX4*, and *VSP2* expression. Alone (*i.e.*, in the absence of LOX6 and LOX2) the expression of the *LOX3/LOX4* pair is barely activated in leaves distal to wounds (note that it is possible that long-distance wound signals activate pre-formed LOX3 and LOX4 enzymes). However, *LOX3* transcript levels increase after mechano-stimulation [29], and LOX3 protein is jasmonate-inducible in the vicinity of wounds [30].

Under the conditions used herein, LOX3 and LOX4 do not strongly activate *LOX2* and *LOX6* gene expression (Figure 2), therefore arrow 2 in Figure 6a is unidirectional. Our studies have not determined the relative contributions of LOX3 and LOX4 to leaf defense, however, differential activities for these LOXs in leaves are possible [31]. Work remains to further refine the models shown in Figure 6. Additionally, the LOX network we propose is likely to extend beyond 13-LOXs and there is already evidence for interaction between 13- and 9-LOX pathways in rice (e.g., [32]). In *Arabidopsis*, transcripts of *LOX1*, a 9-lipoxygenase, are strongly jasmonate-inducible [33], raising the possibility that the 13-LOX network encompasses 9-LOXs.

Lastly, each of the four *Arabidopsis* 13-LOXs should provide unique insights into signalling mechanisms that operate to initiate jasmonate synthesis. Recently, it was proposed that stress-responsive signal pathways involving coupled glutamate receptor-like/lipoxygenase (GLR/ LOX) modules may exist [34]. If so, each GLR-LOX pathway might have unique characteristics.

## 3. Experimental Section

### 3.1. Plant Material and Growth Conditions

*A. thaliana* (L.) Heynh. T-DNA insertion mutants in the Columbia background were obtained from the European Arabidopsis Stock Center (NASC): *LOX3* (At1g17420; *lox3B* = SALK_147830), *LOX4* (At1g72520; *lox4A* = SALK_071732), and *LOX6* (At1g67560; *lox6A* = SALK_138907). The *lox2-1* (At3g45140) mutant, double, triple, and quadruple mutants all based on the *lox2-1, lox3B, lox4A*, and *lox6A* alleles have been described previously [15,17,35]. The *aos* mutant was from [36]. Plants were grown individually in 7 cm diameter pots at 21 °C, 10 h light $d^{-1}$ (100 $\mu mol \cdot s^{-1} \cdot m^{-2}$) photoperiod, and 70% humidity. Leaves of five-week-old plants were numbered from oldest to youngest. All wounding experiments (crushing 50% of leaf 8 with metal forceps) were performed between 12 a.m. and 16.30 p.m. Leaves were snap-frozen in liquid $N_2$ and stored at $-80$ °C before extractions for RT-qPCR.

### 3.2. Insect Feeding Assays

Eleven pots, each with a 4.5 week-old plant, were isolated in eleven plexiglass boxes. Newly hatched *Spodoptera littoralis* (Boisduval; Noctuidae; Lepidoptera) caterpillars were placed on each plant (four larvae per pot). Larvae were harvested after 11 days. The weight of larvae from one box per number of recovered larvae was considered as one biological replicate [15].

### 3.3. Gene Expression

RT-qPCR was performed as described in [15] with a PCR program detailed in [30]. Data were standardized to the *UBC21* ubiquitin conjugase reference gene. The following primers were used: *UBC21* (At5g25760): 5'-CAGTCTGTGTGTAGAGCTATCATAGCAT-3', 5'-AGAAGATTCCCTGAG TCGCAGTT-3'. *VSP2* (At5g24770): 5'-CATCATAGAGCTCGGGATTGAACCC-3', 5'-AGATGCTTC CAGTAGGTCACGC-3'. *LOX2*: 5'-ATTACGGTAGAAGACTACGCACAAC-3', 5'-GTAATTTAA GCTCTACCCCCTTGAG-3'. *LOX3*: 5'-AACACAACCACATGGTCTTAAACTC-3', 5'-GGAGCT CAGAGTCTGTTTTGATAAG-3'. *LOX4*: 5'-ATAGAACGAGTCACGTCTTATCCAC-3', 5'-CATAAA CAAACGGTTCGTCTCTAAC-3'.

### 3.4. GUS Staining and Light Microscopy

Promoter/GUS fusion plants for *LOX2* (At3g45140), *LOX3* (At1g17420), and *LOX4* (At1g72520) were from ([9]; see supplemental material in that paper for LOX3pro:GUS and LOX4pro:GUS). The LOX6pro:GUS fusion plants were from [15]. 26 individual $T_1$ plants from each construct were stained and at least six selected to produce $T_2$ independent lines from which at least six for each independent line were stained for in-depth analysis. For staining and embedding [37], 3.5 week-old rosettes were prefixed in acetone:water (90:10, v:v) for 45 min on ice, washed twice in 50 mM sodium phosphate buffer (pH 7.2), and vacuum-infiltrated for 10 min (for rosette staining) or 2 h (for leaf sectioning), then left 16 h at 37 °C in the dark in 10 mM $Na_2$EDTA; 50 mM sodium phosphate buffer, pH 7.2; 2 mM (for rosettes) or 3 mM (for thin sections) $K_4Fe(CN)_6$; 2 mM (for rosettes) or 3 mM (for thin sections) $K_3Fe(CN)_6$; 0.1% (v/v) Triton X-100; 0.6 mg·$mL^{-1}$ 5-bromo-4-chloro-3-indolyl-β-D-glucuronide (X-Gluc). Rosettes were then either transferred to ethanol: water (70:30, v/v) for photography or fixed (glutaraldehyde:formaldehyde:50 mM sodium phosphate (pH 7.2) buffer = 2:5:43, v:v:v). For embedding, stained leaves were dehydrated in an ethanol gradient (10%, 30%, 50%, 70%, 90%, and twice absolute) for 30 min. Embedding in Technovit 7100 resin (Haslab GmbH, Ostermundigen, Switzerland) was carried out according to the manufacturer's instructions. Sections (5 μm) were made with a RM2255 microtome (Leica, Muttenz, Switzerland) using disposable Leica TC-65 blades.

## 4. Conclusions

Our results highlight the fact that three of the four *13-LOXs* in *Arabidopsis* are expressed primarily in the vasculature. The exception, *LOX2*, is expressed in soft tissues; this *LOX* has only low basal expression in mature veins. Although they have different roles in the wound response, all four *13-LOXs* operate together in a regulatory network. This network can be seen as being comprised of two pairs of *LOXs*; the first pair (LOX6 and LOX2) produces precursors of jasmonates in response to wounding. This helps to control the expression of the second *LOX* pair: *LOX3* and *LOX4*, and the enzymes encoded by these genes also contribute to jasmonate synthesis. We raise the possibility that other plant genomes may encode pairs of *LOXs*, and that these may also act hierarchically to control each other's expression.

**Acknowledgments:** We thank A.-S. Fiorucci and I. Acosta for valuable comments, N. Geldner and E. Dohmann for plasmids, C. Bonnet and P. Reymond for insect eggs, and S. Stolz for excellent technical help. Supported by Swiss National Science Foundation grants 31003A-138235 (to E.E.F.) and 200020-146200 (to J.-L.W. and E.E.F.).

**Author Contributions:** A.C. and A.L. performed experiments; A.C., A.L., J.L.W. and E.E.F. designed research; E.E.F. wrote the paper.

**Conflicts of Interest:** The authors declare no conflict of interest.

## Appendix

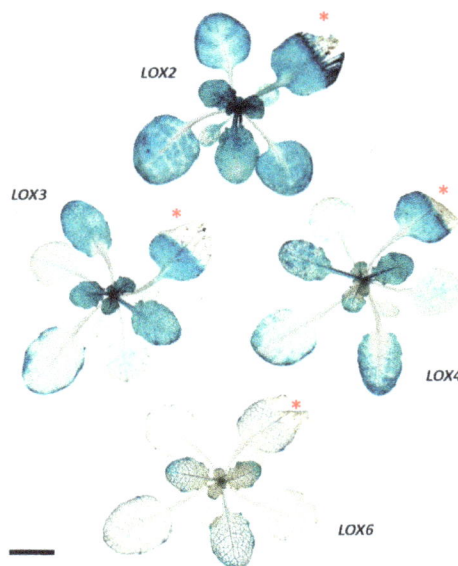

**Figure A1.** Expression of *LOX* promoter::GUS reporters in wounded rosettes. Plants were wounded on a single leaf (orange asterisk) and harvested 6 h later prior to staining for GUS activity. Scale bar = 1 cm.

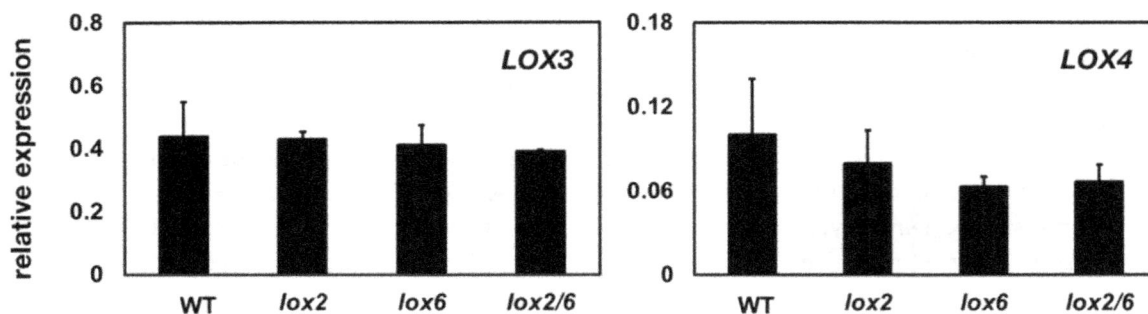

**Figure A2.** Basal *LOX3* and *LOX4* expression is not affected in the *lox2* and *lox6* backgrounds. Samples were from leaf 8 of five-week-old plants. *LOX3* and *LOX4* transcript levels are relative to those of *UBC21*.

## References

1.   Haeggstrom, J.Z.; Funk, C.D. Lipoxygenase and leukotriene pathways: Biochemistry, biology, and roles in disease. *Chem. Rev.* **2011**, *111*, 5866–5898. [CrossRef] [PubMed]

2.   Andreou, A.; Feussner, I. Lipoxygenases—Structure and reaction mechanism. *Phytochemistry* **2009**, *70*, 1504–1510. [CrossRef] [PubMed]

3.   Bannenberg, G.; Martinez, M.; Hamberg, M.; Castresana, C. Diversity of the enzymatic activity in the lipoxygenase gene family of *Arabidopsis thaliana*. *Lipids* **2009**, *44*, 85–95. [CrossRef] [PubMed]

4.   Browse, J. Jasmonate passes muster: A receptor and targets for the defense hormone. *Ann. Rev. Plant Biol.* **2009**, *60*, 183–205. [CrossRef] [PubMed]

5.   Wasternack, C.; Hause, B. Jasmonates: Biosynthesis, perception, signal transduction and action in plant stress response, growth and development. An update to the 2007 review in Annals of Botany. *Ann. Bot.* **2013**, *111*, 1021–1058. [CrossRef] [PubMed]

6.   Dong-Sun, L.; Nioche, P.; Hanberg, M.; Raman, C.S. Structural insights into the evolutionary paths of oxylipin biosynthetic enzymes. *Nature* **2008**, *455*, 363–368.

7.   Schaller, A.; Stintzi, A. Enzymes in jasmonate biosynthesis—Structure, function, regulation. *Phytochemistry* **2009**, *70*, 1532–1538. [CrossRef] [PubMed]

8.   Fonseca, S.; Chini, A.; Hamberg, M.; Adie, B.; Porzel, A.; Kramell, R.; Miersch, O.; Wasternack, C.; Solano, R. (+)-7-iso-Jasmonoyl-L-isoleucine is the endogenous bioactive jasmonate. *Nat. Chem. Biol.* **2009**, *5*, 344–350. [CrossRef] [PubMed]

9.   Gasperini, D.; Chauvin, A.; Acosta, I.F.; Kurenda, A.; Wolfender, J.-L.; Stolz, S.; Chételat, A.; Farmer, E.E. Axial and radial oxylipin transport. *Plant Physiol.* **2015**. [CrossRef] [PubMed]

10.  Royo, J.; Leon, J.; Vancanneyt, G.; Albar, J.P.; Rosahl, S.; Ortego, F.; Castanera, P.; Sanchez-Serrano, J.J. Antisense-mediated depletion of a potato lipoxygenase reduces wound induction of proteinase inhibitors and increases weight gain of insect pests. *Proc. Natl. Acad. Sci. USA* **1999**, *96*, 1146–1151. [CrossRef] [PubMed]

11.  Halitschke, R.; Baldwin, I.T. Antisense *LOX* expression increases herbivore performance by decreasing defense responses and inhibiting growth-related transcriptional reorganization in *Nicotiana attenuata*. *Plant J.* **2003**, *36*, 794–807. [CrossRef] [PubMed]

12.  Yan, L.; Zhai, Q.; Wei, J.; Li, S.; Wang, B.; Huang, T.; Du, M.; Sun, J.; Kang, L.; Li, C.-B.; *et al*. Role of tomato lipoxygenase D in wound-induced jasmonate biosynthesis and plant immunity to insect herbivores. *PLoS Genet.* **2013**, e1003964. [CrossRef] [PubMed]

13.  Zhou, G.; Qi, J.; Ren, N.; Cheng, J.; Erb, M.; Mao, B.; Lou, Y. Silencing *OsHI-LOX* makes rice more susceptible to chewing herbivores, but enhances resistance to a phloem feeder. *Plant J.* **2009**, *60*, 638–648. [CrossRef] [PubMed]

14. Christensen, S.A.; Nemchenko, A.; Borrego, E.; Murray, I.; Sobhy, I.S.; Bosak, L.; DeBlasio, S.; Erb, M.; Robert, C.A.M.; Vaughn, K.A.; *et al.* The maize lipoxygenase, *ZmLOX10*, mediates green leaf volatile, jasmonate and herbivore-induced plant volatile production for defense against insect attack. *Plant J.* **2013**, *74*, 59–73. [CrossRef] [PubMed]

15. Chauvin, A.; Caldelari, D.; Wolfender, J.-L.; Farmer, E.E. Four 13-lipoxygenases contribute to rapid jasmonate synthesis in wounded *Arabidopsis thaliana* leaves: A role for lipoxygenase 6 in responses to long-distance wound signals. *New Phytol.* **2013**, *197*, 566–575. [CrossRef] [PubMed]

16. Bell, E.; Creelman, R.A.; Mullet, J.E. A chloroplast lipoxygenase is required for wound-induced jasmonic acid accumulation in *Arabidopsis*. *Proc. Natl. Acad. Sci. USA* **1995**, *92*, 8675–8679. [CrossRef] [PubMed]

17. Glauser, G.; Dubugnon, L.; Mousavi, S.A.R.; Rudaz, S.; Wolfender, J.-L.; Farmer, E.E. Velocity estimates for signal propagation leading to systemic jasmonic acid accumulation in wounded *Arabidopsis*. *J. Biol. Chem.* **2009**, *284*, 34506–34513. [CrossRef] [PubMed]

18. Glauser, G.; Grata, E.; Dubugnon, L.; Rudaz, S.; Farmer, E.E.; Wolfender, J.-L. Spatial and temporal dynamics of jasmonate synthesis and accumulation in *Arabidopsis* in response to wounding. *J. Biol. Chem.* **2008**, *283*, 16400–16407. [CrossRef] [PubMed]

19. Seltmann, M.A.; Stingl, N.E.; Lautenschlaeger, J.K.; Krischke, M.; Mueller, M.J.; Berger, S. Differential impact of lipoxygenase 2 and jasmonates on natural and stress-induced senescence in *Arabidopsis*. *Plant Physiol.* **2010**, *152*, 1940–1950. [CrossRef] [PubMed]

20. Zoeller, M.; Stingl, N.; Krischke, M.; Fekete, A.; Walker, F.; Berger, S.; Mueller, M.J. Lipid profiling of the *Arabidopsis* hypersensitive response reveals specific lipid peroxidation and fragmentation processes: Biogenesis of pimelic and azelaic acid. *Plant Physiol.* **2012**, *160*, 365–378. [CrossRef] [PubMed]

21. Nilsson, A.K.; Fahlberg, P.; Ellerstrom, M.; Andersson, M.X. Oxo-phytodienoic acid (OPDA) is formed on fatty acids esterified to galactolipids after tissue disruption in *Arabidopsis thaliana*. *FEBS Lett.* **2012**, *586*, 2483–2487. [CrossRef] [PubMed]

22. Vellosillo, T.; Martinez, M.; Lopez, M.A.; Vicente, J.; Cascon, T.; Dolan, L.; Hamberg, M.; Castresana, C. Oxylipins produced by the 9-lipoxygenase pathway in *Arabidopsis* regulate lateral root development and defense responses through a specific signaling cascade. *Plant Cell.* **2007**, *19*, 831–846. [CrossRef] [PubMed]

23. Rodríguez, V.M.; Chételat, A.; Majcherczyk, P.; Farmer, E.E. Chloroplastic phosphoadenosine phosphosulfate (PAPS) metabolism regulates basal levels of the prohormone jasmonic acid in Arabidopsis leaves. *Plant Physiol.* **2010**, *152*, 1335–1345. [CrossRef] [PubMed]

24. Hause, B.; Hause, G.; Kutter, C.; Miersch, O.; Wasternack, C. Enzymes of jasmonate biosynthesis occur in tomato sieve elements. *Plant Cell Physiol.* **2003**, *44*, 643–648. [CrossRef] [PubMed]

25. Grebner, W.; Stingl, N.E.; Oenel, A.; Mueller, M.J.; Berger, S. Lipoxygenase6-dependent oxylipin synthesis in roots is required for abiotic and biotic stress resistance of *Arabidopsis*. *Plant Physiol.* **2013**, *161*, 2159–2170. [CrossRef] [PubMed]

26. Berger, S.; Mitchell-Olds, T.; Stotz, H.U. Local and differential control of vegetative storage protein expression in response to herbivore damage in *Arabidopsis thaliana*. *Physiol. Plant* **2002**, *114*, 85–91. [CrossRef] [PubMed]

27. Allmann, S.; Halitschke, R.; Schuurink, R.C.; Baldwin, I.T. Oxylipin channelling in *Nicotiana attenuata*: Lipoxygenase 2 supplies substrates for green leaf volatile production. *Plant Cell Environ.* **2010**, *33*, 2028–2040. [CrossRef] [PubMed]

28. Duan, H.; Huang, M.-Y.; Palacio, K.; Schuler, M.A. Variations in *CYP74B2* (*Hydroperoxide lyase*) gene expression differentially affect hexenal signaling in the Columbia and Landsberg *erecta* ecotypes of Arabidopsis. *Plant Physiol.* **2005**, *139*, 1529–1544. [CrossRef] [PubMed]

29. Chehab, E.W.; Yoa, C.; Henderson, Z.; Kim, S.; Braam, J. *Arabidopsis* touch-induced morphogenesis is jasmonate-mediated and protects against pests. *Curr. Biol.* **2012**, *22*, 701–706. [CrossRef] [PubMed]

30. Gfeller, A.; Bearenfaller, K.; Loscos, J.; Chetelat, A.; Baginsky, S.; Farmer, E.E. Jasmonate controls polypeptide patterning in undamaged tissue in wounded *Arabidopsis* leaves. *Plant Physiol.* **2011**, *156*, 1797–1807. [CrossRef] [PubMed]

31. Ozalvo, R.; Cabrere, J.; Escobar, C.; Christensen, S.A.; Borrega, E.J.; Kolomiets, M.V.; Castresana, C.; Iberkleid, I.; Horowitz, S.B. Two closely related members of *Arabidopsis* 13-lipoxygenases (13-LOXs), LOX3 and LOX4, reveal distinct functions in response to plant-parasitic nematode infection. *Mol. Plant Pathol.* **2014**, *15*, 319–332. [CrossRef] [PubMed]

32. Zhou, G.; Ren, N.; Qi, J.; Lu, J.; Xiang, C.; Ju, H.; Cheng, J.; Lou, Y. The 9-lipoxygenase Osr9-LOX1 interacts with the 13-lipoxygenase-mediated pathway to regulate resistance to chewing and piercing-sucking herbivores in rice. *Physiol. Plant* **2014**, *152*, 59–69. [CrossRef] [PubMed]

33. Melan, M.A.; Dong, X.; Endara, M.E.; Davis, K.R.; Ausubel, F.; Peterman, T.K. An *Arabidopsis thaliana* lipoxygenase gene can be induced by pathogens, abscisic acid, and methyl jasmonate. *Plant Physiol.* **1993**, *1001*, 441–450. [CrossRef]

34. Farmer, E.E.; Gasperini, D.; Acosta, I. The squeeze cell hypothesis for the activation of jasmonate synthesis in response to wounding. *New Phytol.* **2014**, *204*, 282–288. [CrossRef] [PubMed]

35. Caldelari, D.; Wang, G.; Farmer, E.E.; Dong, X. Arabidopsis *lox3 lox4* double mutants are male sterile and defective in global proliferative arrest. *Plant Mol. Biol.* **2011**, *75*, 25–33. [CrossRef] [PubMed]

36. Park, J.H.; Halitschlke, R.; Kim, H.B.; Baldwin, I.T.; Feldmann, K.A.; Feyereisen, R. A knock-out mutation in allene oxide synthase results in male sterility and defective wound signal transduction in *Arabidopsis* due to a block in jasmonic acid biosynthesis. *Plant J.* **2002**, *31*, 1–12. [CrossRef] [PubMed]

37. Scheres, B.; Wolkenfelt, H.; Terlouw, M.; Lawson, E.; Dean, C.; Weisbeek, P. Embryonic Origin of the *Arabidopsis* Primary Root and Root-Meristem Initials. *Development* **1994**, *120*, 2475–2487.

# Differential Mechanisms of Photosynthetic Acclimation to Light and Low Temperature in *Arabidopsis* and the Extremophile *Eutrema salsugineum*

Nityananda Khanal [1,†], Geoffrey E. Bray [2,‡,§], Anna Grisnich [1,‡], Barbara A. Moffatt [3] and  Gordon R. Gray [1,2,*]

[1] Department of Plant Sciences, University of Saskatchewan, Saskatoon, SK S7N 5A8, Canada; nityananda.khanal@agr.gc.ca (N.K.); anniethansen@icloud.com (A.G.)

[2] Department of Biochemistry, University of Saskatchewan, Saskatoon, SK S7N 5E5, Canada; gbray@usk.edu

[3] Department of Biology, University of Waterloo, Waterloo, ON N2L 3G1, Canada; moffatt@uwaterloo.ca

[*] Correspondence: gr.gray@usask.ca

[†] Present Address: Agriculture and Agri-Food Canada, Lacombe Research and Development Centre, 1  Research Road, P.O. Box 29, Beaverlodge, AB T0H 0C0, Canada.

[‡] These authors contributed equally to this work.

[§] Present Address: University of Saint Katherine, 1637 Capalina Road, San Marcos, CA 92069, USA.

**Abstract:** Photosynthetic organisms are able to sense energy imbalances brought about by the overexcitation of photosystem II (PSII) through the redox state of the photosynthetic electron transport chain, estimated as the chlorophyll fluorescence parameter $1-q_L$, also known as PSII excitation pressure.  Plants employ a wide array of photoprotective processes that modulate photosynthesis to correct these energy imbalances. Low temperature and light are well established in their ability to modulate PSII excitation pressure. The acquisition of freezing tolerance requires growth and development a low temperature (cold acclimation) which predisposes the plant to photoinhibition. Thus, photosynthetic acclimation is essential for proper energy balancing during the cold acclimation process. *Eutrema salsugineum* (*Thellungiella salsuginea*) is an extremophile, a  close relative of *Arabidopsis thaliana*, but possessing much higher constitutive levels of tolerance to abiotic stress. This comparative study aimed to characterize the photosynthetic properties of *Arabidopsis* (Columbia accession) and two accessions of *Eutrema* (Yukon and Shandong) isolated from contrasting geographical locations at cold acclimating and non-acclimating conditions.  In addition, three different growth regimes were utilized that varied in temperature, photoperiod and irradiance which resulted in different levels of PSII excitation pressure.  This study has shown that these accessions interact differentially to instantaneous (measuring) and long-term (acclimation) changes in PSII excitation pressure with regard to their photosynthetic behaviour. *Eutrema* accessions contained a higher amount of photosynthetic pigments, showed higher oxidation of P700 and possessed more resilient photoprotective mechanisms than that of *Arabidopsis*, perhaps through the prevention of PSI acceptor-limitation. Upon comparison of the two *Eutrema* accessions, Shandong demonstrated the greatest PSII operating efficiency ($\Phi_{PSII}$) and P700 oxidizing capacity, while Yukon showed greater growth plasticity to irradiance. Both of these *Eutrema* accessions are able to photosynthetically acclimate but do so by different mechanisms. The Shandong accessions demonstrate a stable response, favouring energy partitioning to photochemistry while the Yukon accession shows a more rapid response with partitioning to other (non-photochemical) strategies.

**Keywords:** adaptive (phenotypic) plasticity; *Arabidopsis thaliana*; cold acclimation; *Eutrema salsugineum*; low temperature; photoinhibition; photosynthesis; photosynthetic acclimation

## 1. Introduction

Photosynthesis is a highly coordinated and environmentally sensitive metabolic process. Photoautrophs modulate the structure and function of their photosynthetic apparatus to changes in the environment to maintain cellular energy balance called photostasis. Photostasis results in the balancing of the light energy absorbed by the photosystems with the energy consumed by the metabolic sinks of the plants [1–3]. Imbalances in cellular energy are sensed through changes in chloroplastic excitation pressure (or PSII excitation pressure), reflected in the redox state of the photosynthetic electron transport chain and estimated as the in vivo chlorophyll fluorescence parameter 1-$q_P$ or 1-$q_L$. It is well established that modulation of light and/or temperature cause similar energy imbalances and thus, a change in PSII excitation pressure [2–6].

Cold acclimation refers to a process whereby plants acquire the ability to tolerate freezing (freezing tolerance). Exposure to low, non-freezing temperatures, for a period of days to weeks, triggers a series of alterations resulting in a complex reconfiguration of cellular processes at multiple levels of organization [7–10]. The level of freezing tolerance attained during cold acclimation is a coordinated response to environmental cues dependent on the genotype of the plant. The changes occurring in leaves during the transition to low temperature are thought to represent transient stress responses whereas leaves that develop at low temperature establish a new metabolic homeostasis that represents the true cold acclimated state [11,12]. It has also been demonstrated that photosynthesis interacts with other processes during cold acclimation involving crosstalk between photosynthetic redox, cold acclimation and sugar-signalling pathways to regulate plant acclimation to low temperatures [3,6]. The process of cold acclimation results in an interesting dilemma for the plant. The exposure of leaves to low temperature creates an energy imbalance between the capacity to harvest light energy and the consumption capacity of photosynthesis resulting in excess PSII excitation pressure [3–5]. This, in turn, can potentially result in generation of reactive oxygen species (ROS) which can lead to photoinhibition and photooxidative damage of photosystem II (PSII) and photosystem I (PSI) [13–15]. This predisposition to photoinhibition has made it necessary for cold acclimated plants to develop a wide array of both short- and long-term photoprotective strategies to deal with excessive light [16–18].

In response to increased excitation pressure, cold acclimated cereals increase their photosynthetic capacity by increasing the RuBP-regeneration and subsequently electron flux through the Calvin cycle [6,11,19]. This results in enhanced photochemical quenching through increased $CO_2$ assimilation [5,6]. Alternative mechanisms for the utilization/dissipation of the excitation energy through other photochemical reactions which sustain the photosynthetic reduction of $CO_2$ and/or $O_2$ are also possible strategies [20,21]. Another major PSII photoprotective mechanism is the $\Delta$pH- and zeaxanthin-dependent non-photochemical quenching (NPQ), which dissipates any excess energy not used in photosynthesis as heat to protect the PSII reaction centre from overexcitation [16,17,22]. The thermal deactivation of excess light energy occurring within the PSII reaction centre (reaction centre quenching) has also been proposed as an effective photoprotective mechanism in cold acclimated plants [18].

Regardless of the mechanism employed, which depends on the species and genera examined, plants that actively grow and develop at low temperatures (high excitation pressure) are characterized by an increased tolerance to photoinhibition, that is, these organisms are less susceptible to the light-dependent inhibition of photosynthesis as a consequence of the re-establishment of photostasis. The sensing of energy imbalances brought about by changes in temperature, irradiance or any environmental stress that alters the redox state of the photosynthetic electron transport chain and modulates excitation pressure appears to be a fundamental feature of various taxonomic groups of photosynthetic organisms including cyanobacteria, green algae, and herbaceous plants [5,6,23]. It has also been suggested that the concepts of photostasis and excitation pressure provide the context to explain phenotypic plasticity and photosynthetic performance associated with cold acclimation and photoacclimation [3,6].

Many studies have extensively characterized plant stress tolerance using *Arabidopsis thaliana*. However, more stress tolerant species may possess different and/or additional protective mechanisms that cannot be found in the commonly studied *Arabidopsis* accessions [24–26]. *Eutrema salsugineum* (*Thellungiella salsuginea*) is an alternative plant model species particularly well suited for the examination of stress tolerance as this genus is also part of the Brassicaceae family and therefore closely related to *Arabidopsis* [27]. However, *Eutrema* is often referred to as an extremophile, owing, in part, to its high capacity to withstand various abiotic stresses such as freezing, water deficit, nutrient-deficiency and salinity [28–35] Primarily, two accessions from stress-prone, geographically diverse locations have been examined; the Shandong accession originating from Shandong Province, China and the Yukon accession, native to the Yukon Territory, Canada. These two accessions grow under contrasting natural habitats with Yukon being subjected to a subarctic and semi-arid climate and Shandong growing under warm temperate regions in high-salinity coastal areas with more frequent precipitation [28,34]. Plant populations evolved in different ecological niches are known to have differential adaptive specificities of photosynthetic properties [36,37]. Therefore, we would anticipate that the Shandong and Yukon accessions would perform, at least in part, based on their ecological backgrounds.

*Eutrema* has been used as a model for the study of cold acclimation and freezing tolerance [31,38,39]. While the study of [39] demonstrated that *Eutrema* and *Arabidopsis* did not differ in their constitutive level of freezing tolerance or short-term cold acclimation capacity, *Eutrema* outperformed *Arabidopsis* in long-term acclimation capacity suggesting a wider phenotypic plasticity for the trait of freezing tolerance. Furthermore, it was demonstrated that growth conditions, specifically irradiance, were determinants of the level of freezing tolerance attained during cold acclimation suggesting a role for photosynthetic processes in adaptive stress responses. Another study examining the cold-induced proteome showed nearly half of the identified cold-responsive proteins were associated with various aspects of chloroplast metabolic processes [40]. However, the role of photosynthetic acclimation in this process remains unknown for *Eutrema*, despite the fact transcript profiling and stress induced gene expression studies are almost always enriched by the differential expression of photosynthesis-related genes [29,30,33,34,41–44]. Previous studies have focused on the role of photosynthesis under salinity stress and photoinhibitory responses [45–47].

This study reports the results of comparative experiments aimed to characterize basic photosynthetic properties of the Yukon and Shandong accessions of *Eutrema* as well as *Arabidopsis* under non-acclimating and cold acclimated conditions. Furthermore, at both non-acclimating and cold acclimating conditions, three different growth regimes were examined which varied in day/night temperatures, photoperiod and irradiance. In addition to an examination of photosynthetic pigments, chlorophyll *a* fluorometry and photosystem I (PSI) spectroscopy were utilized to examine photosynthetic parameters with an underlying hypothesis that Yukon, Shandong and *Arabidopsis* possess differential capacities to modulate photosynthetic responses to light and temperature based on their contrasting ecophysiological backgrounds.

## 2. Results

### 2.1. Growth Regimes

It is well established that changes in light and temperature are sensed by the redox state of the photosynthetic electron transport chain and reflected in PSII excitation pressure [3–5]. Thus, different growth conditions will result in differential PSII excitation pressures. By manipulating growth parameters one can also manipulate PSII excitation pressure. Three different growth regimes were utilized in this study which varied in day/night temperatures, photoperiod and irradiance (Table 1). In addition, both non-acclimating and cold acclimating conditions were examined (Table 1). These conditions were chosen so as to be representative of growth conditions typically used to propagate *Eutrema* and *Arabidopsis* plant material and are referred to as the Yukon, Shandong or *Arabidopsis* growth regimes (see [24] and references contained within).

**Table 1.** Growth regimes used in this study.

| Growth Regime | Acclimation Status | | | | | | | | | |
| --- | --- | --- | --- | --- | --- | --- | --- | --- | --- | --- |
| | Non-Acclimated | | | | | Cold Acclimated | | | | |
| | Temperature (Light/Dark) | Photoperiod (Light/Dark) | PPFD (µmol photons m$^{-2}$ s$^{-1}$) | DPR (µmol photons m$^{-2}$) | DAT (°C) | Temperature (Light/Dark) | Photoperiod (Light/Dark) | PPFD (µmol photons m$^{-2}$ s$^{-1}$) | DPR (µmol photons m$^{-2}$) | DAT (°C) |
| Yukon | 22/10 °C | 21/3 h | 250 | 18.90 | 20.5 | 5/4 °C | 21/3 h | 250 | 18.90 | 4.9 |
| Shandong | 22/19 °C | 16/8 h | 250 | 14.40 | 21.0 | 5/4 °C | 16/8 h | 250 | 14.40 | 4.7 |
| *Arabidopsis* | 20/20 °C | 16/8 h | 100 | 5.75 | 20.0 | 5/4 °C | 16/8 h | 100 | 5.75 | 4.7 |

See Section 4.1. Plant Material and Growth Conditions for calculations of DPR and DAT. DAT = daily average temperature; DPR = daily photon receipt; PPFD = photosynthetic photon flux density.

The daily average temperature (DAT) varied by 1.0 and 0.2 °C at non-acclimating and cold acclimating growth conditions respectively (Table 1). In contrast, the daily photon receipt (DPR) was highest in the Yukon, followed by the Shandong and finally the *Arabidopsis* growth regimes (Table 1). The Yukon and Shandong growth regimes exhibited a 3.3- and 2.5-fold increase in DPR in comparison to the *Arabidopsis* growth regime. In addition, the Yukon growth regime showed a 1.3-fold increase in DAT compared to the Shandong growth regime (Table 1). In this study it is clear that DPR plays a much more prominent role in contributing to PSII excitation pressure under the various growth regimes. The use of multiple growth regimes was not to mimic natural conditions, but rather to demonstrate increased PSII excitation pressure ($1-q_L$) with different combinations of growth parameters (irradiance, temperature, photoperiod) that are reflective of a wide variety of controlled environment studies using *Arabidopsis* or *Eutrema*.

## 2.2. Photosynthetic Pigmentation

Growth conditions affect variables such as leaf area and thickness that in turn has an effect on the leaf pigmentation. Moreover, leaf pigment composition can also reflect the acclimation and adaptation strategies of plants in response to various environmental factors that affect photosynthesis.

### 2.2.1. Analysis of Chlorophyll and Carotenoid Contents

Combined analysis showed a significant difference in chlorophyll content per unit leaf fresh weight (ChlFW) between the accessions ($P < 0.001$), between the growth regimes ($P = 0.037$) and between the cold acclimation status ($P < 0.001$) of the plants (Table 2). For this parameter, significant two or three-way interactions were found between the genotypic and environmental factors: accessions by growth regime ($P = 0.014$), accessions by acclimation ($P < 0.001$), growth regime by cold acclimation ($P = 0.04$) and accession by growth regime by cold acclimation ($P = 0.005$). Chlorophyll content per unit leaf area (ChlLA) also varied significantly between the accessions ($P < 0.001$), across the growth regimes ($P < 0.001$) and cold acclimation status ($P = 0.036$) along with a significant two and three way interactions between accessions, growth regime and acclimation status ($P < 0.001$ to 0.05; Table 2).

Similarly, carotenoid content per unit leaf weight (CarFW) differed significantly between the accessions ($P < 0.001$) and between the acclimation status ($P < 0.001$). However, no significant difference was observed for this parameter across the growth regimes ($P = 0.188$; Table 2). CarFW was affected significantly by the interactions between the accessions and growth regime ($P = 0.004$), between accessions and acclimation status ($P < 0.001$), between growth regime and cold acclimation ($P = 0.008$) and between accession, growth regime and acclimation status ($P = 0.039$). Carotenoid content per unit leaf area (CarLA) also displayed highly significant difference between the accessions, growth regimes and acclimation status ($P < 0.001$ for all factors; Table 2) with significant two and three way interactions between the genotypic and environmental factors ($P < 0.001$ to 0.015).

In the overall analysis, the accessions did not vary significantly in the chlorophyll *a:b* ratio (Chl *a:b*; $P = 0.91$), but significant alterations in the values were observed across the growth regimes ($P < 0.001$) and acclimation status ($P = 0.004$). For this parameter, significant interactions were found between accessions and growth regime ($P < 0.001$), between accession and acclimation status ($P < 0.001$) and between accession, growth regime and acclimation status ($P = 0.002$). However, no significant interaction was observed between growth regime and acclimation status ($P = 0.099$) in Chl *a:b* (Table 2). Unlike other pigmentation parameters, the chlorophyll:carotenoid ratio (Chl:Car) did not differ significantly between the accessions ($P = 0.389$) and across growth regimes ($P = 0.06$). However, acclimation status altered the ratio significantly ($P < 0.001$), with significant interactions between the accession and acclimation ($P = 0.018$), between accessions and growth regimes ($P = 0.019$), and between accessions, growth regimes and acclimation status ($P = 0.001$; Table 2). The significant interactions between the accessions with growth regimes and measurement temperatures imply that the accessions adopt differential photosynthetic adjustment strategies in response to changes in growth regimes and measurement conditions.

**Table 2.** Comparison of photosynthetic leaf pigmentation in the Yukon and Shandong accessions of *Eutrema* and *Arabidopsis* at three different growth regimes under non-acclimating and cold acclimating conditions. Values followed by small case letters along columns are significant within growth regime and the capital letters along the rows denote the significant difference of an accession across the acclimation state.

| Parameter | Taxa (Accession) | Growth Regime | | | | | |
| --- | --- | --- | --- | --- | --- | --- | --- |
| | | Yukon | | Shandong | | Arabidopsis | |
| | | Non-Acclimated | Cold Acclimated | Non-Acclimated | Cold Acclimated | Non-Acclimated | Cold Acclimated |
| ChlFW (µg mg⁻¹) | *Eutrema* (Yukon) | 2.4 ± 0.07 a,A | 1.9 ± 0.08 a,B | 2.4 ± 0.10 a,A | 1.9 ± 0.05 a,B | 2.4 ± 0.29 a,A | 2.2 ± 0.04 a,A |
| | *Eutrema* (Shandong) | 2.2 ± 0.02 a,A | 1.6 ± 0.03 b,B | 2.2 ± 0.11 a,A | 1.6 ± 0.02 b,B | 2.5 ± 0.03 a,A | 1.1 ± 0.21 c,B |
| | *Arabidopsis* (Columbia) | 1.3 ± 0.05 b,A | 1.3 ± 0.06 c,A | 1.2 ± 0.07 b,A | 1.5 ± 0.03 c,A | 1.7 ± 0.05 b,A | 1.6 ± 0.07 b,A |
| CarFW (µg mg⁻¹) | *Eutrema* (Yukon) | 0.36 ± 0.009 a,B | 0.46 ± 0.010 a,A | 0.40 ± 0.005 a,B | 0.49 ± 0.012 a,A | 0.41 ± 0.047 a,A | 0.48 ± 0.022 a,A |
| | *Eutrema* (Shandong) | 0.43 ± 0.003 a,A | 0.35 ± 0.010 b,B | 0.34 ± 0.016 b,A | 0.37 ± 0.008 b,A | 0.38 ± 0.008 a,A | 0.30 ± 0.053 b,A |
| | *Arabidopsis* (Columbia) | 0.23 ± 0.010 b,B | 0.33 ± 0.010 b,A | 0.20 ± 0.013 c,B | 0.29 ± 0.012 c,A | 0.29 ± 0.021 b,A | 0.33 ± 0.013 b,A |
| ChlLA (µg cm⁻²) | *Eutrema* (Yukon) | 64.5 ± 5.22 a,A | 63.3 ± 1.35 b,A | 64.1 ± 2.45 a,A | 65.7 ± 4.08 a,A | 39.0 ± 2.2 b,B | 62.0 ± 2.48 a,A |
| | *Eutrema* (Shandong) | 72.2 ± 2.69 a,A | 71.3 ± 2.67 a,A | 71.3 ± 4.05 a,A | 78.1 ± 4.82 a,A | 61.7 ± 1.84 a,A | 32.8 ± 5.80 b,B |
| | *Arabidopsis* (Columbia) | 36.1 ± 2.38 b,A | 38.3 ± 2.35 c,A | 26.8 ± 1.26 b,B | 46.8 ± 1.46 b,A | 32.9 ± 1.26 c,B | 42.6 ± 1.30 b,A |
| CarLA (µg cm⁻²) | *Eutrema* (Yukon) | 9.9 ± 0.64 b,B | 15.4 ± 0.58 a,A | 10.8 ± 0.64 a,B | 17.3 ± 0.57 a,A | 6.7 ± 0.32 b,B | 13.5 ± 0.36 a,A |
| | *Eutrema* (Shandong) | 14.1 ± 0.44 a,A | 15.4 ± 0.43 a,A | 11.2 ± 0.73 a,B | 18.1 ± 1.16 a,A | 9.4 ± 0.14 a,A | 8.9 ± 1.47 b,A |
| | *Arabidopsis* (Columbia) | 6.5 ± 0.34 c,B | 9.5 ± 0.43 b,A | 4.4 ± 0.24 b,B | 9.0 ± 0.27 b,A | 5.5 ± 0.19 c,B | 8.7 ± 0.23 b,A |
| Chl *a*:*b* | *Eutrema* (Yukon) | 3.1 ± 0.13 c,B | 4.9 ± 0.50 a,A | 3.3 ± 0.29 a,B | 4.8 ± 0.06 a,A | 2.8 ± 0.15 b,A | 3.8 ± 0.39 a,A |
| | *Eutrema* (Shandong) | 4.7 ± 0.12 a,A | 4.2 ± 0.10a b,A | 3.6 ± 0.20 a,B | 4.8 ± 0.22 a,A | 3.2 ± 0.14 b,A | 2.2 ± 0.30 b,B |
| | *Arabidopsis* (Columbia) | 3.9 ± 0.05 b,A | 3.7 ± 0.37 b,A | 4.0 ± 0.08 a,A | 3.3 ± 0.24 b,B | 3.9 ± 0.22 a,A | 4.1 ± 0.11 a,A |
| Chl:Car | *Eutrema* (Yukon) | 6.5 ± 0.21 a,A | 4.1 ± 0.16 a,B | 6.0 ± 0.28 a,A | 3.8 ± 0.14 b,B | 5.9 ± 0.41 a,A | 4.6 ± 0.29 a,A |
| | *Eutrema* (Shandong) | 5.1 ± 0.04 b,A | 4.6 ± 0.12 a,A | 6.4 ± 0.27 a,A | 4.3 ± 0.10 b,B | 6.6 ± 0.16 a,A | 3.7 ± 0.20 b,B |
| | *Arabidopsis* (Columbia) | 5.5 ± 0.23 b,A | 4.0 ± 0.24 a,B | 6.1 ± 0.05 a,A | 5.2 ± 0.28 a,A | 6.0 ± 0.28 a,A | 4.9 ± 0.08 a,B |

Values represent means ± SE ($n$ = 3 to 5). The data were analysed using ANOVA and the means were separated using Fisher's individual error rate at significance level of 0.05. Data are expressed on a leaf FW (FW) or a leaf area basis (LA). ANOVA = analysis of variance; Chl = chlorophyll; Car = carotenoid; SE = standard error.

## 2.2.2. Effect of Cold Acclimation on Pigmentation across Growth Regimes

Cold acclimation had differential effects on photosynthetic pigmentation between the accessions. Upon cold acclimation for 3 weeks, Yukon plants showed a significant decrease in ChlFW under the Yukon and Shandong growth regimes, but there was no significant change in this parameter in the *Arabidopsis* growth regime. Shandong plants consistently displayed a significant decrease in ChlFW upon cold acclimation across all three growth regimes. On the other hand, *Arabidopsis* plants showed no significant change in ChlFW upon cold acclimation under all three growth regimes. The ChlLA remained more or less stable without any significant change due to cold acclimation in all three accessions across growth regimes (Table 2).

Yukon and *Arabidopsis* plants showed similar trends in CarFW due to cold acclimation, with significant increases in CarFW in the Yukon and Shandong growth regimes, while having no significant change in *Arabidopsis* growth regime. Contrarily, Shandong plants did not show significant changes in CarFW upon cold acclimation across all regimes (Table 2).

The accessions did not show any definite trends in Chl *a:b* upon cold acclimation. Yukon plants showed a significant increase in Chl *a:b* in the Yukon and Shandong growth regimes while having no significant change in the *Arabidopsis* growth regime. Contrarily, upon cold acclimation, Shandong plants underwent a significant increase in Chl *a:b* in the Shandong growth regime, while displaying no significant change in this parameter in the Yukon growth regime and a significant decrease in the *Arabidopsis* growth regime. On the other hand, cold acclimation brought about no significant change in Chl *a:b* in *Arabidopsis* plants in the Yukon and *Arabidopsis* growth regimes and surprisingly decreased the value of this parameter in the Shandong growth regime. A generalized observation of cold acclimation was a decrease in the Chl:Car ratio in all accessions across all three growth regimes. However, in some of the cases, the change in the ratio was not statically significant (Table 2).

With only a few exceptions, the *Eutrema* accessions showed a significantly higher content of both chlorophyll and carotenoid pigments both on a per unit fresh weight and per unit leaf area basis across all growth regimes under both acclimated and non-acclimated conditions. Again with few exceptions, Yukon plants seemed to be significantly higher than or at par with Shandong plants for the pigment content considered. Except for the *Arabidopsis* growth regime, Shandong plant had either a significantly higher or similar Chl *a:b* ratio across all regimes, while the relationship was opposite in the *Arabidopsis* growth regime. Contrarily, for the Chl:Car ratio, Shandong plants appeared to be significantly lower than or equivalent to the Yukon and *Arabidopsis* plants across all growth regimes. In most cases, *Arabidopsis* and Yukon plants displayed similar levels of the Chl:Car ratio across all growth regimes (Table 2). Some exceptions in the relative content of the photosynthetic pigment parameters suggest that there was an interaction between the accessions and the growth regimes. The above results show that photosynthetic pigmentation parameters differed across growth regimes with significant interactions between the accessions and the growth regimes. However, due to the multifactor variation in the growth regimes including the temperature, irradiance and photoperiod, the results did not differentiate the effects of individual factors on the pigmentation parameters.

## 2.3. Comparative PSII Photochemistry

A comparison of the maximum quantum efficiencies of PSII ($F_v/F_m$) suggests that growth regime and acclimation status had minor but differential effects on Yukon, Shandong and *Arabidopsis* plants (Figure 1). The accessions differed significantly in $F_v/F_m$ values ($P < 0.0001$). The acclimation status had significant effects on $F_v/F_m$ ($P < 0.0001$), while the growth regimes per se had no significant effects ($P = 0.64$). Significant two-way (accessions by acclimation, accessions by growth regime, and acclimation by growth regime) and three-way interactions (accessions by growth regime by acclimation status) ($P < 0.0001$) suggest a differential responses of accessions to the environmental conditions. The *Eutrema* accessions displayed a more consistent $F_v/F_m$ trend than that of *Arabidopsis* across the growth regimes and acclimating conditions. However, differences in *Eutrema* accessions were evident by consistently higher $F_v/F_m$ values of Shandong (0.81 to 0.84) than those of Yukon (0.79 to 0.82)

(Figure 1). Unlike *Eutrema*, *Arabidopsis* plants underwent a consistent decrease in $F_v/F_m$ values upon cold acclimation across all growth regimes. The cold-acclimated values of *Arabidopsis* (0.77 to 0.80) remained lower than the non-acclimated control values (0.82 to 0.83) (Figure 1). Under non-acclimated conditions, the $F_v/F_m$ values of *Arabidopsis* were comparable to those of Shandong plants (Figure 1).

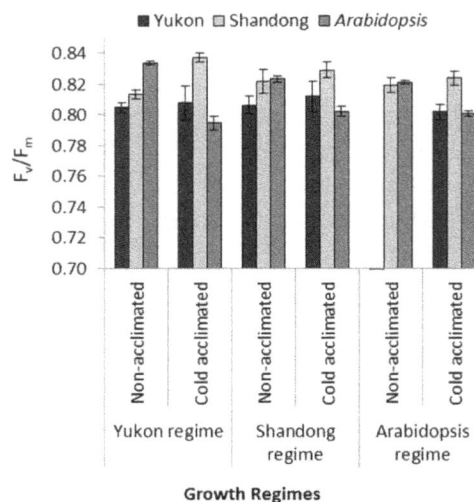

**Figure 1.** Maximum quantum efficiency of PSII ($F_v/F_m$) of non-acclimated and cold acclimated *Eutrema* accessions and *Arabidopsis* developed under three different growth regimes. Yukon accession (■); Shandong accession (■); *Arabidopsis* (■). No results were obtained for the Yukon accession in the non-acclimated *Arabidopsis* regime due to poor growth. Values represent means ± SE ($n$ = 3 to 6). PSII = photosystem II; SE = standard error. Means were grouped by Fisher's individual error rate at significance level of 0.05.

## 2.4. Photoinhibition and Recovery of PSII

Exposure of non-acclimated and cold acclimated *Arabidopsis* and *Eutrema* plants to a high irradiance of 1750 μmol photons $m^{-2}$ $s^{-1}$ coupled with low temperature (7 °C) for 4 h resulted in differential levels of photoinhibition between the accessions across the growth regimes (Figure 2). Cold acclimation significantly enhanced the tolerance to photoinhibition. The reduction in $F_v/F_m$ was significantly ($P < 0.001$) higher (32 to 54%) in non-acclimated plants than that of cold acclimated plants (12 to 31%; Figure 2). Under non-acclimating conditions, plants grown with an irradiance of 250 μmol photons $m^{-2}$ $s^{-1}$ exhibited a reduction of $F_v/F_m$ in the range of 32 to 36%, while those values for plants grown with an irradiance of 100 μmol photons $m^{-2}$ $s^{-1}$ were 35 to 54%. The cold acclimated plants from a growth irradiance of 250 μmol photons $m^{-2}$ $s^{-1}$ had a lower extent of photoinhibition (12 to 25%) than those from the lower growth irradiance (17 to 31%; Figure 2). The interactions of experimental accessions with growth regimes and acclimation status resulted in variation of the extent of photoinhibition between the accessions. Yukon and Shandong plants displayed relatively more stable values of $F_v/F_m$ with a lower extent of photoinhibition than *Arabidopsis* (Figure 2). Under higher growth irradiance (250 μmol photons $m^{-2}$ $s^{-1}$), photoinhibitory responses of *Arabidopsis* plants were at par with *Eutrema*. However, when grown under lower irradiance (100 μmol photons $m^{-2}$ $s^{-1}$), *Arabidopsis* plants showed greater susceptibility to photoinhibition than the *Eutrema* accessions. For instance, with a growth irradiance of 250 μmol photons $m^{-2}$ $s^{-1}$, the non-acclimated plants of all three experimental accessions underwent photoinhibition by 32 to 36% and the cold acclimated plants by 12 to 24%. On the other hand, *Arabidopsis* underwent photoinhibition by 54% and 31% in non-acclimated and cold-acclimated plants respectively, while the corresponding values for *Eutrema* were 35 to 38% in non-acclimated plants and 17 to 22% in cold acclimated plants under lower growth irradiance (100 μmol photons $m^{-2}$ $s^{-1}$ at *Arabidopsis* growth regime). After the release of photoinhibitory treatments, plants were kept at room temperature (22 °C) with a low irradiance of approximately 30 μmol photons m$^{-2}$ s$^{-1}$. After 24 h of releasing the photoinhibitory stress, all plants fully recovered their $F_v/F_m$ to the equivalent level of the control plants (Figure 2). This recovery suggests that the photoinhibition in all accessions was reversible, suggesting no differences in the effectiveness of repair of the photosynthetic apparatus or permanent photooxidative damage.

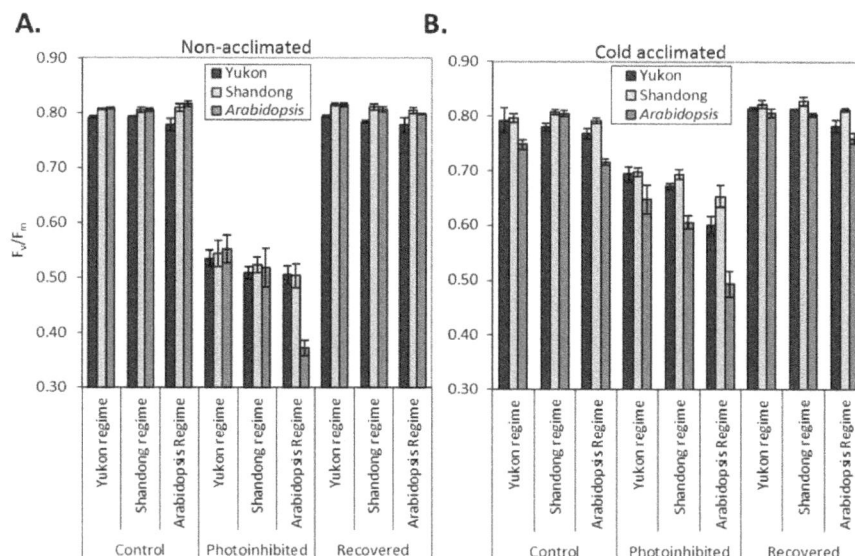

**Figure 2.** Photoinhibition and recovery measured as changes in maximum quantum efficiency of PSII ($F_v/F_m$) for (**A**) non-acclimated and (**B**) cold acclimated *Eutrema* accessions and *Arabidopsis* developed under three different growth regimes. Yukon accession (■); Shandong accession (■); *Arabidopsis* (■). Photoinhibition occurred for 4 h with a photosynthetic photon flux density (PPFD) of 1750 μmol photons m$^{-2}$ s$^{-1}$ at 7 °C. Recovery occurred at 22 °C under dim light (30 μmol photons m$^{-2}$ s$^{-1}$) for 24 h. Values represent means ± SE ($n$ = 3 to 6). PSII = photosystem II; SE = standard error. Fisher's individual error rate at significance level of 0.05 was used for inter-specific means comparison.

## 2.5. Chlorophyll Fluorescence Quenching Analyses

### 2.5.1. Excitation Pressure (1-q$_L$)

The parameter 1-q$_L$ reflects the redox poise of the primary quinone electron acceptor of PSII (Q$_A$) and is an estimate of PSII excitation pressure. In all accessions under various growth regimes, 1-q$_L$ increased non-linearly with the increase in measuring photosynthetic photon flux densities (PPFDs) (Figure 3). With respect to measurement at respective growth temperatures, the 1-q$_L$ light response curves of non-acclimated and cold acclimated plants clustered distinctly, displaying higher 1-q$_L$ in cold acclimated plants (Figure 3A–C). This indicates that the temperature-dependence of 1-q$_L$ is not fully compensated through cold acclimation. On the other hand, measurement of non-acclimated and cold acclimated plants at reciprocal temperatures across the range of PPFDs resulted in contrasting interactions of measuring irradiance and temperature (Figure 3D–F). With reference to the control (non-acclimated warm measured; NAWM), the non-acclimated cold measured (NACM; cold shock) resulted in the acceleration of 1-q$_L$, while the cold acclimated warm measured (CAWM) gave rise to a relaxation of 1-q$_L$.

Similarly, the 1-q$_L$ trend was also affected by the growth conditions. This is substantiated by the fact that plants grown in *Arabidopsis* growth regime displayed a higher 1-q$_L$ than those in the Yukon and Shandong growth regimes across the range of measuring PPFDs under different measuring temperatures and acclimation status (Figure 3). Similarly, in response to increasing PPFDs, the plants grown in Shandong growth regime showed a slightly lower 1-q$_L$ than those in Yukon growth regime.

The taxonomic differences in $1\text{-}q_L$ were negligible under all experimental conditions, except for those grown in *Arabidopsis* growth regime (Figure 3). The measurement of non-acclimated plants at low temperature (NACM) triggered an acceleration of $1\text{-}q_L$ that was substantially lower in Shandong (42%) than that of *Arabidopsis* and Yukon plants that had identical values of 51% (Table 3). Similarly, cold acclimation increased relaxation in $1\text{-}q_L$, and was higher in plants of *Arabidopsis* (20%) followed by Yukon (17%) and then Shandong (13%) (Table 3). When cold acclimated plants were exposed to non-acclimated (control) growth temperature (CAWM), there was substantial relaxation in $1\text{-}q_L$ and that was estimated to be 58%, 48% and 52% on the average for Yukon, Shandong and *Arabidopsis* plants respectively (Table 3).

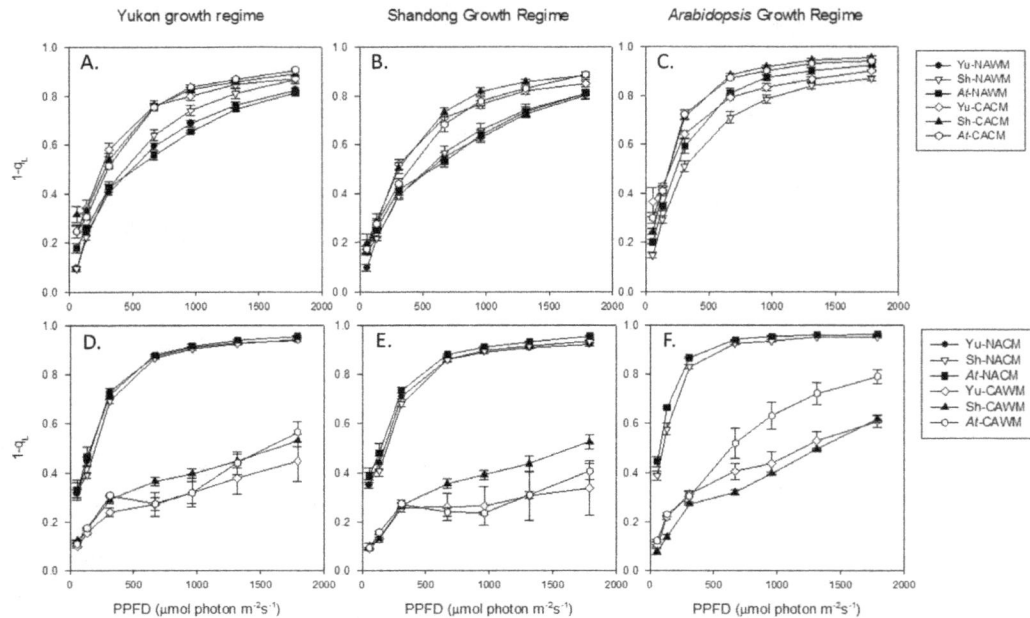

**Figure 3.** Light response curves of $1\text{-}q_L$ for *Eutrema* accessions and *Arabidopsis* developed under three different growth regimes. Non-acclimated and cold acclimated plants of Yukon (Yu), Shandong (Sh) and *Arabidopsis* (At) were subjected to their respective growth temperatures (**A–C**) as well as reciprocal temperature measurements (**D–F**). No results were obtained for the Yukon accession in the non-acclimated *Arabidopsis* regime due to poor growth. Values represent means ± SE ($n$ = 3 to 6). CACM = cold acclimated cold-measured; CAWM = cold acclimated warm-measured; NACM = non-acclimated cold measured; NAWM = non-acclimated warm-measured; PSII = photosystem II; $1\text{-}q_L$ = PSII excitation pressure; SE = standard error.

**Table 3.** Effect of cold shock, cold acclimation and thermal relaxation/augmentation as the fraction of corresponding values of photosynthetic parameters for *Eutrema* accessions and *Arabidopsis*.

| Parameters | Cold Shock Effect (NAWM-NACM)/NAWM | | | Cold Acclimative Effect (CACM-NACM)/NACM | | | Relaxation/Augmentation Effect (CAWM-CACM)/CACM | | |
|---|---|---|---|---|---|---|---|---|---|
| | Yukon | Shandong | *Arabidopsis* | Yukon | Shandong | *Arabidopsis* | Yukon | Shandong | *Arabidopsis* |
| $1\text{-}q_L$ | −0.51 | −0.42 | −0.51 | −0.17 | −0.13 | −0.20 | −0.58 | −0.48 | −0.52 |
| $RETR_{PSII}$ | 0.54 | 0.49 | 0.57 | 0.34 | 0.32 | 0.40 | 1.05 | 1.18 | 0.94 |
| $q_O$ | −0.39 | −0.26 | −0.99 | 0.16 | 0.13 | 0.30 | −0.50 | −0.52 | −0.41 |

Values are averages from the Yukon and Shandong growth regimes. Results from the *Arabidopsis* growth regime were excluded due to incomplete data sets for the Yukon accession. Negative signs before the values indicate the direction of the treatment effect on the specific parameters. Magnitudes can be interpreted in the absolute terms. CACM = cold acclimated cold measured; CAWM = cold acclimated warm-measured; $RETR_{PSII}$ = relative non-cyclic electron transport rate through PSII; NACM = non-acclimated cold-measured; NAWM = non-acclimated warm-measured; PSII = photosystem II; $q_O$ = basal fluorescence quenching coefficient; $1\text{-}q_L$ = PSII excitation pressure.

## 2.5.2. Relative Electron Transport Rate (RETR$_{PSII}$)

Yukon, Shandong and *Arabidopsis* plants exhibited comparable effects of measurement temperature, cold acclimation and growth regimes on the light response curves of RETR$_{PSII}$ (Figure 4). Under the respective growth temperatures, cold acclimated plants showed less RETR$_{PSII}$ in the range of experimental PPFDs along with light-saturation of RETR$_{PSII}$ at lower PPFDs, compared to that of non-acclimated plants (Figure 4A–C). Measurement of non-acclimated plants at low-temperature (NACM) displayed a further depression of RETR$_{PSII}$ as the indication of an inhibitory effect of low temperature on photosynthesis of non-acclimated plants (Figure 4D–F). On the other hand, measurement of cold acclimated plants at warm temperature (CAWM) augmented the RETR$_{PSII}$ that substantially exceeded the control values of RETR$_{PSII}$ from NAWM plants. Similarly, compared to the RETR$_{PSII}$ of NACM, the cold acclimated cold measured (CACM) values were higher in the light response curves. While the experimental accessions displayed common trends across the various growth regimes, quantitative differences between their responses were also evident. Shandong plants generally outperformed both of its counterparts in that *Arabidopsis* remained mostly at the lower scale, while Yukon was in an intermediate position between the Shandong and *Arabidopsis* plants in the light response curves. However, there are a few exceptions to this generalization indicating differential interactions between the genotype and environmental conditions (Figure 4). The NACM depression of RETR$_{PSII}$ was highest in *Arabidopsis* plants and lowest in Shandong plants across all growth regimes (Table 3). These amounted to 49% in Shandong, 54% in Yukon and 57% in *Arabidopsis* plants. On the other hand, the increases in RETR$_{PSII}$ due to cold acclimation were higher in *Arabidopsis* (40%) followed by Yukon (34%) and then Shandong (32%). When cold acclimated plants were exposed to non-acclimating (control) growth temperatures (CAWM), there was a rapid escalation of RETR$_{PSII}$ that was estimated to be 118%, 105% and 94%, for Shandong, Yukon and *Arabidopsis* plants respectively (Table 3).

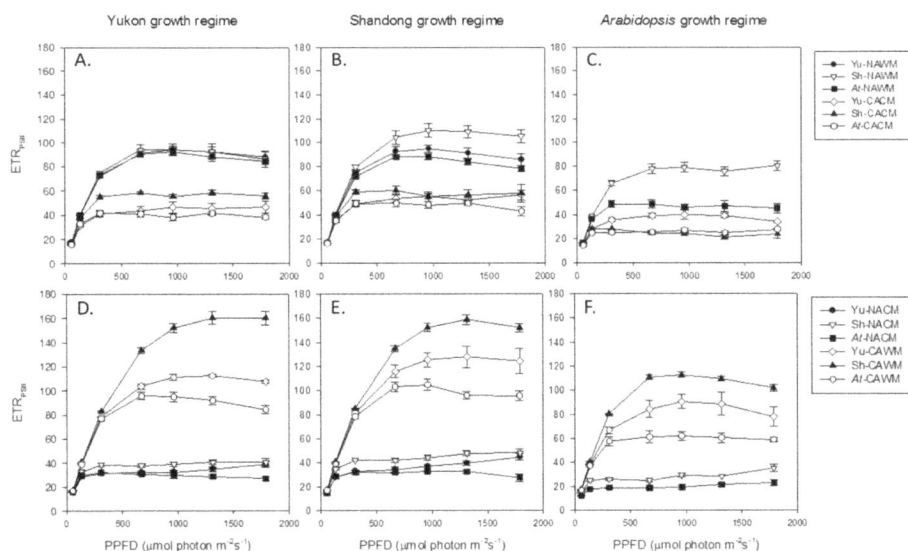

**Figure 4.** Light response curves of RETR$_{PSII}$ for *Eutrema* accessions and *Arabidopsis* developed under three different growth regimes. Non-acclimated and cold acclimated plants of Yukon (Yu), Shandong (Sh) and *Arabidopsis* (At) subjected to their respective growth temperatures (**A–C**) as well as reciprocal temperature measurements (**D–F**). No results were obtained for the Yukon accession in the non-acclimated *Arabidopsis* regime due to poor growth. Values represent means ± SE (*n* = 3 to 6). CACM = cold acclimated cold measured; CAWM = cold acclimated warm-measured; RETR$_{PSII}$, non-cyclic electron transport rate through PSII; NACM = non-acclimated cold measured; NAWM = non-acclimated warm-measured; PSII = photosystem II; SE = standard error.

### 2.5.3. Basal Fluorescence Quenching ($q_O$)

The coefficient $q_O$ is an indicator of dissipation of excitation energy from light-harvesting antenna of PSII (antenna quenching). Excluding the results of the *Arabidopsis* growth regime where a complete set of comparative data were lacking, cold acclimated plants had consistently higher values of $q_O$ than non-acclimated plants when measured at their respective growth temperatures (Figure 5A–C). The results of reciprocal measurement temperatures were virtually opposite from those of growth temperatures where light response curves of non-acclimated plants remained consistently above those of cold acclimated plants (Figure 5D–F). All experimental accessions exhibited similar responses of $q_O$ upon cold acclimation. On the other hand, under non-acclimated conditions the $q_O$ of *Arabidopsis* plants remained quantitatively lower than *Eutrema* accessions (Figure 5A–C). The experimental accessions exhibited differential effects of cold acclimation and measurement temperature on $q_O$. These responses were well discernible at PPFDs higher than 310 $\mu$mol photons m$^{-2}$ s$^{-1}$ (Figure 5). Low measuring temperature of the non-acclimated plants (NACM) triggered a rise in $q_O$ which was calculated for the irradiance levels higher than 310 $\mu$mol photons m$^{-2}$ s$^{-1}$. This rise in $q_O$ was significantly higher in *Arabidopsis* plants (99% on the average) followed distantly by Yukon (39%) and Shandong (26%) (Table 3). Similarly, cold acclimation resulted in a gain in $q_O$ with increases in PPFDs beyond 310 $\mu$mol photons m$^{-2}$ s$^{-1}$. This increase was significantly higher in *Arabidopsis* plants (30% on the average) followed by Yukon (16%) and Shandong (13%). When cold acclimated plants were exposed to non-acclimating (control) growth temperatures (CAWM) there was substantial down-shift in the light response curve of $q_O$. Such a subsidence in $q_O$ due to warm measuring temperature was estimated for the PPFD levels above 310 $\mu$mol photons m$^{-2}$ s$^{-1}$. This was higher in plants of *Eutrema* (50% in Yukon and 52% in Shandong) than that in *Arabidopsis* (41%) (Table 3). These results show that the effects of low measuring temperature and cold acclimation on $q_O$ was more pronounced in *Arabidopsis* plants than in *Eutrema*.

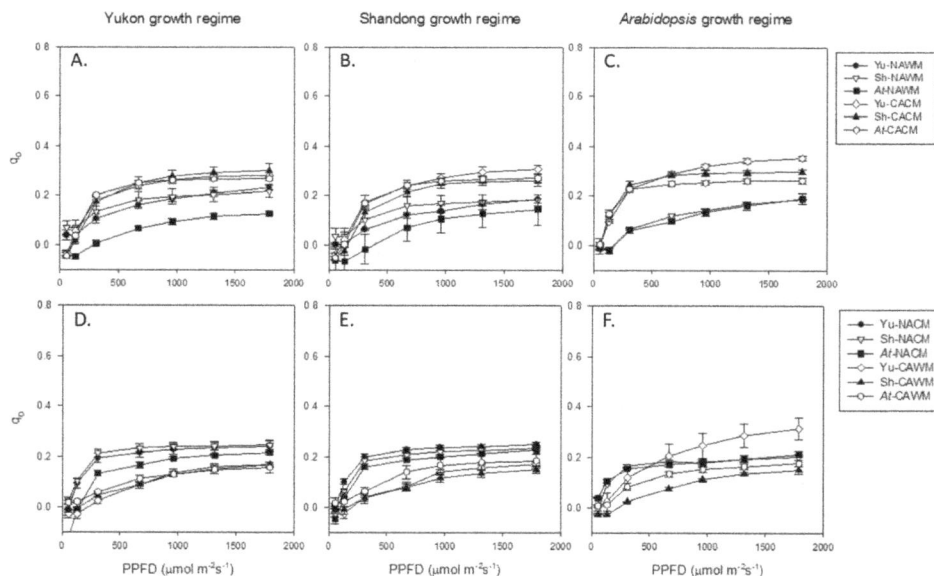

**Figure 5.** Light response curves of $q_O$ for *Eutrema* accessions and *Arabidopsis* developed under three different growth regimes. Non-acclimated and cold acclimated plants of Yukon (Yu), Shandong (Sh) and *Arabidopsis* (*At*) subjected to their respective growth temperatures (**A–C**) as well as reciprocal temperature measurements (**D–F**). No results were obtained for the Yukon accession in the non-acclimated *Arabidopsis* regime due to poor growth. Values represent means $\pm$ SE ($n$ = 3 to 6). CACM = cold acclimated cold measured; CAWM = cold acclimated warm-measured; NACM = non-acclimated cold measured; NAWM = non-acclimated warm-measured; $q_O$ = coefficient of basal fluorescence quenching; SE = standard error.

### 2.5.4. Excitation Energy Partitioning with Increasing Irradiance

Increasing measurement PPFD resulted in the non-linear decrease in the efficiency of PSII photochemistry ($\Phi_{PSII}$) with the concomitant increase in the non-photochemical dissipation (NPQ) of the excitation energy in all growth regimes and measurement temperatures (Figures 6–8). The fraction of dissipated energy was discerned into two components: the first component being the light independent, constitutive non-photochemical energy dissipation and fluorescence ($\Phi_{NO}$), and the second component being the light regulated, predominantly ΔpH- and/or zeaxanthin-dependent non-photochemical dissipation within the PSII antenna ($\Phi_{NPQ}$). The $\Phi_{NO}$ exhibited a negligible effect of experimental accessions, growth regimes and measurement temperature, remaining more or less constant (approximately 0.2). It was the $\Phi_{NPQ}$ that competed with $\Phi_{PSII}$ in the excitation energy partitioning in response to changes in the measurement temperature, measurement irradiance and the growth regimes (Figures 6–8).

The acclimation status of plants and measurement temperature triggered a marked effect, while the growth regimes and plant accessions had only subtle effects on the partitioning of the excitation energy. When measured at respective growth temperatures, cold acclimated plants responded to increasing PPFD with a more rapid down-regulation of $\Phi_{PSII}$ with a proportionate increase in $\Phi_{NPQ}$ (Figures 6–8). Generally, the NAWM plants displayed higher $\Phi_{PSII}$ than $\Phi_{NPQ}$ depending on the measurement PPFD and accessions from the three different growth regimes. In contrast, the CACM plants showed more competitive $\Phi_{NPQ}$ that surpassed $\Phi_{PSII}$ depending on the growth regimes and accessions (Figures 6–8). Measurement at reciprocal temperatures showed a more contrasting trend of energy partitioning in non-acclimated and cold acclimated plants (Figures 6–8). The $\Phi_{PSII}$ was surpassed by $\Phi_{NPQ}$ at much lower PPFDs in NACM plants compared to the CAWM counterparts.

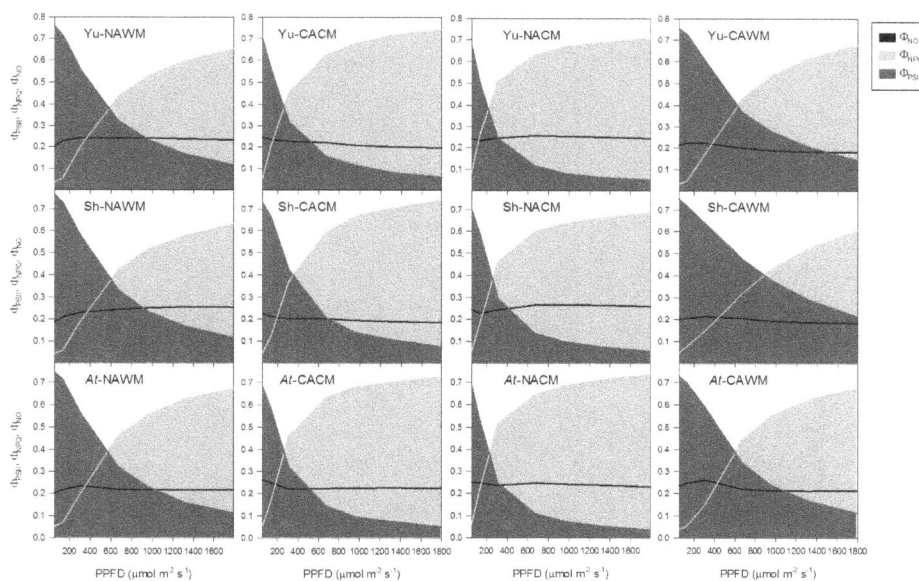

**Figure 6.** Partitioning of excitation energy as a function of irradiance for *Eutrema* accessions and *Arabidopsis*. The fraction of excitation energy flow via PSII photochemistry ($\Phi_{PSII}$) and non-photochemical dissipation pathways ($\Phi_{NO}$ and $\Phi_{NPQ}$) in Yukon (Yu), Shandong (Sh) and *Arabidopsis* (At) were estimated for plants developed under a Yukon growth regime. Non-acclimated and cold acclimated plants were subjected to their respective growth temperatures as well as reciprocal temperature measurements. CACM = cold acclimated cold measured; CAWM = cold acclimated warm-measured; NACM = non-acclimated cold measured; NAWM = non-acclimated warm-measured; NPQ = non-photochemical quenching; PSII = photosystem II; $\Phi_{NO}$ = efficiency of constitutive non-photochemical energy dissipation and fluorescence; $\Phi_{NPQ}$ = efficiency of light dependent NPQ; $\Phi_{PSII}$ = PSII operating efficiency.

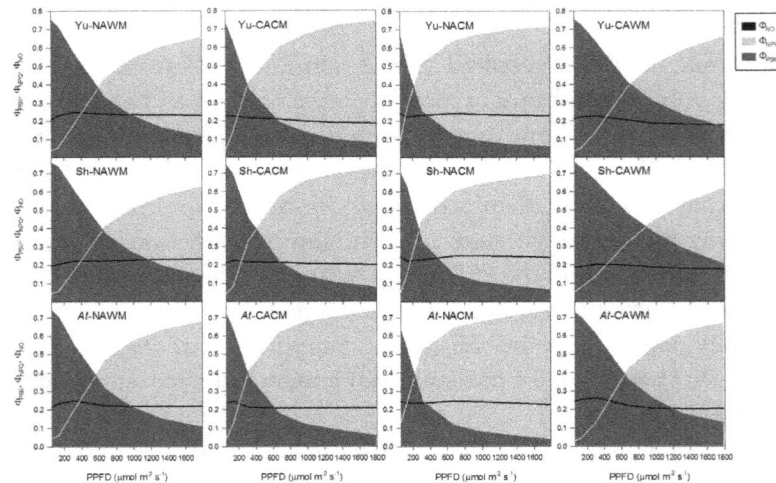

**Figure 7.** Partitioning of excitation energy as a function of irradiance for *Eutrema* accessions and *Arabidopsis*. The fraction of excitation energy flow via PSII photochemistry ($\Phi_{PSII}$) and non-photochemical dissipation pathways ($\Phi_{NO}$ and $\Phi_{NPQ}$) in Yukon (Yu), Shandong (Sh) and *Arabidopsis* (*At*) were estimated for plants developed under a Shandong growth regime. Non-acclimated and cold acclimated plants were subjected to their respective growth temperatures as well as reciprocal temperature measurements. CACM = cold acclimated cold measured; CAWM = cold acclimated warm-measured; NACM = non-acclimated cold measured; NAWM = non-acclimated warm-measured; NPQ = non-photochemical quenching; PSII = photosystem II; $\Phi_{NO}$ = efficiency of constitutive non-photochemical energy dissipation and fluorescence; $\Phi_{NPQ}$ = efficiency of light dependent NPQ; $\Phi_{PSII}$ = PSII operating efficiency.

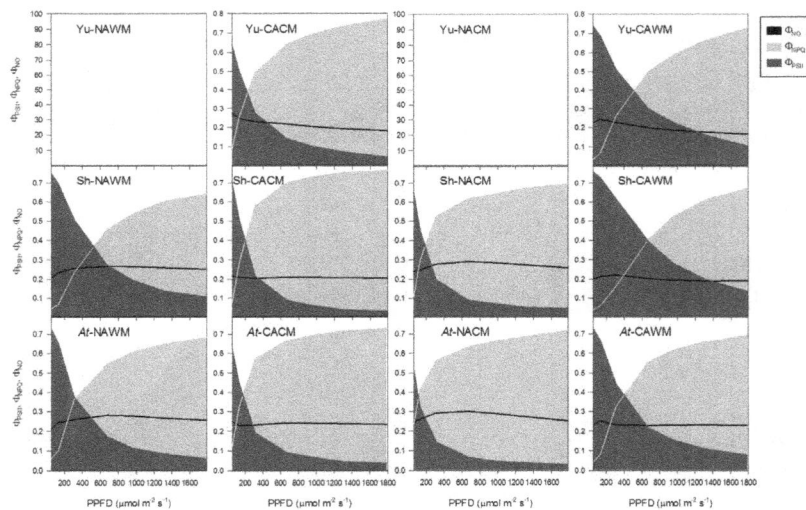

**Figure 8.** Partitioning of excitation energy as a function of irradiance for *Eutrema* accessions and *Arabidopsis*. The fraction of excitation energy flow via PSII photochemistry ($\Phi_{PSII}$) and non-photochemical dissipation pathways ($\Phi_{NO}$ and $\Phi_{NPQ}$) in Yukon (Yu), Shandong (Sh) and *Arabidopsis* (*At*) were estimated for plants developed under an *Arabidopsis* growth regime. Non-acclimated and cold acclimated plants were subjected to their respective growth temperatures as well as reciprocal temperature measurements. The blank graphs of the Yukon accession denote no results obtained due to poor growth of the accession in that condition. CACM = cold acclimated cold measured; CAWM = cold acclimated warm-measured; NACM = non-acclimated cold measured; NAWM = non-acclimated warm-measured; NPQ = non-photochemical quenching; PSII = photosystem II; $\Phi_{NO}$ = efficiency of constitutive non-photochemical energy dissipation and fluorescence; $\Phi_{NPQ}$ = efficiency of light dependent NPQ; $\Phi_{PSII}$ = PSII operating efficiency.

The effect of growth regimes on energy partitioning was also evident. In general, at given PPFD levels and experimental conditions the *Arabidopsis* growth regime had lower $\Phi_{PSII}$ and higher $\Phi_{NPQ}$ than the Yukon and Shandong growth regimes (Figures 6–8).

### 2.5.5. Excitation Energy Partitioning with Increasing Excitation Pressure

It is intriguing to examine whether the key photosynthetic correlates of *Eutrema* accessions and *Arabidopsis* respond similarly to excitation pressure. Separate regression results of Shandong, Yukon and *Arabidopsis* plants display that changes in excitation pressure explain the variation in photochemical and non-photochemical correlates of photosynthesis (Figure S1). Though the trends of all experimental accessions were fairly similar, some differences in slopes were distinguishable. In general, Shandong is quantitatively more responsive to excitation pressure for $\Phi_{PSII}$ and $\Phi_{NPQ}$ and *Arabidopsis* for $q_O$ (Figure S1).

The high correlation of excitation pressure with photochemical and NPQ parameters led to the consideration that excitation pressure may serve as a unifying determinant of energy partitioning. To examine the pattern of energy partitioning in response to excitation pressure, the results of energy fractionation from different growth regimes and measurement conditions were plotted against the excitation pressure (Figures S2–S4). When measured at growth irradiance, *Arabidopsis* plants displayed relatively lower $\Phi_{PSII}$ or higher $\Phi_{NPQ}$ at a given excitation pressure than did *Eutrema* accessions. At the reciprocal measurement temperatures, cold acclimated plants displayed higher sensitivity of energy partitioning pattern to the changes in excitation pressure than the non-acclimated plants (Figures S2–S4).

### 2.6. *Redox State of PSI and the Intersystem Electron Pool*

Exposure of leaves to far-red (FR) light results in an absorbance change at 820 nm ($\Delta A_{820}$), an indicator of the oxidation of P700. The P700$^+$ is transiently reduced with the application of saturating single-turnover (ST) flash or multiple turnover (MT) flashes in the presence of background FR light. The ratio of the extent of reduction of P700 triggered by the MT flash to the ST flash is an indicator of the number of electrons stored in the intersystem electron transport chain (e$^-$/P700) or intersystem electron pool size.

### 2.6.1. PSI Oxidation

Yukon, Shandong and *Arabidopsis* plants differed significantly ($P < 0.001$) for P700 oxidation within and across the growth regimes, acclimation status and measurement temperatures. The accessions also displayed significant interactions ($P < 0.001$) with growth regime and cold acclimation for the extent of P700 oxidation (Figure 9). In all experimental accessions, the effect of measurement temperature on P700 oxidation was consistently similar. The generalized effect was that all three experimental accessions showed a consistently higher extent of P700 oxidation at cold temperature measurement than at warm temperature measurement (Figure 9). The differences between the corresponding values were significant, though the magnitudes of differences were not very high (2.9% to 13.78% on average). Unlike the low temperature measurement, the effect of cold acclimation on P700 oxidation was variable between the experimental accessions (Figure 9). Cold acclimation enhanced the capacity of P700 oxidation in Yukon (3% to 68% increase) and *Arabidopsis* plants (22% to 98% increase) grown across all growth regimes. Shandong plants grown at 100 µmol photons m$^{-2}$ s$^{-1}$ (*Arabidopsis* growth regime) also displayed a significant increase (61% to 77%) in P700 oxidation due to cold acclimation. However, when Shandong plants were grown under a growth irradiance of 250 µmol photons m$^{-2}$ s$^{-1}$ (Shandong and Yukon growth regimes) the modulating effect of cold acclimation on P700 oxidation disappeared ($-9.5\%$ to 0.8% change).

In all experimental accessions, an increase in growth irradiance from 100 (*Arabidopsis* growth regime) to 250 µmol photons m$^{-2}$ s$^{-1}$ (Yukon and Shandong growth regimes) significantly increased P700 oxidation under non-acclimated conditions (Figure 9). However, Shandong and Yukon plants had a better response to growth irradiance in relation to P700 oxidation with a greater magnitude of increase (82% to 112% increases) than that of *Arabidopsis* (28% to 37% increases) under non-acclimated conditions. Similarly, the cold acclimated Yukon accession distinguished itself with the highest response to growth irradiance with a 142% to 197% increase in P700 oxidation due to increasing growth irradiance. The Shandong accession followed the trend with a moderate response (4% to 17% increases) in the amount of oxidized P700 with the increase in irradiance. However, cold acclimated *Arabidopsis* plants showed contrasting responses in that the P700 oxidation was 7% to 15% lower in plants grown under higher growth irradiances. Non-acclimated Shandong plants consistently displayed higher amounts of oxidized P700 than both Yukon and *Arabidopsis* plants. Cold acclimated values of P700 oxidation of the Yukon and Shandong accessions were quite similar (Figure 9).

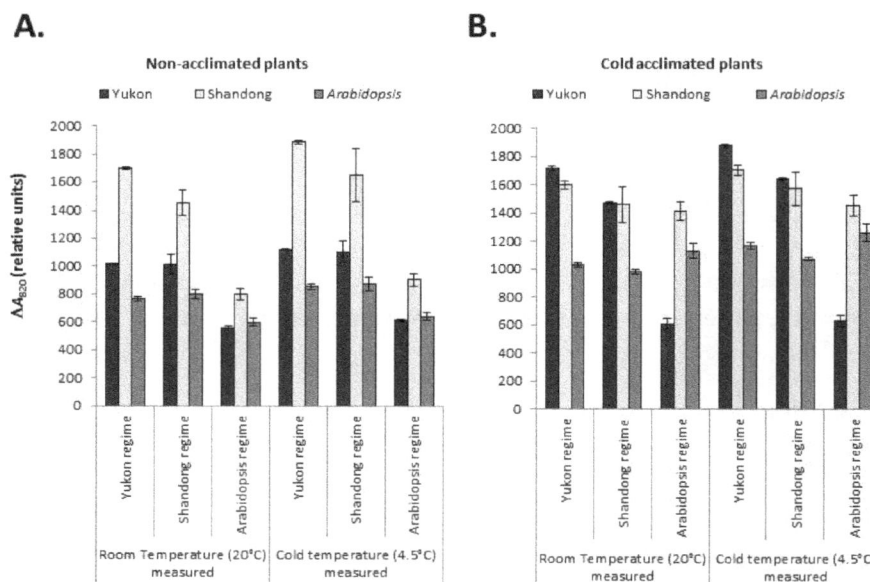

**Figure 9.** P700 oxidation measured as $\Delta A_{820}$ for (**A**) non-acclimated and (**B**) cold acclimated *Eutrema* accessions and *Arabidopsis* developed under three different growth regimes. Yukon accession (■); Shandong accession (■); *Arabidopsis* (■). Plants were measured at room (20 °C) and cold (4.5 °C) temperatures. Values represent means ± SE ($n$ = 4 to 9). SE = standard error.

## 2.6.2. Intersystem Electron Pool

The $e^-$/P700 was found to be the product of complex interactions between the experimental accessions, growth regimes or measurement conditions. There were no consistent responses of the experimental accessions to measurement temperature, cold acclimation or growth regime for this parameter (Figure 10). Exceptionally higher values of $e^-$/P700 were detected in NAWM Yukon plants in the *Arabidopsis* growth regime and CACM measured values of *Arabidopsis* plants in the Yukon growth regime. These observations were associated with apparent growth abnormalities. In the former case, Yukon plant growth was arrested due to a limitation of growth irradiance, while in the latter case, *Arabidopsis* plants displayed pale and stunted foliage as a combined effect of cold, longer photoperiod and high irradiance. A generalized scenario of combined analysis displayed significant effects of cold acclimation ($P < 0.001$) and measurement temperature ($P = 0.041$). However, these trends were complicated by the significant interaction of accessions with growth regime and measurement temperature. In general, cold acclimation resulted in an increase of the intersystem electron pool in Shandong (7% to 86% increase) and *Arabidopsis* (31% to 126% increases), while for Yukon plants this trend was not amenable for generalization due to interacting effect of growth regime and measurement temperature. Barring few exceptional observations, measurement at low temperature caused the lowering of $e^-$/P700 in all three accessions across all growth regimes. The differences in the low temperature-measured values of the Yukon accession between the non-acclimated and cold acclimated plants were relatively smaller. The trend of the effect of growth irradiance on $e^-/P700$ is variable between the experimental accessions. Yukon and *Arabidopsis* plants displayed interactions between growth regime, acclimation and measurement temperature, making it difficult to discern the effect of growth irradiance. However, for Shandong plants, $e^-/P700$ values showed an increasing trend (10% to 90% increases) as the result of increased growth irradiance (Figure 10).

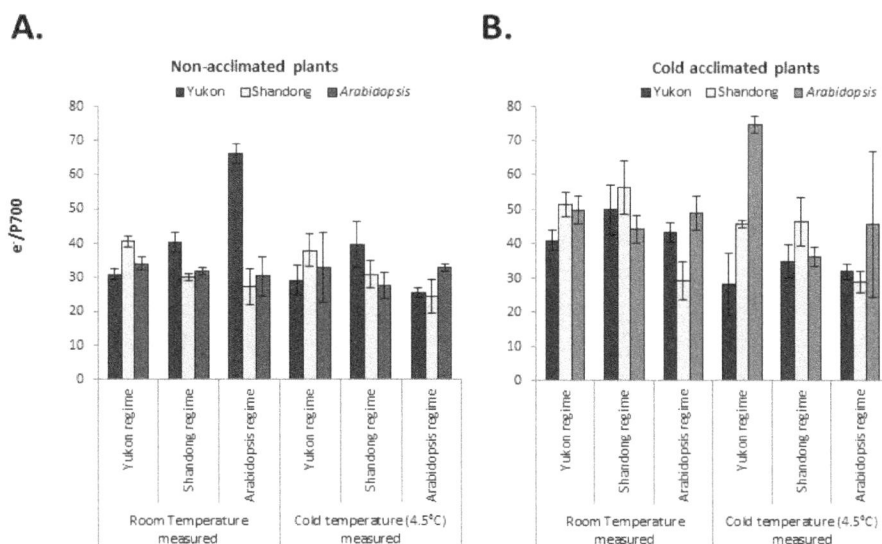

**Figure 10.** Pool size of electrons in the intersystem chain ($e^-/P700$) for (**A**) non-acclimated and (**B**) cold acclimated *Eutrema* accessions and *Arabidopsis* developed under three different growth regimes. Yukon accession (■); Shandong accession (■); *Arabidopsis* (■). Plants were measured at room (20 °C) and cold (4.5 °C) temperatures. Values represent means ± SE ($n$ = 4 to 9). SE = standard error.

## 3. Discussion

### 3.1. Differential Photoprotective Stratagies Indicated by Pigmentation

Chlorophyll and carotenoids are integral components of the photosynthetic machinery. Chlorophyll content is the proxy indicator of photosynthetic competence of plants, while carotenoid content reflects photoprotective processes such as the xanthophyll cycle, resulting in the formation of NPQ. Therefore, leaf pigment composition can reflect the acclimation and adaptation strategies of plants in response to environmental factors that affect photosynthesis [48].

A combined analysis of various photosynthetic pigmentation parameters showed significant differences between the accessions across growth conditions shaped by individual environmental factors and their interplay. Significant two- or three-way interactions between the accessions, growth regimes and acclimation status indicated that Yukon, Shandong and *Arabidopsis* plants differentially respond to environmental variables. With a few exceptions arising from the interactions between the experimental factors, *Eutrema* accessions showed a significantly higher content of both chlorophyll and carotenoid pigments than *Arabidopsis* both on a per unit FW and unit LA basis across all growth regimes under both cold acclimated and non-acclimated conditions. All three accessions acclimated to low temperature by lowering the Chl:Car ratio, but *Eutrema* accessions reduced the chlorophyll content, while *Arabidopsis* increased the carotenoid content. Previous studies have shown that leaves of *Eutrema* contained approximately a 30% higher chlorophyll content than that of *Arabidopsis* and the pigment disparity between the accessions increased upon salinity treatment [45,46,49]. These results indicate that *Eutrema* accessions not only maintain greater photosynthetic and photoprotective potential than *Arabidopsis*, they also modulate photosynthetic and photoprotective strategies more dynamically in response to environmental conditions. These observations correspond well with the significantly higher capacity of *Eutrema* to oxidize P700 and have higher light saturation points with respect to $\Phi_{PSII}$, $RETR_{PSII}$, $1\text{-}q_L$ and $\Phi_{NPQ}$ (see below). The increase in light saturation levels of PSII performance parameters due to cold acclimation (acclimation capacity) was also higher in *Eutrema*. These features reflect the extremophillic ecological background of *Eutrema* that requires a more dynamic mechanism to balance photosynthetic and photoprotective potentials, in contrast to the glycophytic adaptation of *Arabidopsis*.

Yukon and Shandong plants also showed differential pigment modulating properties in response to environmental variables. Shandong plants seemed to have the strategy of more abundant light interception and energy transformation, with a higher content of both chlorophyll and carotenoids per unit LA than that of Yukon plants. The higher pigment content of Shandong relates to the significantly higher $RETR_{PSII}$ of this accession compared to that of Yukon accession. These accessions also differed in cold acclimation strategies in that Yukon had a significant increase Chl *a:b* due to cold acclimation across all growth regimes, but Shandong showed variable trends with no appreciable alteration in Chl *a:b*. This suggests that the Yukon accession tends to cold acclimate by reducing photosynthetic light harvesting, while the Shandong accession undergoes a proportionate decrease in both pigments (Chl *a+b*) during cold acclimation. These differences in cold acclimation strategies are also reflected in the differential trends of P700 oxidation in Yukon and Shandong plants (see below).

### 3.2. Similar Trends but Quantitative Differences in PSII Fluorescence Parameters

*Eutrema* and *Arabidopsis* plants grown under various conditions showed comparable trends in PSII performance indicators. As was anticipated, a generalized pattern of each accession across all growth conditions was that measurements at low temperature caused significant increases in excitation pressure with a concomitant down-regulation of $\Phi_{PSII}$ and $RETR_{PSII}$, and consequently saturation of photosynthesis at lower irradiance levels. At the same time, the photosynthetic down-regulation was also associated with upregulation of NPQ parameters.

An examination of electron transport rate and pattern of energy partitioning revealed that photochemical down-regulation with concomitant upregulation of NPQ was a common manifestation of cold acclimation in both *Eutrema* accessions and *Arabidopsis*. At respective growth temperatures, the light response curves of cold acclimated plants positioned invariably at lower levels and light saturation occurred at lower PPFDs than that of non-acclimated control plants.

When plants were exposed to temperatures that contrasted with their respective growth temperatures, cold acclimated plants out-performed the non-acclimated counterparts with regard to PSII photochemistry. Exposure of cold acclimated plants to warm temperature (CAWM) resulted in the thermal augmentation of PSII performance characterized by the relaxation of excitation pressure, a greater fraction of excitation energy partitioned to photochemistry and a concomitant reduction of NPQ over a wide range of PPFDs. On the other hand, non-acclimated plants upon exposure to low measuring temperature (NACM) displayed a significant depression in the light response curves of photochemical indicators, suggesting the inhibition of photosynthesis. Growth regimes also triggered conspicuous effects on photosynthetic light response curves. Compared to the plants grown at low irradiance (*Arabidopsis* growth regime), the plants grown under higher irradiance (Yukon and Shandong growth regimes) underwent a down-shift in the trend of PSII excitation pressure coupled with upward shifts in light response curves for $\Phi_{PSII}$ and $RETR_{PSII}$.

Amidst the common general trends of taxonomic responses to environmental variables, *Eutrema* and *Arabidopsis* plants differed from each other in the relative magnitude of PSII performance parameters. In most instances of the experimental settings, Shandong plants stood superior to the other plants, followed by the Yukon accession and then closely by *Arabidopsis*. This generalization was more applicable after the exclusion of results from the *Arabidopsis* growth regime that was proven to be a sub-optimal growth condition for the Yukon accession. Both the Yukon and Shandong accessions of *Eutrema* were able to cold acclimate without affecting their potential quantum efficiencies, while

*Arabidopsis* plants acclimated to low growth temperature by lowering the $F_v/F_m$ by about 5% to 10%. These observations corroborate with an earlier finding that showed incomplete recovery of photosynthetic capacity of *Arabidopsis* upon cold acclimation [50]. In the range of experimental PPFDs, Shandong plants underwent a relatively larger fraction of excitation energy partitioning towards PSII photochemistry compared to that of Yukon and *Arabidopsis* plants, while the latter displayed more or less similar responses to each other. Compared to Yukon and *Arabidopsis* plants, the Shandong accession showed higher rates of $RETR_{PSII}$ on the light response curve under all experimental conditions, with less intensity of $RETR_{PSII}$ depression of non-acclimated plants due to cold shock (NACM) and a greater magnitude of thermal augmentation of $RETR_{PSII}$ for cold acclimated plants due to exposure to warm measuring temperatures (CAWM). A recent comparative study also showed that Shandong had higher $RETR_{PSII}$ than *Arabidopsis* under both normal and salt-stressed conditions [45]. Shandong plants also displayed less intensity of the effects of cold shock on PSII excitation pressure and NPQ parameters. On the other hand, the acclimative gain from the cold shock level (NACM) and thermal relaxation due to the warming effect (CAWM) were relatively lower in Shandong than that of *Arabidopsis*. These are the indication that Shandong has better photosynthetic stability. The Yukon and *Arabidopsis* exhibited contrasting effects of cold shock on NPQ parameters. The Yukon accession displayed highest sensitivity of overall NPQ while *Arabidopsis* showed highest sensitivity of $q_O$ to the cold shock.

*Eutrema* and *Arabidopsis* plants also differed in photoinhibition of PSII. Yukon and Shandong accessions displayed relatively more stable values of $F_v/F_m$ in response to photoinhibitory treatments than *Arabidopsis* plants. When grown under higher irradiance non-acclimating conditions (250 µmol photons m$^{-2}$ s$^{-1}$), photoinhibitory responses of *Arabidopsis* were at par with *Eutrema*. However, when grown under lower irradiance non-acclimating conditions (100 µmol photons m$^{-2}$ s$^{-1}$), *Arabidopsis* showed greater susceptibility to photoinhibition than the *Eutrema* accessions. When grown under cold acclimating conditions, *Eutrema* accessions consistently displayed significantly less photoinhibition than *Arabidopsis*. Evidently *Eutrema* possess greater cold-acclimation capacity of photosynthetic machinery than that of *Arabidopsis*. The relative resistance to photoinhibition was found to be associated with carotenoid content in the leaves since these parameters were negatively correlated.

### 3.3. Differential Responses of Eutrema and Arabidopsis Plants to PSII Excitation Pressure

Photochemistry and the non-photochemical dissipation of energy are competing processes. A high degree of determination of regression of the photochemical and NPQ parameters on 1-$q_L$ suggested the later to be an important link in photosynthetic processes. The scatter plots and regression parameters of Shandong, Yukon and *Arabidopsis* plants showed that Shandong is quantitatively more responsive to excitation pressure for $\Phi_{PSII}$ and $\Phi_{NPQ}$, while in *Arabidopsis* $q_O$ shows more involvement. It has already been postulated that the energy partitioning through these competing pathways is regulated by the environmental signals perceived by photosynthesis itself through the redox state of photosynthetic electron transport components [3,51].

The above observations led to the consideration that excitation pressure may serve as the unifying determinant of the energy partitioning. To examine the pattern of energy partitioning in response to excitation pressure, the results of energy fractionation from different experimental conditions were plotted against the excitation pressure. The portrait of energy partitioning against excitation pressure revealed a complex nature of energy fractionation that cannot be ascribed to short-term changes in excitation pressure. However, the results suggested that an excitation pressure of around 0.50 is critical for relative predominance of $\Phi_{PSII}$ or $\Phi_{NPQ}$. When measured at growth irradiance, the intersection of $\Phi_{PSII}$ or $\Phi_{NPQ}$ curves occurred between excitation pressures of 0.49 and 0.62 in non-acclimated plants and between 0.42 and 0.56 in cold acclimated plants. *Arabidopsis* plants displayed relatively lower $\Phi_{PSII}$ or higher $\Phi_{NPQ}$ at a given excitation pressure than *Eutrema* accessions. At reciprocal measurement temperatures, cold acclimated plants displayed a higher sensitivity of energy partitioning patterns to the changes in excitation pressure than the non-acclimated plants. This suggests that cold acclimation results in the development of more responsive mechanisms to detect changes in environmental conditions, thereby enabling them to adjust the pattern of excitation energy partitioning. Considering the extremophillic adaptation of *Eutrema* in contrast with the glycophytic adaptation of *Arabidopsis*, it was anticipated that *Eutrema* would possess better resiliency of PSII performance especially under low temperature conditions. However, Shandong plants only displayed minor quantitative differences from *Arabidopsis*, while Yukon appeared fairly similar to *Arabidopsis*. This may be due to the fact that the treatments imposed in the experiments were within the adaptive range of *Arabidopsis*.

*3.4. Yukon, Shandong and Arabidopsis Plants Show Divergent Trends in PSI Performance*

PSI has crucial role in balancing the phosphorylating and reducing potentials of cellular metabolism [52]. Oxidation of P700, an indicator of PSI activity was measured as the far-red light induced absorbance change at 820 nm ($\Delta A_{820}$). The redox state of P700 reflects the metabolic condition of chloroplast including the availability of the electron acceptors such as $NADP^+$, the extent of alternative electron transfer pathways around PSI, the redox state of the ferredoxin pool and electron transfer from PSII [53]. These properties, in turn, are a function of genotype, environmental variables and their interactions. Oxidized P700 is reduced by electrons mainly originating from the photooxidation of water which are conveyed through linear electron transport via PSII, the plastoquinone pool and plastocyanin. The intersystem electron pool is also fed by other pathways [54]. These properties constitute components of photosynthetic adjustment strategies of plants in response to environmental conditions [55].

Yukon, Shandong and *Arabidopsis* plants displayed divergent patterns of P700 oxidation with interactions between the accessions and environmental variables. The difference in the P700 oxidation was accompanied with active electron transport from PSII, as evidenced by a full reduction of oxidized P700 from MT flashes. Except for the case of Yukon plants in the light-limited condition of the *Arabidopsis* growth regime, *Eutrema* accessions displayed a significantly higher amount of oxidizable P700 than *Arabidopsis* under all conditions and acclimation status. Under non-acclimated conditions, both accessions of *Eutrema* and *Arabidopsis* showed a growth irradiance-dependent response of P700 oxidation. However, both accessions of *Eutrema* contrasted with *Arabidopsis* in having a higher magnitude of growth irradiance response. With the increase in irradiance from 100 μmol photons $m^{-2}$ $s^{-1}$ (*Arabidopsis* growth regime) to 250 μmol photons $m^{-2}$ $s^{-1}$ (Yukon and Shandong growth regimes), the *Eutrema* accessions responded with a minimum of an 82% increase in P700 oxidation whereas the corresponding increase in P700 oxidation of *Arabidopsis* was approximately 28%. Non-acclimated Shandong plants consistently displayed higher amounts of oxidized P700 than both Yukon and *Arabidopsis* plants. The Yukon accession significantly outperformed *Arabidopsis* under a growth irradiance of 250 μmol photons $m^{-2}$ $s^{-1}$. However, it is notable that Yukon had the lowest oxidizable P700 in the *Arabidopsis* growth regime that presumably barely met the light compensation point of this accession, resulting in stagnant growth and eventual collapse of plants after 4 weeks of germination (data not shown). This shows an interesting relationship between P700 oxidation and plant growth. In fact, an earlier study has shown the relationship between the oxidizable P700 and $CO_2$ assimilation in plants [53]. Yukon and Shandong plants also contrasted each other and with *Arabidopsis* plants in the cold acclimated response of P700. Although cold acclimated *Arabidopsis* had lower oxidizable P700 than *Eutrema*, it displayed the greatest acclimatory change in this parameter due to cold acclimation. The Yukon accession also underwent a significant increase in P700 oxidation as a cold acclimatory response. However, Shandong showed a growth irradiance-dependent acclimation pattern. In this accession, there was marked increase in oxidizable P700 as a result of acclimation in the *Arabidopsis* growth regime (low irradiance), but no acclimatory response in the Yukon and Shandong growth regimes (higher irradiance). This is presumably due to enhancement of plastid terminal oxidase (PTOX) activity of Shandong at higher irradiance, rendering acclimatory adjustment in PSI unnecessary in the given environmental conditions. This argument can be supported by a previous study that demonstrated salinity treatment triggered a significant up-regulation of PTOX with a concomitant increase in $RETR_{PSII}$ in Shandong, while in *Arabidopsis* there was no such up-regulation of PTOX and RETR$_{PSII}$ was actually reduced [45]. In the current study, higher irradiance could have triggered such a response in the Shandong accession. Presumably, the irradiance was not high enough for triggering PTOX or other alternative mechanisms in Yukon, as this accession had more sensitive growth plasticity to irradiance suggesting a higher irradiance requirement for optimal growth than to that of Shandong and *Arabidopsis* plants.

For the pool size of electrons in the intersystem chain (e$^{-}$/P700), there were no consistent responses of the experimental accessions to measurement temperature, cold acclimation or growth regime. In general, Shandong plants showed a trend of increasing e$^{-}$/P700 with increase in irradiance and also due to cold acclimation. In the Shandong accession, there was some correspondence between e$^{-}$/P700 and RETR$_{PSII}$ that explains partly what results in the increased e$^{-}$/P700 size. However, Yukon plants did not display any definite trend of e$^{-}$/P700 in response to environmental variables, while *Arabidopsis* responded to cold acclimation with an increase in e$^{-}$/P700, without any definite trend across the growth regimes. Moreover, Yukon and *Arabidopsis* plants that had perturbed phenotypes observed in two contrasting growth regimes and acclimation status showed an exceptional escalation of e$^{-}$/P700 for contrasting measurement temperatures. In Yukon plants, those observations were associated with warm temperature measurements of the plants from the non-acclimated *Arabidopsis* growth regime where plant growth was severely arrested due presumably to a deficit of irradiance to meet the metabolic demand of the plants. In *Arabidopsis*, on the other hand, the exceptional results were associated with low temperature measurement of cold acclimated plants from the Yukon growth regime, where plant phenotypes were chlorotic in nature. In fact, cold acclimated *Arabidopsis* had significantly lower chlorophyll contents in the Yukon growth regime compared to other growth regimes. These results showed that there was a complex interaction between the accessions and environmental variables that determine the relative predominance of electron flux in the intersystem pool of electrons.

## 4. Materials and Methods

### 4.1. Plant Material and Growth Conditions

Seeds of *Arabidopsis thaliana* (L.) Heynh. (accession Columbia, Col-0, stock no. CS60000) and *Eutrema salsugineum* (Pall.) Al-Shehbaz Shehbaz & Warwick (Shandong accession, stock no. CS22504 and Yukon accession, stock no. CS22664) were obtained from the Arabidopsis Biological Resource Center (ABRC, The Ohio State University, Columbus, OH, USA) were germinated and maintained in controlled environment growth chambers (Model PGR15; Conviron, Winnipeg, MB, Canada) using the non-acclimating growth conditions described in Table 1 [39]. These are referred to as Yukon, Shandong and *Arabidopsis* growth regimes.

The calculation of daily photon receipt (DPR) was performed as follows: the instantaneous units of irradiance as μmol photons m$^{-2}$ s$^{-1}$ were converted to mol photons m$^{-2}$ d$^{-1}$, where d denotes day.

$$DPR = (\text{μmol photons m}^{-2} \text{ s}^{-1})/1{,}000{,}000 \times \text{photoperiod} \times 60 \times 60$$

The cumulative daily average temperature (DAT) was estimated as:

$$DAT = [(\text{Light time temperature} \times \text{light period})/24] + [(\text{dark time temperature} \times \text{dark period})/24]$$

For cold acclimation, separate flats of plants maintained under these conditions were shifted to a 5/4 °C (day/night) temperature with the same irradiance and photoperiod as the non-acclimated plants in all growth regimes (Table 1). Photosynthetic parameters of non-acclimated *Eutrema* and *Arabidopsis* were measured after 4 weeks and 3 weeks of sowing respectively and those of cold acclimated plants were measured after 3-weeks of cold acclimation.

It is noteworthy that the poor growth of the Yukon accession under the non-acclimating, *Arabidopsis* growth regime resulted in a lack of suitably sized leaves for certain chlorophyll fluorescence measurements. This was circumvented in some cases by using fluorescence imaging or the construction of leaf discs comprised of multiple leaves. However, upon cold acclimation for 3 weeks, the Yukon leaves attained measurable growth allowing for subsequent analyses.

## 4.2. Chlorophyll a Fluorescence

### 4.2.1. Steady-State Fluorescence Quenching

Chlorophyll steady-state fluorescence quenching characteristics were determined in vivo using detached leaves under saturated $CO_2$ conditions using a XE-PAM xenon-pulse amplitude modulation fluorometer (Heinz Walz GmbH, Effeltrich, Germany) as described in detail previously [56]. Measurements were made using a Hansatech leaf-disc chamber (LD2/3; Hansatech Instruments Ltd., King's Lynn, Norfolk, UK) modified with an adapter (LD/FA; Hansatech) to accept the PAM fibreoptic. The fluorescence nomenclature and derivation of parameters were adopted from [57–59].

The temperature inside the chamber was maintained by a refrigerated circulating water bath (model RC6 CS; Lauda Dr. R. Wobser GmbH and Co., KG, Lauda-Königshofen, Germany) and matched that of the growth temperature for non-acclimated and cold acclimated conditions in each growth regime. Measurements were made at reciprocal temperatures so that there were four measuring conditions in each treatment; they are: non-acclimated plants measured at the growth temperature (non-acclimated warm-measured, NAWM; control), non-acclimated plants measured at low temperature (non-acclimated cold-measured, NACM; cold shock), cold acclimated plants measured at the growth temperature (cold acclimated cold-measured, CACM; cold acclimative), and cold acclimated plants measured at non-acclimating temperature (cold acclimated warm-measured, CAWM; thermal relaxation/augmentation). This design allowed for the dissection of the effects of cold shock, cold acclimation and thermal relaxation/augmentation of photosynthetic parameters by the comparison of measurements as follows:

$$\text{Cold shock effect with respect to corresponding control} = (NAWM - NACM)/NAWM$$

$$\text{Cold acclimative effect with respect to corresponding cold shock} = (CACM - NACM)/NACM$$

$$\text{Thermal relaxation/augmentation effect} = (CAWM - CACM)/CACM$$

These calculations were performed over the range of actinic irradiance where the tissue had equilibrated to the cuvette temperature to experience the cold shock or warming effect. These were 133 to 1790 $\mu$mol photons m$^{-2}$ s$^{-1}$ for 1-$q_L$ and RETR$_{PSII}$, and 670 to 1790 $\mu$mol photons m$^{-2}$ s$^{-1}$ for $q_O$.

Leaves were dark-adapted for 15 min prior to the onset of measurement which was determined empirically. Further increases in dark adaptation time did not result in any changes in fluorescence parameters. Minimal fluorescence in the dark-adapted state ($F_o$) was determined by subjecting the leaf sample to ms probe flashes from a xenon-arc measuring beam. A saturating pulse (6500 $\mu$mol photons m$^{-2}$ s$^{-1}$) of light for 800 ms was used to determine the maximal fluorescence ($F_m$) in the dark-adapted state. The leaves were then exposed to a series of actinic PPFDs of various intensities from 55 to 1790 $\mu$mol photons m$^{-2}$ s$^{-1}$ until a stable, steady-state level of fluorescence ($F'$) was achieved, approximately 10 min after switching to the next higher light level. Application of another saturating pulse gave the maximal fluorescence ($F_m'$) in light-adapted state. Minimal fluorescence in the light-adapted state ($F_o'$) was determined immediately after turning off the actinic source in the presence of far-red (FR) background light for 4 s to ensure maximal oxidation of PSII electron acceptors. WinControl software (ver 1.93; Heinz Walz) was used in conjunction with the PAM-data acquisition system (PDA-100; Heinz Walz) to control the timing, settings and trigger signals for the various actinic and saturating pulse light sources. Fluorescence traces were captured and analysed using the WinControl software.

The maximum quantum efficiency of PSII photochemistry ($F_v/F_m = (F_m - F_o)/F_m$) and the coefficient of basal fluorescence quenching ($q_O = (F_o - F_o')/F_o$) were calculated according to [60] and [61] respectively. PSII excitation pressure was expressed as 1-$q_L$ and calculated as $1 - (F_q'/F_v')$

$\times$ ($F_o'/F'$), where $F_q' = F_m' - F'$ [58]. The relative non-cyclic electron transport rate through PSII (RETR$_{PSII}$) was determined as $\Phi_{PSII} \times I \times 0.42$, where I is the incident PPFD on the leaf and 0.42 is the product of the spectral absorbance of the leaf (84%) and the fraction of incident photons that are absorbed by PSII (50%) [62].

The partitioning of absorbed light energy into PSII photochemistry and non-photochemical processes was estimated according to the model proposed by [57]. In this model, the fraction of energy utilized to drive PSII photochemistry (PSII operating efficiency) is estimated as $\Phi_{PSII}$ (($F_m - F'$)/$F_m' = F_q'/F_m'$), the fraction of energy dissipated as light-dependent non-photochemical quenching is estimated as $\Phi_{NPQ}$ ((($F_m - F_m'$)/$F_m$) $\times$ ($F'/F_m'$) = ($F'/F_m'$) − ($F'/F_m$)) and the efficiency of constitutive non-photochemical energy dissipation and fluorescence is estimated as $\Phi_{NO}$ ($F'/F_m$), hence $\Phi_{PSII} + \Phi_{NPQ} + \Phi_{NO} = 1$.

### 4.2.2. Chlorophyll Fluorescence Imaging

Chlorophyll fluorescence imaging was used to monitor photoinhibition and recovery in whole plants (see below) using a commercially available modulated imaging fluorometer (FluorCam; Photon System Instruments, Brno, Czech Republic) as described in detail by [56] except that only numerical data was acquired. Plants were dark-adapted for 15 min prior to measurement and the $F_v/F_m$ ratio with values integrated for the entire plant.

### 4.3. Photoinhibitory Treatments and Recovery

For photoinhibitory treatments, whole plants were exposed to 1750 μmol photons m$^{-2}$ s$^{-1}$ in a cold room set at 2 °C for 4 h using high pressure sodium bulbs (Sylvania Lumalux LU400/Eco, Osram Sylvania Products Inc., Manchester, NH, USA). Temperature at the leaf surface never exceeded 7 °C. The plants were allowed to recover at low light (30 μmol photons m$^{-2}$ s$^{-1}$) at 22 °C for 24 h. Photoinhibition and recovery were quantified by assessment of changes in $F_v/F_m$ as described above.

### 4.4. Photosystem I Spectroscopy

The relative redox state of P700 was estimated in vivo as the light induced change in absorbance at 820 nm ($\Delta A_{820}$) under ambient $CO_2$ conditions using a PAM-101 modulated fluorometer (Heinz Walz) as described in detail previously by [63]. Plants for each treatment were measured at room (20 °C) and cold (4.5 °C) temperatures to represent the growth temperatures for non-acclimated and cold acclimated conditions in each growth regime. Measurements were also done at reciprocal temperatures. The temperature was maintained by a refrigerated circulating water bath (Lauda Dr. R. Wobser).

Briefly, FR light ($\lambda_{max}$ 735 nm) at an intensity of approximately 70 μmol photons m$^{-2}$ s$^{-1}$ was applied on the adaxial side of the leaf in a custom designed cuvette modified to accept the PAM fibre optic. The redox state of P700 was evaluated as $\Delta A_{820}$ due to the formation of the cation radical (P700$^+$). After the P700$^+$ signal reached a steady-state level, single turnover (ST) saturating flashes (half peak width 14 μs) and multiple turnover (MT) saturating flashes (50 ms) were applied for the transient reduction of P700$^+$ in the presence of the background FR irradiance. WinControl software (version 1.93; Heinz Walz) was used in conjunction with the PAM-data acquisition system (PDA-100; Heinz Walz) to control the timing, settings and trigger signals for the various light sources. Traces were captured using the WinControl software and data files were exported to MicoCal Origin (version 6.0; MicroCal Software Inc., Northampton, MA, USA) for plotting, smoothing and integration of the areas under the curves representing the reduction of oxidized P700 due to the ST and MT turnover flashes. The apparent size of the intersystem electron donor pool to PSI (e$^-$/P700) was estimated in vivo as a ratio of the area associated with the MT and ST flashes (MT area/ST area) as described by [54].

### 4.5. Photosynthetic Pigment Determination

Plant pigments were extracted from leaves in 80% ($v/v$) acetone by grinding in a mortar and pestle followed by centrifugation for 5 minutes at 4500 rpm. Pigments were quantified spectrophotometrically

(SmartSpec Plus; Bio-Rad Laboratories, Hercules, CA, USA) according to the equations of [64] and expressed on a leaf fresh weight (FW) or leaf area (LA) basis. Measurements of leaf FW and leaf area were carried out using an analytical balance (Mettler Toledo, Columbus, OH, USA) and leaf area meter (LI-COR Inc., Lincoln, NE, USA).

### 4.6. Experimental Design and Statistical Analyses

The experiments were conducted in completely randomized design and measurements were replicated three or more times. A planting tray containing 36 cells constituted a block in which three experimental accessions were randomly assigned to the group of 12 cells. One tray was considered one replicate. Data were analysed by using descriptive statistics, correlation and analysis of variance (ANOVA) techniques with the aid of Microsoft Excel 2007 (Microsoft Corporation, Redmond, Washington, DC, USA), SigmaPlot 12 (Systat Software, Inc., San Jose, CA, USA), and Minitab version 15 (Minitab Inc., State College, PA, USA). Means of the experimental factors were grouped by Fisher's individual error rate at significance level of 0.05.

## 5. Conclusions

Photosynthesis is highly sensitive to environmental variability and a prime target of abiotic stress. Adaptation to prevailing growth conditions requires dynamic and flexible modulation of the photosynthetic machinery in response to environmental cues, including a tightly regulated energy balancing mechanism of photosynthetic metabolism. In this study, we have performed comparative experiments aimed to characterize basic photosynthetic properties of the Yukon and Shandong accessions of *Eutrema* as well as *Arabidopsis* grown under different growth regimes. Because of their extremophillic nature, it was expected that *Eutrema* would possess a differential capacity to modulate photosynthetic responses to PSII excitation pressure (light and low temperature) in comparison to *Arabidopsis*.

### 5.1. Comparisons between Eutrema and Arabidopsis

Our analyses revealed that photosynthetic parameters differed intraspecifically between the two accessions of *Eutrema* and *Arabidopsis*. *Eutrema* accessions showed a significantly higher content of both chlorophyll and carotenoid pigments across all growth regimes under both acclimated and non-acclimated conditions in comparision to *Arabidopsis*. Significant interactions between accessions, growth regime and acclimation status for photosynthetic pigments content and ratios suggests differential photosynthetic acclimation responses of the accessions to growth temperature, irradiance and photoperiod. *Eutrema* accessions were able to cold acclimate without an apparent effect on $F_v/F_m$, while *Arabidopsis* cold acclimates with a down-regulation of $F_v/F_m$. The processes modulating $F_v/F_m$, such as growth conditions, are prominent strategies of cold acclimation in both *Eutrema* and *Arabidopsis*. These accessions showed differential physiological plasticity to growth temperature and irradiance. However, growth irradiance seemed to be an important determinant of the tolerance of plants to photoinhibition of photosynthesis. In general, plants grown with higher irradiance experienced less photoinhibition than those grown under lower irradiance. *Eutrema* accessions possessed greater constitutive tolerance to photoinhibition and were better able to attained energy balancing during cold acclimation than *Arabidopsis* plants. Photoinhibition at low temperature was reversible in all accessions, suggesting no differences in the effectiveness of repair of the photosynthetic apparatus. All accessions demonstrated a full recovery from photoinhibition after the removal of the stress.

Fluorescence quenching analyses demonstrated a similar trend in responses, although quantitative differences were evident. Plants grown under a higher DPR presented lower values of $1-q_L$ when exposed to momentary high irradiance and/or cold temperature. In the range of low to saturating irradiance levels, *Eutrema* plants showed more stable $1-q_L$ than *Arabidopsis* when exposed to cold shock and/or thermally relaxing conditions. Both *Eutrema* and *Arabidopsis* plants acclimated under low temperatures by down-regulating photosynthesis. However, *Eutrema* showed greater stability in RETR$_{PSII}$ than *Arabidopsis* in the range of low to saturating irradiance levels upon the exposure to cold shock and/or thermally relaxing conditions. In all accessions, cold acclimation elicited a greater dissipation of excitation energy from the light-harvesting antenna of PSII (q$_O$). Such dissipation of excitation energy was higher in non-acclimated *Eutrema* than in non-acclimated *Arabidopsis*, and contrastingly lower in cold acclimated *Eutrema* than in cold acclimated *Arabidopsis*. Growth conditions exerting higher excitation pressure (high irradiance and low temperature) resulted in the energy partitioning in favour of NPQ at the expense of PSII photochemistry. However, growth conditions with higher irradiance proved more favorable for the development of more resilient photochemistry and generally, energy partitioning in *Eutrema* was more favourable for photochemistry than that of *Arabidopsis* plants. Moreover, cold acclimation of photosynthesis resulted in only partial recovery of photochemistry in both *Arabidopsis* and *Eutrema* plants.

A property that both *Eutrema* accessions shared while contrasting with *Arabidopsis* was that of significantly higher PSI activity in all growth regimes. Both accessions of *Eutrema* outperformed *Arabidopsis* plants for P700 oxidation in all environmental conditions that meet the minimal growth requirement of the plants. Cold acclimation under higher DPR conditions (higher excitation pressure) significantly enhanced the P700 oxidizing capacity in *Eutrema*, but *Arabidopsis* underwent a depression in P700 oxidizing capacity due to cold acclimation. Thus, it appears mitigation of PSI limitation may play a prominent role in the mechanism of photosynthetic acclimation to high light and low temperature, ensuring PSI is not acceptor-limited and conferring protection to PSII by oxidation of the plastoquinone pool. While not examined in this study, the role of alternative electron sinks may also contribute to the acclimatory adjustments enabling adaptation to harsh environmental conditions as have been shown previously for both *Eutrema* and *Arabidopsis* [21,45].

## 5.2. Comparisons between Eutrema Accessions

Numerous lines of evidence substantiate that geographical separation of a population leads to the development of photosynthetic strategies as dictated by the local environmental conditions [36,37]. Having evolved in contrasting ecophysiological backgrounds, the Yukon and Shandong accessions of *Eutrema* as well as *Arabidopsis* presumably respond differentially to environmental variability.

Surprisingly, it was the Shandong accession of *Eutrema* which exhibited better photosynthetic plasticity and not the Yukon accession as we would have predicted based on its natural habitat which is much harsher than that of Shandong [34]. Both *Eutrema* accessions acclimated under low temperatures with the down-regulation of photosynthesis. In the range of low to saturating irradiance, Shandong plants showed a greater stability in 1-q$_L$ and RETR$_{PSII}$ than Yukon plants when exposed to cold shock and thermally relaxing conditions. Energy partitioning in the Shandong accession of *Eutrema* was more favourable for photochemistry than that of Yukon accession. While P700 oxidation was greater in *Eutrema* overall, non-acclimated Shandong plants consistently displayed higher amounts of oxidized P700 in comparison to Yukon plants. Cold acclimation under higher DPR conditions (higher 1-q$_L$) significantly enhanced the P700 oxidizing capacity in Yukon followed distantly by Shandong.

While both accessions respond in a manner that allows more efficient photosynthetic acclimation than *Arabidopsis*, the mechanisms utilized are quite different and likely a result of adaptation to the natural conditions of these accessions. For the Yukon accession, we propose that as a result of growth in an environment where a plant must deal with multiple stress simultaneously (high 1-q$_L$) a mechanism of rapid response is required. This is activated only in response to stress conditions and along with already higher levels of constitutive tolerance facilitate survival. In the Yukon accession this results in excitation partitioning to more protective processes which likely involve photochemical and non-photochemical strategies. In contrast, the Shandong accession is much more stable in its photosynthetic response to stress and responds more slowly, partitioning excess excitation to photochemistry. Thus, it is possible that at least for photosynthetic acclimation, the Yukon accession has 'locked in' mechanisms of adjustment that permit rapid responses to environmental change that ensure survival, albeit at the expense of the enhanced flexibility exhibited by the Shandong accession.

**Acknowledgments:** This project was supported, in part, from Discovery funding provided by the Natural Sciences and Engineering Research Council (NSERC) of Canada and by funding from the Agriculture and Agri-Food Canada (AAFC)/NSERC Research Partnerships Program (with Performance Plants Inc.).

**Author Contributions:** Nityananda Khanal performed all the experiments, analysed the data and prepared a first draft of the manuscript. Geoffrey E. Bray and Anna Grisnich performed initial fluorescence and spectroscopy experiments. Barbara A. Moffatt and Gordon R. Gray provided funding. Gordon R. Gray designed the research, analysed the data and prepared the final draft of the manuscript. All authors have read and approved the final manuscript.

**Conflicts of Interest:** The authors declare no conflict of interest.

## References

1.  Melis, A. *Photostasis in Plants*; Plenum Press: New York, NY, USA, 1998; pp. 207–220.
2.  Hüner, N.P.A.; Öquist, G.; Melis, A. *Photostasis in Plants, Green Algae and Cyanobacteria: The Role of Light Harvesting Antenna Complexes*; Kluwer Academic Publishers: Dordrecht, The Netherlands, 2003; Volume 13, pp. 401–421.
3.  Ensminger, I.; Busch, F.; Huner, N.P.A. Photostasis and cold acclimation: Sensing low temperature through photosynthesis. *Physiol. Plant.* **2006**, *126*, 28–44. [CrossRef]
4.  Huner, N.P.A.; Maxwell, D.P.; Gray, G.R.; Savitch, L.V.; Krol, M.; Ivanov, A.G.; Falk, S. Sensing environmental change: PSII excitation pressure and redox signalling. *Physiol. Plant.* **1996**, *98*, 358–364. [CrossRef]
5.  Huner, N.P.A.; Öquist, G.; Sarhan, F. Energy balance and acclimation to light and cold. *Trends Plant Sci.* **1998**, *3*, 224–230. [CrossRef]
6.  Hüner, N.P.A.; Bode, R.; Dahal, K.; Busch, F.A.; Possmayer, M.; Szyszka, B.; Rosso, D.; Ensminger, I.; Krol, M.; Ivanov, A.G.; et al. Shedding some light on cold acclimation, cold adaptation, and phenotypic plasticity. *Can. J. Bot.* **2013**, *91*, 127–136. [CrossRef]
7.  Guy, C.L. Cold acclimation and freezing stress tolerance: Role of protein metabolism. *Annu. Rev. Plant Physiol. Plant Mol. Biol.* **1990**, *41*, 187–223. [CrossRef]
8.  Guy, C.L.; Kaplan, F.; Kopka, J.; Selbig, J.; Hincha, D.K. Metabolomics of temperature stress. *Physiol. Plant.* **2008**, *132*, 220–235. [CrossRef] [PubMed]
9.  Thomashow, M.F. Molecular basis of plant cold acclimation: Insights gained from studying the CBF cold response pathway. *Plant Physiol.* **2010**, *154*, 571–577. [CrossRef] [PubMed]
10. Theocharis, A.; Clément, C.; Barka, E.A. Physiological and molecular changes in plants grown at low temperatures. *Planta* **2012**, *235*, 1091–1105. [CrossRef] [PubMed]
11. Huner, N.P.A.; Öquist, G.; Hurry, V.M.; Krol, M.; Falk, S.; Griffith, M. Photosynthesis, photoinhibition and low temperature acclimation in cold tolerant plants. *Photosynth. Res.* **1993**, *317*, 9–39. [CrossRef] [PubMed]
12. Gray, G.R.; Heath, D. A global reorganization of the metabolome in *Arabidopsis* during cold acclimation is revealed by metabolic fingerprinting. *Physiol. Plant.* **2005**, *124*, 236–248. [CrossRef]
13. Powles, S.B. Photoinhibition of photosynthesis induced by visible light. *Annu. Rev. Plant Physiol.* **1984**, *35*, 15–44. [CrossRef]
14. Aro, E.M.; Virgin, I.; Andersson, B. Photoinhibition of photosystem II. Inactivation, protein damage and turnover. *Biochim. Biophys. Acta* **1993**, *1143*, 113–134. [CrossRef]
15. Scheller, H.; Haldrup, A. Photoinhibition of photosystem I. *Planta* **2005**, *221*, 5–8. [CrossRef] [PubMed]
16. Demmig-Adams, B.; Adams, W.W. Photoprotection and other responses of plants to high light stress. *Annu. Rev. Plant Physiol. Plant Mol. Biol.* **1992**, *43*, 599–626. [CrossRef]
17. Horton, P.; Ruban, A.V.; Walters, R.G. Regulation of light harvesting in green plants. *Annu. Rev. Plant Physiol. Plant Mol. Biol.* **1996**, *47*, 655–684. [CrossRef] [PubMed]
18. Ivanov, A.G.; Sane, P.V.; Hurry, V.; Öquist, G.; Huner, N.P.A. Photosystem II reaction centre quenching: Mechanisms and physiological role. *Photosynth. Res.* **2008**, *98*, 565–574. [CrossRef] [PubMed]
19. Stitt, M.; Hurry, V. A plant for all seasons: Alterations in photosynthetic carbon metabolism during cold acclimation in *Arabidopsis*. *Curr. Opin. Plant Biol.* **2002**, *5*, 199–206. [CrossRef]
20. Osmond, B.; Badger, M.; Maxwell, K.; Björkman, O.; Leegood, R. Too many photons: Photorespiration, photoinhibition and photooxidation. *Trends Plant Sci.* **1997**, *2*, 119–121. [CrossRef]

21.   Ivanov, A.G.; Rosso, D.; Savitch, L.V.; Stachula, P.; Rosembert, M.; Öquist, G.; Hurry, V.; Hüner, N.P.A. Implications of alternative electron sinks in increased resistance of PSII and PSI photochemistry to high light stress in cold acclimated *Arabidopsis thaliana*. *Photosynth. Res.* **2012**, *113*, 191–206. [CrossRef] [PubMed]

22.   Ort, D.R. When there is too much light. *Plant Physiol.* **2001**, *125*, 29–32. [CrossRef] [PubMed]

23.   Öquist, G.; Hüner, N.P.A. Photosynthesis of overwintering evergreen plants. *Annu. Rev. Plant Biol.* **2003**, *54*, 329–355. [CrossRef] [PubMed]

24.   Amtmann, A. Learning from evolution: *Thellungiella* generates new knowledge on essential and critical components of abiotic stress tolerance in plants. *Mol. Plant* **2009**, *2*, 3–12. [CrossRef] [PubMed]

25.   Oh, D.H.; Dassanayake, M.; Bohnert, H.J.; Cheeseman, J.M. Life at the extreme: Lessons from the genome. *Genome Biol.* **2012**, *13*, 241. [CrossRef] [PubMed]

26.   Bressan, R.A.; Park, H.C.; Orsini, F.; Oh, D.H.; Dassanayake, M.; Inan, G.; Yun, D.J.; Bohnert, H.J.; Maggio, A. Biotechnology for mechanisms that counteract salt stress in extremophile species: A genome-based view. *Plant Biotechnol. Rep.* **2013**, *7*, 27–37. [CrossRef]

27.   Koch, M.A.; German, D.A. Taxonomy and systematics are key to biological information: *Arabidopsis*, *Eutrema* (*Thellungiella*), *Noccaea* and *Schrenkiella* (Brassicaceae) as examples. *Front. Plant Sci.* **2013**, *4*, 267. [CrossRef] [PubMed]

28.   Inan, G.; Zhang, Q.; Li, P.; Wang, Z.; Cao, Z.; Zhang, H.; Zhang, C.; Quist, T.M.; Goodwin, S.M.; Zhu, J.; et al. Salt cress: A halophyte and cryophyte Arabidopsis relative model system and it applicability to molecular genetic analyses of growth and development of extremophiles. *Plant Physiol.* **2004**, *135*, 1718–1737. [CrossRef] [PubMed]

29.   Taji, T.; Seki, M.; Satou, M.; Sakurai, T.; Kobayashi, M.; Ishiyama, K.; Narusaka, Y.; Narusaka, M.; Zhu, J.K.; Shinozaki, K. Comparative genomics in salt tolerance between Arabidopsis and Arabidopsis-related halophyte salt cress using Arabidopsis microarray. *Plant Physiol.* **2004**, *135*, 1697–1709. [CrossRef] [PubMed]

30.   Gong, Q.Q.; Li, P.H.; Ma, S.S.; Rupassara, S.I.; Bohnert, H.J. Salinity stress adaptation competence in the extremophile *Thellungiella halophila* in comparison with its relative Arabidopsis thaliana. *Plant J.* **2005**, *44*, 826–839. [CrossRef] [PubMed]

31.   Griffith, M.; Timonin, M.; Wong, C.E.; Gray, G.R.; Akhter, S.R.; Saldanha, M.; Rogers, M.A.; Weretilnyk, E.A.; Moffatt, B. *Thellungiella*: An *Arabidopsis*-related model plant adapted to cold temperatures. *Plant Cell Environ.* **2007**, *30*, 529–538. [CrossRef] [PubMed]

32.   Kant, S.; Bi, Y.M.; Weretilnyk, E.; Barak, S.; Rothstein, S.J. The Arabidopsis halophytic relative *Thellungiella halophila* tolerates nitrogen-limiting conditions by maintaining growth, nitrogen uptake, and assimilation. *Plant Physiol.* **2008**, *147*, 1168–1180. [CrossRef] [PubMed]

33.   Lugan, R.; Niogret, M.F.; Leport, L.; Guegan, J.P.; Larher, F.R.; Savouré, A.; Kopka, J.; Bouchereau, A. Metabolome and water homeostasis analysis of *Thellungiella salsuginea* suggests that dehydration tolerance is a key response to osmotic stress in this halophyte. *Plant J.* **2010**, *64*, 215–229. [CrossRef] [PubMed]

34.   Guevara, D.R.; Champigny, M.J.; Tattersall, A.; Dedrick, J.; Wong, C.E.; Li, Y.; Labbe, A.; Ping, C.L.; Wang, Y.; Nuin, P.; et al. Transcriptomic and metabolomic analysis of Yukon *Thellungiella* plants grown in cabinets and their natural habitat show phenotypic plasticity. *BMC Plant Biol.* **2012**, *12*, 175. [CrossRef] [PubMed]

35.   Lee, Y.P.; Funk, C.; Erban, A.; Kopka, J.; Köhl, K.I.; Zuther, E.; Hincha, D.K. Salt stress responses in a geographically diverse collection of *Eutrema/Thellungiella* spp. accessions. *Funct. Plant Biol.* **2016**, *43*, 590–606.

36.   Bravo, L.A.; Saavedra-Mella, F.A.; Vera, F.; Guerra, A.; Cavieres, L.A.; Ivanov, A.G.; Huner, N.P.A.; Corcuera, L.J. Effect of cold acclimation on the photosynthetic performance of two ecotypes of *Colobanthus quitensis* (Kunth) Bartl. *J. Exp. Bot.* **2007**, *58*, 3581–3590. [CrossRef] [PubMed]

37.   Yamori, W.; Noguchi, K.; Hikosaka, K.; Terashima, I. Phenotypic plasticity in photosynthetic temperature acclimation among crop species with different cold tolerances. *Plant Physiol.* **2010**, *152*, 388–399. [CrossRef] [PubMed]

38.   Lee, Y.P.; Babakov, A.; de Boer, B.; Zuther, E.; Hincha, D.K. Comparison of freezing tolerance, compatible solutes and polyamines in geographically diverse collections of *Thellungiella* sp. and *Arabidopsis thaliana* accessions. *BMC Plant Biol.* **2012**, *12*, 131. [CrossRef] [PubMed]

39.   Khanal, N.; Moffatt, B.A.; Gray, G.R. Acquisition of freezing tolerance in *Arabidopsis* and two contrasting ecotypes of the extremophile *Eutrema salsugineum* (*Thellungiella salsuginea*). *J. Plant Physiol.* **2015**, *180*, 35–44. [CrossRef] [PubMed]

40.  Gao, F.; Zhou, Y.; Zhu, W.; Li, X.; Fan, L.; Zhang, G. Proteomic analysis of cold stress-responsive proteins in *Thellungiella* rosette leaves. *Planta* **2009**, *230*, 1033–1046. [CrossRef] [PubMed]

41.  Wong, C.E.; Li, Y.; Whitty, B.; Akhter, S.; Diaz, C.; Brandle, J.; Golding, B.; Weretinylk, E.; Moffatt, B.A.; Griffith, M. Expressed sequence tags from the Yukon ecotype of *Thellungiella salsuginea* reveal that gene expression in response to cold, drought and salinity shows little overlap. *Plant Mol. Biol.* **2005**, *58*, 561–574. [CrossRef] [PubMed]

42.  Wong, C.E.; Li, Y.; Labbe, A.; Guevara, D.; Nuin, P.; Whitty, B.; Diaz, C.; Golding, G.B.; Gray, G.R.; Weretilnyk, E.A.; et al. Transcriptional profiling implicates novel interactions between abiotic stress and hormonal responses in *Thellungiella*, a close relative of Arabidopsis. *Plant Physiol.* **2006**, *140*, 1437–1450. [CrossRef] [PubMed]

43.  Champigny, M.J.; Sung, W.W.L.; Catana, V.; Salwan, R.; Summers, P.S.; Dudley, S.A.; Provart, N.J.; Cameron, R.K.; Golding, G.B.; Weretilnyk, E.A. RNA-Seq effectively monitors gene expression in *Eutrema salsugineum* plants growing in an extreme natural habitat and in controlled growth cabinet conditions. *BMC Genom.* **2013**, *14*, 578. [CrossRef] [PubMed]

44.  Lee, Y.P.; Giorgi, F.M.; Lohse, M.; Kvederaviciute, K.; Klages, S.; Usadel, B.; Meskiene, I.; Reinhardt, R.; Hincha, D.K. Transcriptome sequencing and microarray design for functional genomics in the extremophile Arabidopsis relative *Thellungiella salsuginea* (*Eutrema salsugineum*). *BMC Genom.* **2013**, *14*, 793. [CrossRef] [PubMed]

45.  Stepien, P.; Johnson, G.N. Contrasting responses of photosynthesis to salt stress in the glycophytic Arabidopsis and the halophytic *Thellungiella*: Role of the plastid terminal oxidase as an alternative electron sink. *Plant Physiol.* **2009**, *149*, 1154–1165. [CrossRef] [PubMed]

46.  Sui, N.; Han, G. Salt-induced photoinhibition of PSII is alleviated in halophyte *Thellungiella halophila* by increases of unsaturated fatty acids in membrane lipids. *Acta Physiol. Plant.* **2014**, *36*, 983–992. [CrossRef]

47.  Malik, V.M.; Lobo, J.M.; Stewart, C.; Irani, S.; Todd, C.D.; Gray, G.R. Growth irradiance affects ureide accumulation and tolerance to photoinhibition in *Eutrema salsugineum* (*Thellungiella salsuginea*). *Photosynthetica* **2016**, *54*, 93–100. [CrossRef]

48.  Nishio, J.N. Why are higher plants green? Evolution of the higher plant photosynthetic pigment complement. *Plant Cell Environ.* **2000**, *6*, 539–548. [CrossRef]

49.  M'rah, S.; Ouerghi, Z.; Berthomieu, C.; Havaux, M.; Jungas, C.; Hajji, M.; Grignon, C.; Lachaal, M. Effects of NaCl on the growth, ion accumulation and photosynthetic parameters of *Thellungiella halophila*. *J. Plant Physiol.* **2006**, *163*, 1022–1031. [CrossRef] [PubMed]

50.  Savitch, L.V.; Barker-Astrom, J.; Ivanov, A.G.; Hurry, V.; Öquist, G.; Huner, N.P.A.; Gardeström, P. Cold acclimation of *Arabidopsis thaliana* results in incomplete recovery of photosynthetic capacity, associated with an increased reduction of the chloroplast stroma. *Planta* **2001**, *214*, 295–303. [CrossRef] [PubMed]

51.  Bräutigam, K.; Dietzel, L.; Kleine, T.; Ströher, E.; Wormuth, D.; Dietz, K.J.; Radke, D.; Wirtz, M.; Hell, R.; Dörmann, P.; et al. Dynamic plastid redox signals integrate gene expression and metabolism to induce distinct metabolic states in photosynthetic acclimation in Arabidopsis. *Plant Cell* **2009**, *21*, 2715–2732. [CrossRef] [PubMed]

52.  Eberhard, S.; Finazzi, G.; Wollman, F.A. The dynamics of photosynthesis. *Annu. Rev. Genet.* **2008**, *42*, 463–515. [CrossRef] [PubMed]

53.  Harbinson, J.; Hedley, C.L. Changes in P700 oxidation during the early stages of the induction of photosynthesis. *Plant Physiol.* **1993**, *103*, 649–660. [CrossRef] [PubMed]

54.  Asada, K.; Heber, U.; Schreiber, U. Pool size of electrons that can be donated to P700$^+$, as determined in intact leaves: Donation to P700$^+$ from stromal components via the intersystem chain. *Plant Cell Physiol.* **1992**, *33*, 927–932.

55.  Bukhov, N.G.; Samson, G.; Carpentier, R. Non-photosynthetic reduction of the intersystem electron transport chain of chloroplasts following heat stress: The pool size of stromal reductants. *Photochem. Photobiol.* **2001**, *74*, 438–443. [CrossRef]

56.  Gray, G.R.; Hope, B.J.; Qin, X.Q.; Tayler, B.G.; Whitehead, C.L. The characterization of photoinhibition and recovery during cold acclimation in *Arabidopsis thaliana* using chlorophyll fluorescence imaging. *Physiol. Plant.* **2003**, *119*, 365–375. [CrossRef]

57.  Hendrickson, L.; Furbank, R.T.; Chow, W.S. A simple alternative approach to assessing the fate of absorbed light energy using chlorophyll fluorescence. *Photosynth. Res.* **2004**, *82*, 73–81. [CrossRef] [PubMed]

58.	Kramer, D.M.; Johnson, G.; Kiirats, O.; Edwards, G.E. New fluorescence parameters for the determination of $Q_A$ redox state and excitation energy fluxes. *Photosynth. Res.* **2004**, *79*, 209–218. [CrossRef] [PubMed]

59.	Baker, N.R. Chlorophyll fluorescence: A probe of photosynthesis in vivo. *Annu. Rev. Plant Biol.* **2008**, *59*, 89–113. [CrossRef] [PubMed]

60.	Genty, B.; Briantais, J.M.; Baker, N.R. The relationship between the quantum yield of photosynthetic electron transport and quenching of chlorophyll fluorescence. *Biochim. Biophys. Acta* **1989**, *990*, 87–92. [CrossRef]

61.	Bilger, W.; Schreiber, U. Energy-dependent quenching of dark-level chlorophyll fluorescence in intact leaves. *Photosynth. Res.* **1986**, *10*, 303–308. [CrossRef] [PubMed]

62.	Schreiber, U.; Bilger, W.; Neubauer, C. *Chlorophyll Fluorescence as a Nonintrusive Indicator for Rapid Assessment of In Vivo Photosynthesis*; Spinger-Verlag: Berlin, Germany, 1994; pp. 49–70.

63.	Gray, G.R.; Ivanov, A.G.; Krol, M.; Huner, N.P.A. Adjustment of thylakoid plastoquinone content and Photosystem I electron donor pool size in response to growth temperature and growth irradiance in winter rye (*Secale cereale* L.). *Photosynth. Res.* **1998**, *56*, 209–221. [CrossRef]

64.	Lichtenthaler, H.K.; Wellburn, A.R. Determination of total carotenoids and chlorophylls *a* and *b* of leaf extracts in different solvents. *Biochem. Soc. Trans.* **1983**, *11*, 591–592. [CrossRef]

# First Report on the Ethnopharmacological Uses of Medicinal Plants by Monpa Tribe from the Zemithang Region of Arunachal Pradesh, Eastern Himalayas, India

**Tamalika Chakraborty** [1,2], **Somidh Saha** [3,4,5,*] **and Narendra S. Bisht** [3,6]

[1]   Institute of Ethnobiology, School of Studies in Botany, Jiwaji University, Gwalior 474011, India;
      tamalika.chakraborty@waldbau.uni-freiburg.de
[2]   Chair of Site Classification and Vegetation Science, Institute of Forest Sciences, University of Freiburg,
      Tennenbacherstr. 4, D-79106 Freiburg, Germany
[3]   Resource Survey and Management Division, Forest Research Institute, PO New Forest, Dehra Dun 248006,
      India; bishtnsifs@yahoo.com
[4]   Chair of Silviculture, Institute of Forest Sciences, University of Freiburg, Tennenbacherstr. 4,
      D-79106 Freiburg im Breisgau, Germany
[5]   Institute for Technology Assessment and Systems Analysis (ITAS), Karlsruhe Institute of Technology (KIT),
      Karlstr. 11, D-76133 Karlsruhe, Germany
[6]   Directorate of Extension, Indian Council of Forestry Research and Education, PO New Forest,
      Dehra Dun 248006, India
*    Correspondence: somidh.saha@waldbau.uni-freiburg.de or somidhs@gmail.com

Academic Editor: Milan S. Stankovic

**Abstract:** The Himalayas are well known for high diversity and ethnobotanical uses of the region's medicinal plants. However, not all areas of the Himalayan regions are well studied. Studies on ethnobotanical uses of plants from the Eastern Himalayas are still lacking for many tribes. Past studies have primarily focused on listing plants' vernacular names and their traditional medicinal uses. However, studies on traditional ethnopharmacological practices on medicine preparation by mixing multiple plant products of different species has not yet been reported in published literature from the state of Arunachal Pradesh, India, Eastern Himalayas. In this study, we are reporting for the first time the ethnopharmacological uses of 24 medicines and their procedures of preparation, as well as listing 53 plant species used for these medicines by the Monpa tribe. Such documentations are done first time in Arunachal Pradesh region of India as per our knowledge. Our research emphasizes the urgent need to document traditional medicine preparation procedures from local healers before traditional knowledge of tribal people living in remote locations are forgotten in a rapidly transforming country like India.

**Keywords:** medicinal plants; traditional knowledge; Eastern Himalayas; mountain plants; ethnobotany; ethnopharmacology; bioprospecting

## 1. Introduction

The Himalayas are rich in diversity of medicinal plant species [1]. The culture of traditional healing of diseases using these plants is still prevalent among aboriginal mountain communities in the Himalayas. Arunachal Pradesh (approximately 84,000 km$^2$ in size), a state belonging to the Republic of India, is situated in the Eastern Himalayas. The entire state is declared as a "biodiversity hotspot" with 5000 endemic flowering plant species as well as very high faunal diversity [1,2]. Also, this state is the home to 28 major tribes and 110 sub-tribes and is considered to be one of the most splendidly variegated and multilingual tribal areas of the world [3]. The traditional wisdom of healing among mountain tribal communities is orally transferred from one generation to the next generation through traditional healers, spiritual gurus, and elderly or sometimes ordinary people. This traditional wisdom, if not properly documented, can be lost by rapid modernization and religious reformation among mountain communities in Arunachal Pradesh where traditional customary practices are often regarded as a symbol of "*backwardness*" and "*unscientific*" by the educated and younger generations. Nevertheless, plant-based traditional wisdom inherited and carried forward to generation after generation in traditional communities has become a recognized tool in the search for new sources of drugs and pharmaceuticals in modern medicine [4]. Therefore, field based ethnobotanical and ethnopharmacological surveys to list medicinal plants and their uses are still relevant and worth the effort in order to bring out new clues for the development of drugs to treat human diseases [5].

Before coming to our research objectives, we would like to briefly mention the state of the art of ethnopharmacological research in the Himalayas. There are plenty of research works on the listing of the traditional uses of medicinal plants from the Himalayas. A search with the terms "medicinal plants * Himalayas" yielded 163 peer-reviewed articles listed in ISI Web of Knowledge on 20 February 2017. However, out of those 163 articles, 19 articles were found from the Eastern Himalayas and only two were on the Monpa tribe (please see Materials and Methods section for a detailed sociocultural description of the Monpa tribe). Haridasan et al., in the seminal works produced in 1998 and 1990, comprehensively listed medicinal and edible plants of the Monpa tribe and other tribes of Arunachal Pradesh [2]. Recently, Namsa et al. (2011) listed 50 plant species and recorded their ethnobotanical uses among people of the Monpa tribe at the southern range of their habitation (i.e., Kalaktang circle of West Kameng district of Arunachal Pradesh) [6]. These two publications provided general descriptions of the plants, traditional uses of the plants to cure certain diseases, and traditional ways of consumption of these plants or plant parts (e.g., pills, syrups, decoctions, etc.). Nevertheless, no ethnopharmacological studies have yet reported how, and in what proportion, multiple plant parts from different species can be used to prepare specific ethnomedicines for healing of diseases among the Monpa tribes or any other tribes of the Eastern Himalayas as per our literature research as of 20 February 2017. In addition, the traditional knowledges of the people of the Monpa tribe residing at their northern habitation range (i.e., Zemithang circle of Tawang district of Arunachal Pradesh) are still not adequately documented due to the remoteness of the location.

Documentations of traditional ethnopharmacological know-hows are necessary for the preservation of traditional knowledges of Himalayan tribal communities. Such documentations could create interest among professional pharmacologists for the search of new medicines and motivate ethnologists to study high cultural diversity of the Eastern Himalayas of India. Those were the main motivations to carry out this research. This study aims to document traditional ethnopharmacological know-hows of medicinal drug making among Monpa people in the Zemithang region of the state of Arunachal Pradesh.

## 2. Results

Our study was a notable departure from the previous studies from the area that mostly documented and described the use of plant parts in individual plant species. We documented and described 24 ethnomedicines prepared by traditional healers based on 53 species (Table 1). The medicines were comprised of 53 plant species of medicinal plants belonging to 21 families (Table 2). These traditional medicines were most commonly used to heal a wide range of diseases such as arthritis, rheumatic pain, malaria, cough and cold, dysentery, etc. In addition, we recorded descriptions of medicines for the treatment of diseases such as epilepsy (*Pambrey*), herpes (*Bukbukpa-khaksa-chandongbra*), and oedema (*Darshek sheng nye putpoo*) that have rarely been reported in past studies. Our main result is presented in Table 1 which provides a list of ethnomedicines and their preparations by traditional ethnopharmacological techniques.

**Table 1.** List of 24 ethnomedicines used by the Zemithang Monpa people and the associated medicinal plants documented in this study.

| Number | Name of the Ethnomedicines (in Monpa Tribal Language of Zemithang Dialect) | Type of Medicines | Name of Medicinal Plants Used for Ethnomedicines | Proportion of Used Plant Parts (Bray in local language is a Buddhist prayer bowl. It could be made of gold, silver, brass, copper, stone, or wood and is often used for religious offerings. 1 bray can contain approximately 900 g of grain). | Mode of Preparations | Medicinal Uses |
|---|---|---|---|---|---|---|
| 1 | Arkadamasisi | paste | Crawfurdia speciosa Wall. | 1/4 bray of dried root + 1/8 bray of dried flower | dried roots and flower crushed together to prepare powder and then mixed with water to prepare paste | paste is applied externally for healing wounds |
| 2 | Baribama | decoction | Aristolochia griffithii Hook.f. | 1/8 bray of raw washed roots | roots are boiled with water to prepare a decoction | decoction is taken as blood purifier and purgative |
| 3 | Blenga | pills | Hedychium spicatum Buch.-Ham. | 1 bray of washed raw roots | raw roots are crushed and small round pills are prepared and sun dried | pills are taken orally for treatment of dysentery, chest pain, cough and cold |
| 4 | Bomdeng | paste | Cirsium falconeri Hook.f., Cirsium verutum D. Don and Onopordum acanthium L. | 1/2 bray of washed raw root of C. falconeri + 1/2 bray of washed raw root of C. verutum + 3/4 bray of washed raw root of O. acanthium | raw roots are mixed together and crushed to prepare paste | paste is applied externally to treat arthritis |
| 5 | Bragen | syrup | Bergenia stracheyi Hook.f. & Thorns. | 1 bray of washed fresh leaves | clear fresh leaves are crushed to prepare paste and mixed with 1/4 bray local millets wine to prepare syrup | syrup is taken for treating rheumatic pains |
| 6 | Bukbukpa-khaksa-chandongbra | paste | Campanula latifolia Linn., Codonopsis clematidea Schrenk. and Codonopsis viridis Wall. | 1/2 bray of washed fresh leaves and 1/4 bray of fresh flowers of each species + 1/4 bray of conch powder + 1/4 bray of water | leaves and flowers are crushed together and mixed with conch powder and water to prepare paste | paste is applied externally to treat herpes |
| 7 | Chandoo-konghlin-bhor | powder | Aconitum ferox Wall. ex Ser., Aconitum heterophyllum Wall. ex Royle, Aconitum hookeri Stapf., Geranium polyanthes Edgeworth & J. D. Hooker, Geranium wallichianum D. Don and Picrorhiza kurrooa Royle ex Benth. | 1 small dried root from each plants of A. ferox, A. heterophyllum and A. hookeri (total 5 g mixture of three plants) + 3 bray of dried root of G. polyanthes and G. wallichianum + 1 bray of dried root P. kurroa | all ingredients are mixed together and crushed to prepare a powder | powder is taken orally to overcome poisoning effects |
| 8 | Chhalachhusar | syrup | Meconopsis grandis Prain and Meconopsis paniculata D. Don | 1/2 bray of dry leaves from each plant + 1/2 bray of dy flowers from each plant | dried leaves and flowers are mixed together and crushed to prepare powder, and a small amount of powder (5 g) is mixed with 1 bray of water to prepare a syrup | syrup is taken to treat sexually transmitted diseases |

**Table 1.** *Cont.*

| Number | Name of the Ethnomedicines (in Monpa Tribal Language of Zemithang Dialect) | Type of Medicines | Name of Medicinal Plants Used for Ethnomedicines | Proportion of Used Plant Parts (*Bray* in local language is a Buddhist prayer bowl. It could be made of gold, silver, brass, copper, stone, or wood and is often used for religious offerings. 1 *bray* can contain approximately 900 g of grain). | Mode of Preparations | Medicinal Uses |
|---|---|---|---|---|---|---|
| 9 | Chhurchu doho keusheng | pills | *Rheum australe* D. Don, *Rheum nobile* Hook.f. & Thoms. and *Bistorta affinis* D. Don | 1/2 bray of fresh roots from each species + 1/4 bray of dried flowers from each species | fresh roots and dried flowers are crushed together to make a paste, then small round pills are prepared and sun dried | pills are taken orally to overcome poisoning effects |
| 10 | Comrep | syrup | *Rubus ellipticus* Smith and *Rubus paniculatus* Smith | 1/2 bray of fresh ripe fruits from each plant | roots are mixed together and crushed to prepare a thick syrup | syrup is used for treatment of cold and cough |
| 11 | Darshek sheng nye putpoo | decoction | *Pieris formosa* (Wallich) D. Don; *Vaccinium nummularia* Hook.f. & Thoms. | 1/4 bray of fruits of *P. formosa* + 1/4 bray of fruits of *V. numularia* + 1/2 bray of fresh roots of *P. formosa* + 1/2 bray of fresh roots of *V. numularia* | mixture of all fresh fruits and roots along with water is boiled to prepare a decoction | decoction is taken to cure oedema |
| 12 | Dhamrep | paste | *Fragaria nubicola* Lindl., *Geum elatum* Wall. and *Potentilla peduncularis* D. Don. | 1/2 bray of *F. nubicola* fresh fruits + 1/8 bray of dried roots of *G. elatum* + 1/4 bray of leaves of *P. peduncularis* | fresh fruits, leaves, and dried roots are crushed together to prepare a paste | paste is taken orally to treat cold, cough, and fever |
| 13 | Gin sheng | powder | *Panax pseudoginseng* Wall. | 1/4 bray of dried rhizomes | dried rhizomes are crushed to prepare powder, which is taken with water | used for treating depression and fatigue |
| 14 | Karpo Chiito | paste | *Iris clarkei* Baker | 1/4 bray of dried flower, leaves, stem, and root | dried flowers, leaves, stem parts, and roots are crushed together to prepare powder and mixed with local millets wine to prepare paste | paste is used externally to treat muscle pain |
| 15 | Lowa bur | pills | *Lomatogonium carinthiacum* (Wulfen) Rchb. | 1/4 bray of dried roots | dried roots are crushed and small round pills are prepared and sun dried | pills are taken orally to treat cold, cough, and fever |
| 16 | Maraptang | pills | *Houttuynia cordata* Thunb. | 1/4 bray of dried roots | dried roots are crushed and small round pills are prepared and sun dried | pills are taken orally for treatment of piles |
| 17 | Nyasheng jormu | paste and pills | *Viscum articulatum* Burm. f. | 1/4 bray of fresh roots + 1/4 bray of fresh leaves + 1/4 bray of fresh stems | fresh roots, leaves, and parts of stem are crushed together to prepare paste; sometimes paste is used to prepare small round pills and sun dried | paste is used to join broken bones, treating pain from swelling of nerves and healing wounds; pills are used for treatment of infertility among women |

**Table 1.** *Cont.*

| Number | Name of the Ethnomedicines (in Monpa Tribal Language of Zemithang Dialect) | Type of Medicines | Name of Medicinal Plants Used for Ethnomedicines | Proportion of Used Plant Parts (*Bray* in local language is a Buddhist prayer bowl. It could be made of gold, silver, brass, copper, stone, or wood and is often used for religious offerings. 1 *bray* can contain approximately 900 g of grain). | Mode of Preparations | Medicinal Uses |
|---|---|---|---|---|---|---|
| 18 | Pambrey | mixture | *Anaphalis monocephala* DC., *Anaphalis triplinervis* Sims., *Gnaphalium hypoleucum* DC., *Leontopodium himalayanum* DC., *Leontopodium jacotianum* Beauv., *Tanacetum tibeticum* Hook.f. and *Tanacetum gracile* Hook.f. & Thoms. | 1/2 bray of flowers from each of the plants | flowers are kept in a dark place for two days after plucking and then mixed together | used to treat epilepsy, mildly warm mixtures are applied on the bare head of the patient (two times a day) consecutively for 15 to 20 days |
| 19 | Pangen | pills | *Gentiana depressa* D. Don, *Gentiana ornata* Wallich ex G. Don, *Gentiana phyllocalyx* C. B. Clarke and *Gentiana tubiflora* Wallich ex G. Don. | 1/4 bray of dried roots from each of the plants | dried roots are crushed and then mixed with 1/4 bray of local millet wine and 1/2 bray of water and small round pills are prepared and sun dried | pills are used to treat cough, cold, and headache |
| 20 | Rah-nya | decoction | *Smilacina purpurea* *S. oleracea* and *Polygonatum multiflorum* Allem. | 1/4 bray of fresh roots from each of the plants | roots are boiled with water to prepare a decoction | is used for the treatment of malaria |
| 21 | Rambhoo tsarphakur | paste | *Morina longifolia* Wall., *Pterocephalus hookeri* (C. B. Clarke) Hock. | 1/4 bray of dried flowers, 1/2 bray of fresh roots, 1/4 bray of fresh fruits of *M. longifolia* + 1/8 bray of dried flower, 1/2 bray of fresh roots, 1/2 bray of fresh fruits of *P. hookeri* | flowers, roots, and fruits of both plants are mixed together and crushed to prepare paste | paste is applied for healing chest pain |
| 22 | Trahm-Sheng | paste | *Corydalis cashmeriana* Royle. | 1/4 bray of fresh leaves + 1/4 bray of fresh flower | fresh leaves and flowers are crushed to prepare paste | paste is applied for healing wounds |
| 23 | Wang La | powder | *Swertia chirayita* (Roxb. ex Flem.) Karst. and *Swertia hookeri* C. B. Clarke | 1/2 bray of dried whole plants | dried whole plants are crushed to prepare powder, which is taken with water | powder is used to treat malaria, and is also used as a purgative and laxative |
| 24 | Whan | pills | *Lilium nepalense* D. Don | 1/2 bray of dried roots | dried roots are crushed and mixed with water to prepare small round pills which are then sun dried | pills are used for treating gastritis and stomachic |

**Table 2.** List of recorded plants used in Ethnomedicine.

| Serial Number | Botanical Name | Family | Type |
|---|---|---|---|
| 1 | *Aconitum ferox* Wall. ex Ser. | Ranunculaceae | herb |
| 2 | *Aconitum heterophyllum* Wall. ex Royle | Ranunculaceae | herb |
| 3 | *Aconitum hookeri* Stapf. | Ranunculaceae | herb |
| 4 | *Anaphalis monocephala* DC. | Compositae | herb |
| 5 | *Anaphalis triplinervis* Sims. | Compositae | herb |
| 6 | *Aristolochia griffithii* Hook.f. | Aristolochiaceae | vine |
| 7 | *Bergenia stracheyi* Hook.f. & Thorns. | Saxifragaceae | herb |
| 8 | *Bistorta affinis* D. Don | Polygonaceae | herb |
| 9 | *Campanula latifolia* Linn. | Campanulaceae | herb |
| 10 | *Cirsium falconeri* Hook. f. | Asteraceae | herb |
| 11 | *Cirsium verutum* D. Don | Asteraceae | herb |
| 12 | *Codonopsis clematidea* Schrenk. | Campanulaceae | herb |
| 13 | *Codonopsis viridis* Wall. | Campanulaceae | herb |
| 14 | *Corydalis cashmeriana* Royle. | Papaveraceae | herb |
| 15 | *Crawfurdia speciosa* Wall. | Gentianaceae | herb |
| 16 | *Fragaria nubicola* Lindl. | Rosaceae | herb |
| 17 | *Gentiana depressa* D. Don | Gentianaceae | herb |
| 18 | *Gentiana ornata* Wallich ex G. Don | Gentianaceae | herb |
| 19 | *Gentiana phyllocalyx* C. B. Clarke | Gentianaceae | herb |
| 20 | *Gentiana tubiflora* Wallich ex G. Don. | Gentianaceae | herb |
| 21 | *Geranium polyanthes* Edgeworth & J. D. Hooker | Geraniaceae | herb |
| 22 | *Geranium wallichianum* D. Don | Geraniaceae | herb |
| 23 | *Geum elatum* Wall. | Rosaceae | herb |
| 24 | *Gnaphalium hypoleucum* DC. | Asteraceae | herb |
| 25 | *Hedychium spicatum* Buch.-Ham. | Zingiberaceae | herb |
| 26 | *Houttuynia cordata* Thunb. | Saururaceae | herb |
| 27 | *Iris clarkei* Baker | Iridaceae | herb |
| 28 | *Leontopodium himalayanam* DC. | Asteraceae | herb |
| 29 | *Leontopodium jacotianum* Beauv. | Asteraceae | herb |
| 30 | *Lilium nepalense* D. Don | Liliaceae | herb |
| 31 | *Lomatogonium carithiacum* (Wulfen) Rchb. | Gentianaceae | herb |
| 32 | *Meconopsis grandis* Prain | Papaveraceae | herb |
| 33 | *Meconopsis paniculata* D. Don | Papaveraceae | herb |
| 34 | *Morina longifolia* Wall. | Dipsacaceae | herb |
| 35 | *Onopordum acanthium* L. | Asteraceae | herb |
| 36 | *Panax pseudoginseng* Wall. | Araliaceae | herb |
| 37 | *Picrorhiza kurrooa* Royle ex Benth. | Scrophulariaceae | herb |
| 38 | *Pieris formosa* (Wallich) D. Don | Ericaceae | shrub |
| 39 | *Polygonatum multiflorum* Allem. | Convallariaceae | herb |
| 40 | *Potentilla peduncularis* D. Don. | Rosaceae | herb |
| 41 | *Pterocephalus hookeri* (C. B. Clarke) Hock. | Dipsacaceae | herb |
| 42 | *Rheum australe* D. Don | Polygonaceae | herb |
| 43 | *Rheum nobile* Hook.f. & Thoms. | Polygonaceae | herb |
| 44 | *Rubus ellipticus* Smith | Rosaceae | shrub |
| 45 | *Rubus paniculatus* Smith | Rosaceae | shrub |
| 46 | *Swertia chirayita* (Roxb. ex Flem.) Karst. | Gentianaceae | herb |
| 47 | *Smilacina oleracea* (Baker) Hook.f. | Liliaceae | herb |
| 48 | *Smilacina purpurea* (Wall.) H.Hara | Liliaceae | herb |
| 49 | *Swertia hookeri* C. B. Clarke | Gentianaceae | herb |
| 50 | *Tanacetum gracile* Hook.f. & Thoms. | Asteraceae | herb |
| 51 | *Tanacetum tibeticum* Hook.f. | Asteraceae | herb |
| 52 | *Vaccinium nummularia* Hook.f. & Thoms. | Ericaceae | shrub |
| 53 | *Viscum articulatum* Burm. f. | Viscaceae | shrub |

## 3. Materials and Methods

### 3.1. Sociocultural Description of the People from the Monpa Tribe

The Monpa people are a Buddhist tribe belonging to the Mahayana (Tibetan–Lamaist) *Gelukpa* and *Nyngmapa* sect. The Monpa people are inhabitants of the western most districts of the Tawang and West Kameng regions of Arunachal Pradesh, India. Their main centers of habitation are in and around the administrative headquarters of Zemithang, Tawang, Dirang, and Kalaktang. Depending on the place of living and the geographical location of these centers, they are often called as Zemithang-Tawang or "Northern Monpas", Dirang or "Central Monpas", and Kalaktang or "Southern Monpas". The language used by Dirang and Kalaktang Monpa are different from that of Tawang Monpas. Dirang and Kalaktang Monpas use a dialect of Bhutanese Brokpa language, whereas Zemithang-Tawang Monpas use a dialect of Tibetan-Bhutanese Dakpa language. However, many other aspects of their life are quite similar. In Dakpa language, the name "Mon" and "Pa" signify the "Men of the Lower Country" or the inhabitants of southern regions to Tibet.

The Monpa villages are often situated on the slopes of the hills or in the valleys. A striking characteristic of the Monpa villages is the presence of a *"Gompa"* (Buddhist village monastery), often situated on the top of the hill and surrounded by prayer flags (*"phan"*), stone shrines (*"mane"*), and small chapels called *"chorten"* which are often found alongside the roads and foot-lanes. The houses are usually double or triple storeyed, and made mainly of locally sourced stone. Each house has a family chapel with a wooden, stone, or brass statue of the *Lord Buddha*.

The adornments and clothing are diverse and colourful. People cover their whole bodies with a variety of well-designed woolen garments. The women do the traditional spinning and weaving of the garments, as well as carpet making. The Monpa people can be recognized from a long distance owing to the attractive color of their clothing, which is a mellow strawberry red. The Monpa people love this color and dye their clothes themselves using the locally available natural dyes from diverse species of Rhododendrons and other plants. They love music and dance. Their musk-dances are very famous and attract a large number of tourists. The "Losar" or the Buddhist New Year is the most important festival celebrated among them, which is organized in February. Monpa villages could be located at a great distance by their high fluttering Buddhist prayer flags on which is printed in Tibetan script *"Om Mani Pame Hung"* which means *"Hail to him who is born as a Jewel in a Lotus"*.

The Monpa people typically eat various types of locally grown vegetables, which are often cultivated by using tradtional methods [7]. Drinking yak milk, making homemade butter and dry cheese from yak milk (e.g., the famous *"churpi"* dry cheese), eating yak meat, pork, chicken, mutton, cultivation of multiple species of cereal and pulse through sustainable mountain agriculture based on tradtional ecological knowledges without any use of pesticides, herbicides, and chemical fertilizers are common practice [7]. Monogamy appears to be the form of marriage followed by the Tibetan Buddhist traditions. Tattooing is not typically observed among Monpa people, which is a stark contrast to the people from other tribes such as the *Nishi* and *Adi* in nearby districts. Information relating to the origin and migration of the Monpa people to their present habitat in Arunachal Pradesh is largely obscure. This is because written records on the history of Monpa people from the middle ages or beyond are very rare. Thus, it remains a matter of further anthropological and archeological research to find out the route and approximate time of their relationship with either the Tibetans or Bhutanese, or even with the people of Pan-Indian origin. When we visited Namshu village in Dirang region, the *"Gaobura"* or the village headman told us a folklore story about a marriage between a prince from Bhutan and a local Monpa girl from that village. The story indicates the Bhutanese influence among Dirang Monpa. The language of the Eastern Bhutan and Dirang areas are similar. Here we can quote from von Fürer-Haimendorf of Austria who was the most prominent anthropologist that ever worked with the tribes of this region [8]: *"THE REGION, WHICH ADJOINS TO THE WEST OF THE MOUNTAIN KINGDOM OF BHUTAN, DIFFERS FROM THE REST OF ARUNACHAL PRADESH BOTH TOPOGRAPHICALLY AND CULTURALLY. WHEREAS, ELSEWHERE NATURE AND THE TERRAIN HAD PREVENTED THE DEVELOPMENT OF CARAVAN ROUTES SUITABLE FOR PACK ANIMALS IN THE WESTERNMOST PART OF ARUNACHAL PRADESH. HOWEVER, WHERE THE CLIMATE AND GEOGRAPHICAL CONDITION WERE FAVORABLE, THE TRADE ROUTES WERE OPENED LINKING THE TERRITORY BOTH WITH TIBET AND THE PLAINS OF ASSAM OF INDIA. HENCE CONDITIONS ARE SIMILAR TO THOSE PREVAILING IN BHUTAN AND FURTHER WEST TO SIKKIM AND NEPAL. ALONG WITH THESE TRANS-HIMALAYAN TRADE ROUTES, TIBETAN CULTURAL ELEMENTS*

*AND ULTIMATELY BUDDHIST MONKS AND NUNS INFILTRATED INTO THE MOUNTAIN REGION LYING BETWEEN THE EASTERNMOST OF BHUTAN AND THE SOUTHERN BORDER OF TIBET."*

### 3.2. Study Area

The study area is located in the extreme north of the north-western Arunachal Pradesh. The areas of investigations are situated at the Lumla–Zemithang administrative circle of the Tawang district of Arunachal Pradesh (Figure 1). This region is situated along the bank of the river Namshyang Chu that flows through the area. The name, exact locations, and altitude of the three villages where the study took place are as follows: (1) village Kublaitang (27°37′070″ N, 91°41′618″ E, elevation 2224 m); (2) village Shakti (27°36′736″ N, 91°42′970″ E, elevation 2020 m); and (3) village Lumpho (27°43′140″ N, 091°43′069″ E, elevation 2550 m). The research areas fall under the middle Himalayan range of the Eastern Himalayas. The soil on the hills is moderately deep and moist, fertile loamy layer stained with humus. At places, shallow soils are not uncommon with underlying boulders and rocks. The subsoil at lower elevations consists of mostly boulders and pebbles superimposed by a layer of a sandy loam of various depths with layers of humus overtop. The relative humidity of this area varies from 30% to 80%. Southern aspects at low altitude areas are more humid than any other places in the region. The annual temperature in this area varies from −10 degrees Celsius to +15 degrees Celsius. The area typically receives 1500–1800 mm rainfall every year. The dry months are December, January, and February. The pre-monsoon rainfall starts from the end of the March. Highest rainfall is observed in June, July, and August [9]. The forest type of the research area is the Northeastern Himalayan subalpine mixed conifer forests. The top canopy of the forest consists of *Abies densa, Juniperus wallichiana, Illicium griffithi, Pinus wallichiana, Quercus* spp., and *Cupressus torulosa*. The secondary canopy layer mainly consists of *Rhododendron* spp., *Betula utilis, Pyrus aucuparia,* and *Salix wallichiana*. The trees of the forest ground storey are dominated by *Juniperus recurva, Cassiope fastigiata,* and *Rhododendron* spp. [10].

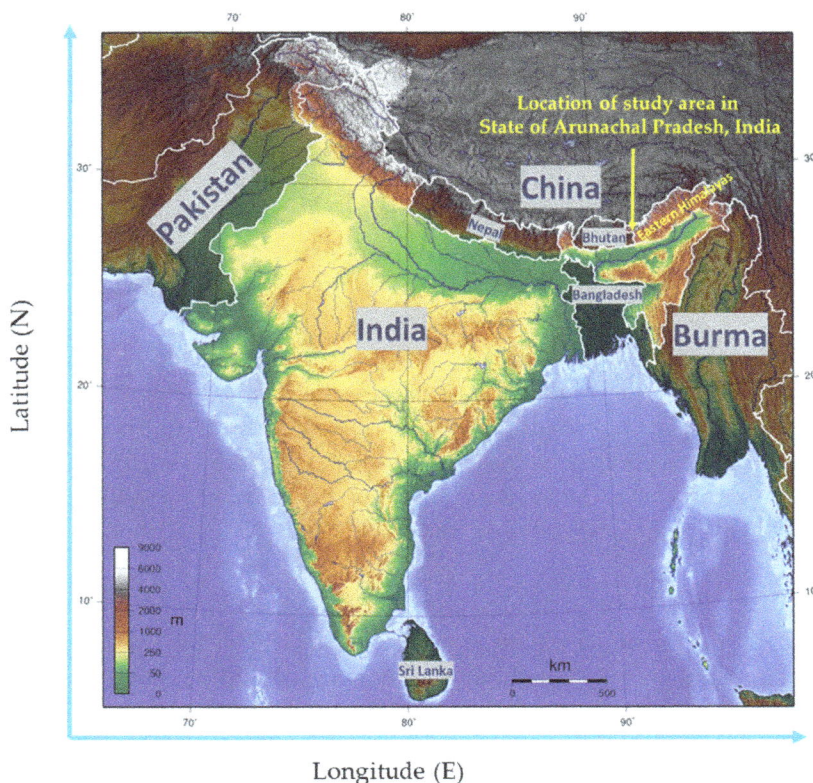

**Figure 1.** Location of the study area at the Zemithang region in the state of Arunachal Pradesh, India pointed by the yellow arrow (map not to scale).

*3.3. Field Surveys*

Field surveys were carried at the three sample villages of the Zemithang region of Tawang district. The research was carried out in three stages. In the first stage, ethnobotanical data were collected from the research area. At the second stage, ethnopharmacological information was collected from the same research area. The herbariums of the collected plant specimens were prepared and verified at the third stage of the research at the Forest Research Institute of Dehradun, India. The field identifications of the plants were mostly done by using field guide with colored photographs of the plants by Polunin and Stainton [11]. Some unidentified and partially identified plants from the field were brought to the specialists at the Forest Research Institute of Dehradun, India for full identification. The participatory transect walk, interview, and discussions with traditional healers were used for ethnobotanical data collection. The total number of participatory transects established were three for every village, resulting in a total of nine across all three villages. The length of each transect was 2 km from the center of the village to three different outward directions, depending on aspects of the village. We used three different groups for the transect walks. These groups were common village people including men and women, hunters, and traditional healers. Two walks with every group with different people were conducted. As such, the total number of transect walks per village was six, thereby totaling 18 transect walks across all three villages. This type of data collection design was followed for the robustness of ethnobotanical information. Apart from this technique for collecting plant specimens with ethnobotanical values, we used a structured questionnaire for interviews and group discussions regarding the ethnopharmacological techniques of medicine preparation for the collected plants. The people who participated in transect walks were not selected for questionnaire surveys in order to avoid repetition and establish a more general idea among larger population groups. The participatory transect walks were mostly carried out in spring and summer when a large flush of herbaceous plants grow in the forest, pasturelands, and meadows after the melting of winter snow. At the second stage of the research, ethnopharmacological information was collected from the high ranked monks and traditional healers who prepare medicine from plants for the healing of the tribal people. In each village, we selected at least three independent healers or monks for this purpose. After gathering the information, we performed a qualitative assessment for reaching a consensus among the respondents and rejected the conflicting responses. The basic information that was collected from these monks and traditional healers were regarding (1) the plants needed to make medicine; (2) the use of plant parts; (3) the different ratios of plant use; (4) the techniques of preparation; (5) the doses and prescription to the patients; and (6) the medicinal uses. The third stage of the research was carried out at the Resource Survey and Management Division of the Forest Research Institute, Dehradun, India. Taxonomical classification was performed with the help of the Botany Division of the Forest Research Institute, and identified plant specimens were confirmed by using the herbaria of the same division for comparison purposes. The specimens with detailed taxonomic information, name of collectors, and place and date of collections were finally deposited to The Course Coordinator of Postgraduate Programs, Forest Research Institute University (Dehra Dun, India) for future references. We had received permission from the local forest authorities in addition to having obtained consents from the traditional healers before doing this survey.

## 4. Discussion

The list of plant species and utilization of plant parts for different diseases documented in this study support a recent study carried out by Namsa et al. (2011) on the southern or Kalaktang Monpas [6]. The list we provided for the medicinal plants is not completely new to ethnobotanists, as it was already listed in old research works on medicinal plants of the Himalayas (see [12–14]). This proves that the plants we listed are already confirmed as "medicinal plants" by past researchers from the other parts of the Himalayas. However, the detailed ethnopharmacological descriptions or traditional ways of preparation for the herbal drugs and medicines were rarely documented. Due to this reason, a search with the terms "ethnopharmacology * Himalayas" yielded only three articles on 20 February 2017 in the ISI Web of Knowledge. For example, Gangwar et al. [15] worked on ethnopharmacological uses of *Mallotus philippinensis* Muell. Arg, and Stobdan et al. [16] did a similar work on *Hippophae rhamnoides* L. We found only one article, a study by Abbasi et al. [17], that was similar to our study and described ethnopharmacological knowledges of medicine preparation from a Himalayan region of the Pakistan Himalayas. Therefore, we emphasize that this is the first documented study on ethnopharmacology of a tribe from Arunachal Pradesh. We assume that in most cases modern pharmacologists and researchers start chemical assessments of the medicinal plants without giving much attention to the traditional ways of drug preparation by the tribal communities. This could be the reason behind the high number of studies on ethnobotany of medicinal plants from the Himalayas, but the comparatively minimal number of studies on ethnopharmacology. In South India, the Kani tribe uses similar approach for traditional medicine making [5], which supports the notion that tribal healers do use certain systematic techniques for drug preparation. The ethnopharmacological knowledge of traditional healers are generally transferred orally to the next generation, thus, making the knowledge vulnerable to being forgotten or lost.

In this context, we would like to provide a few examples of past pharmacological studies that had reported similar utilization of some medicinal plants listed in this study. Ghildiyal et al. (2012) showed that ethanolic extracts from *Hedychium spicatum* can inhibit respiratory as well as gastrointestinal disorders in rats and guinea pigs [18]. We showed in this study that the ethnomedicine *Blenga* prepared from the same plant was used for the treatments of dysentery and chest pain. In 2007, Nazir et al. extracted a drug called "Bergenin" from the species *Bergenia stracheyi* and proved that this drug can be used to treat arthritis in mice [19]. Interestingly, we found that an ethnomedicine named as *Bragen* (prepared from *Bergenia stracheyi* as well) was also used for the treatment of arthritis. Recently in 2014, Kumar et al. reported that the extracts of *Houttuynia cordata* can be used for the healing of hemorrhoids, and this species is frequently used in tradtional Tibetan and Chinese medicines [20]. We found that the ethnomedicine *Maraptang* prepared from *Houttuynia cordata* were also used by the tradtional Monpa healers for the treatment of piles which is a type of hemorrhoid. These examples mentioned above showed that the tradtional ethnomedicines used by the healers of Zemithang Monpa may have some potential to cure or manage some diseases. However, detailed pharmacological studies are needed to evaluate the potential of these medicines. A study by Witt et al. (2009) in Sikkim and Eastern Nepal (also part of the Eastern Himalayas) comprehensively listed 138 species of plants from tropical to alpine regions of the Himalayas used specifically in Tibetan medicine [21]. The majority of the species listed in our study were also reported by Witt et al., but detailed descriptions of the preparations for the ethnomedicines were not provided.

The results of this study should be interpreted very cautiously. The traditional ethnopharmacological knowledge of the Zemithang-Monpa tribe presented here for some diseases must not be treated as a general prescription under any circumstances, as scientific trials have not been undertaken nor the "traditional ethnomedicines" have ever been certified by any governmental authority such as the Central Drugs Standard Control Organization of India. There is also a high probability that the descriptions presented here may not be the same throughout the study region. Nevertheless, our main goal was not to certify or validate traditional medicines, but rather to document the uses and preparation of traditional medicines used by tribal people. The field method applied for data collection (i.e., participatory transect walk) also had some limitations. This method was helpful in remote regions where time and logistics are always a constraint of field work. Nevertheless, future research should establish more sample plots and cover larger regions in order to list more medicinal plants.

## 5. Conclusions

We have documented for the first time the vernacular names combined with ethnopharmacological preparations of ethnomedicines among Monpa tribes from the Zemithang region of Arunachal Pradesh, India. Past studies on ethnobotany in the Arunachal Pradesh, Eastern Himalayas, had listed uses of medicinal plants, however, we found that traditional healers use diverse species and plant parts in specific proportions for drug preparations. Our study illustrates the diversity of medicinal drug preparations and traditional knowledge that has passed through generation after generation of Monpa people. The ethnopharmacological documentation presented in this study should motivate researchers to carry out further scientific work on pharmacology, bioprospecting, and the cultivation of medicinal plants for the socioeconomic development in the region. Under ongoing warming of the Himalayas and mass migration of people from the mountain areas to cities, our study also highlights the need to document the traditional knowledge regarding the use of local flora and to develop strategies to conserve them before the traditional knowledges are lost or forgotten.

**Acknowledgments:** We sincerely thank the Divisional Forest Officers (DFOs) of Tawang Social Forestry Division and West Kameng Forest Division (Suneesh Buxy, M. Sambhu and Adukparon of Indian Forest Service) and R. C. Das (the Range Forest Officer, Lumla Forest Range, Arunachal Pradesh State Forest Service) for giving us permission and providing logistical support to do the field work. Without their support, this study could not have been done in this remote part of India. We thank Prema Khandu (village Lumpho) and Norbu (village Namshu) for all the supports provided during field data collection, and for organizing and participating in long field expeditions. We will never forget the support from Haridasan and Rao (State Forest Research Institute of Arunachal Pradesh, Itanagar), Saroj K. Barik (North Eastern Hill University, Shillong), and Mohammed Latif Khan (Northeastern Regional Institute of Science and Technology, Itanagar) in the identification of herbarium samples. We also acknowledge the cooperation provided by the Pijush K. Datta of WWF-India and Bibhab Talukdar of ATREE-Northeast Program and the IUCN-India during this research work. The article processing charge was funded by the open access publication fund of the Albert-Ludwigs-University of Freiburg.

**Author Contributions:** Tamalika Chakraborty and Somidh Saha conducted the field work, wrote the paper, and equally contributed to this work. Narendra Singh Bisht provided motivation, guidance, and supervision to carry out this research and acted as mentor for the graduate research works of Tamalika Chakraborty and Somidh Saha.

**Conflicts of Interest:** The authors declare no conflict of interest.

## References

1.  Borges, R.M. The frontiers of India's biological diversity. *Biotropica* **2005**, *37*, A1–A3.
2.  Haridasan, K.; Bhuyan, L.R.; Deori, M.L. Wild edible plants of Arunachal Pradesh. *Arunachal For. News* **1990**, *18*, 1–8.
3.  Adak, D.K. A morphometric study of the Thingbu-pa and population comparison with neighbouring Monpa tribes of Arunachal Pradesh, India. *Anthropologischer Anzeiger Bericht uber die Biologisch-Anthropologische Literatur* **2001**, *59*, 365–375. [PubMed]
4.  Sharma, P.P.; Mujundar, A.M. Traditional knowledge on plants from Toranmal Plateau of Maharastra. *Indian J. Tradit. Knowl.* **2003**, *2*, 292–296.
5.  Ayyanar, M.; Ignacimuthu, S. Traditional knowledge of kani tribals in Kouthalai of Tirunelveli hills, Tamil Nadu, India. *J. Ethnopharmacol.* **2005**, *102*, 246–255. [CrossRef] [PubMed]
6.  Namsa, N.D.; Mandal, M.; Tangjang, S.; Mandal, S.C. Ethnobotany of the Monpa ethnic group at Arunachal Pradesh, India. *J. Ethnobiol. Ethnomed.* **2011**, *7*, 31. [CrossRef] [PubMed]
7.  Saha, S.; Bisht, N.S. Role of traditional ecological knowledge in natural resource management among Monpas of north-western Arunachal Pradesh. *Indian For.* **2007**, *133*, 155–164.
8.  Von Fürer-Haimendorf, C. *Tribes of India: The Struggle for Survival*; Oxford University Press: Delhi, India, 1982.
9.  State Government of Arunachal Pradesh. *Working Plan: Tawang Social Forestry Division*; Arunachal Forest Department: Tawang, India, 2001.
10. Champion, S.H.G.; Seth, S.K. *A Revised Survey of the Forest Types of India*; Government of India: New Delhi, India, 1968.
11. Polunin, O.; Stainton, A. *Flowers of the Himalayas*; Oxford University Press: Delhi, India, 1997.
12. Haridasan, K.; Shukla, G.P.; Beniwal, B.S. Medicinal plants of Arunachal Pradesh. *SFRI Inf. Bull.* **1995**, *5*, 32.
13. Kala, C.P. *Medicinal Plants of Indian Trans-Himalaya: Focus on Tibetan Use of Medicinal Resources*; Bishen Singh Mahendra Pal Singh: Dehra Dun, India, 2003.
14. Kala, C.P. Status and conservation of rare and endangered medicinal plants in the Indian trans-Himalaya. *Biol. Conserv.* **2000**, *93*, 371–379. [CrossRef]

15. Gangwar, M.; Goel, R.K.; Nath, G. *Mallotus philippinensis* Muell. Arg (Euphorbiaceae): Ethnopharmacology and phytochemistry review. *BioMed Res. Int.* **2014**, *2014*, 213973. [CrossRef] [PubMed]

16. Stobdan, T.; Targais, K.; Lamo, D.; Srivastava, R.B. Judicious use of natural resources: A case study of traditional uses of seabuckthorn (*Hippophae rhamnoides* L.) in trans-Himalayan Ladakh, India. *Natl. Acad. Sci. Lett.* **2013**, *36*, 609–613. [CrossRef]

17. Abbasi, A.M.; Khan, M.A.; Zafar, M. Ethno-medicinal assessment of some selected wild edible fruits and vegetables of Lesser-Himalayas, Pakistan. *Pak. J. Bot.* **2013**, *45*, 215–222.

18. Ghildiyal, S.; Gautam, M.K.; Joshi, V.K.; Goel, R.K. Pharmacological evaluation of extracts of *Hedychium spicatum* (Ham-ex-Smith) rhizome. *Anc. Sci. Life* **2012**, *31*, 117–122.

19. Nazir, N.; Koul, S.; Qurishi, M.A.; Taneja, S.C.; Ahmad, S.F.; Bani, S.; Qazi, G.N. Immunomodulatory effect of bergenin and norbergenin against adjuvant-induced arthritis—A flow cytometric study. *J. Ethnopharmacol.* **2007**, *112*, 401–405. [CrossRef] [PubMed]

20. Kumar, M.; Prasad, S.K.; Hemalatha, S. A current update on the phytopharmacological aspects of *Houttuynia cordata* Thunb. *Pharmacogn. Rev.* **2014**, *8*, 22–35. [PubMed]

21. Witt, C.M.; Berling, N.E.; Rinpoche, N.T.; Cuomo, M.; Willich, S.N. Evaluation of medicinal plants as part of Tibetan medicine prospective observational study in Sikkim and Nepal. *J. Altern. Complement. Med.* **2009**, *15*, 59–65. [CrossRef]

# Selecting Lentil Accessions for Global Selenium Biofortification

Dil Thavarajah [1,*], Alex Abare [1], Indika Mapa [1], Clarice J. Coyne [2], Pushparajah Thavarajah [3] and Shiv Kumar [4]

[1] Plant and Environmental Sciences, 270 Poole Agricultural Center, Clemson University, Clemson, SC 29634, USA; aabare@g.clemson.edu (A.A.); imapapa@g.clemson.edu (I.M.)

[2] USDA Agriculture Research Service, Western Regional Plant Introduction Station, Washington State University, Pullman, WA 99164-6434, USA; Clarice.Coyne@ars.usda.gov

[3] BOV Solutions Inc., 1105 Garner Bagnal Blvd, Statesville, NC 28677, USA; rajah.thava@gmail.com

[4] Biodiversity and Integrated Gene Management Program, International Centre for Agricultural Research in the Dry Areas (ICARDA), P.O. Box 6299, Rabat-Institute, Rabat, Morocco; SK.Agrawal@cgiar.org

* Correspondence: dthavar@clemson.edu

**Abstract:** The biofortification of lentil (*Lens culinaris* Medikus.) has the potential to provide adequate daily selenium (Se) to human diets. The objectives of this study were to (1) determine how low-dose Se fertilizer application at germination affects seedling biomass, antioxidant activity, and Se uptake of 26 cultivated lentil genotypes; and (2) quantify the seed Se concentration of 191 lentil wild accessions grown in Terbol, Lebanon. A germination study was conducted with two Se treatments [0 (control) and 30 kg of Se/ha] with three replicates. A separate field study was conducted in Lebanon for wild accessions without Se fertilizer. Among cultivated lentil accessions, PI533690 and PI533693 showed >100% biomass increase vs. controls. Se addition significantly increased seedling Se uptake, with the greatest uptake (6.2 $\mu$g g$^{-1}$) by PI320937 and the least uptake (1.1 $\mu$g g$^{-1}$) by W627780. Seed Se concentrations of wild accessions ranged from 0 to 2.5 $\mu$g g$^{-1}$; accessions originating from Syria (0–2.5 $\mu$g g$^{-1}$) and Turkey (0–2.4 $\mu$g g$^{-1}$) had the highest seed Se. Frequency distribution analysis revealed that seed Se for 63% of accessions was between 0.25 and 0.75 $\mu$g g$^{-1}$, and thus a single 50 g serving of lentil has the potential to provide adequate dietary Se (20–60% of daily recommended daily allowance). As such, Se application during plant growth for certain lentil genotypes grown in low Se soils may be a sustainable Se biofortification solution to increase seed Se concentration. Incorporating a diverse panel of lentil wild germplasm into Se biofortification programs will increase genetic diversity for effective genetic mapping for increased lentil seed Se nutrition and plant productivity.

**Keywords:** lentil; selenium; biofortificaiton; wild germplasm

## 1. Introduction

Selenium deficiency is a global public health concern. Recent estimates indicate 15 to 20% of children and adults around the world are Se deficient [1,2]. This means an estimated 30–100 million people are Se deficient, mainly due to low concentrations of bioavailable Se in commonly eaten foods. Biofortification, i.e., enriching staple foods with Se through conventional plant breeding, is considered a sustainable way to increase Se intake and support good general health [3]. Globally, pulses are becoming popular as they are a nutritionally superior, medium-energy food that is low in fat, high in protein, and a good source of micronutrients. For example, a single 50 g serving of lentil (*Lens culinaris* Medikus) provides 3.7–4.5 mg of iron (Fe), 2.2–2.7 mg of zinc (Zn), and 22–34 $\mu$g of Se with very low levels of phytic acid (2.5–4.4 mg g$^{-1}$) [4]. Further, lentil is an excellent source of folate [5] and contains a range of low digestible carbohydrates and resistant starch that could modulate human gut microbiome to reduce obesity, overweight, and related non-communicable diseases [6].

Selenium is an essential element for mammals but the physiological requirement of Se for higher plants is not well understood [7]. Lower plants, such as algae, require Se for normal growth and development, and green algae contain a Se-dependent antioxidant enzyme, glutathione peroxidase [8]. Lobanov et al. (2009) indicated that sequences for selenoproteins have not been revealed for higher plants, but this conclusion was questioned after a recent study involving de novo assembly and annotation of the complete mitochondrial genome of American cranberry [9]. Specifically, American cranberry was found to contain two copies of tRNA-Sec and a selenocysteine insertion sequence element that were lost in higher plants during evolution [10]. The presence of a selenocysteine insertion in the cranberry mitochondrial genome supports the hypothesis that American cranberry has higher antioxidant activity and high production capability in acidic, low nutrient glacial lake bottom soils with deposits of heavy metals including Se. Although seleno genes were lost during plant evolution, other physiological mechanisms may exist to upregulate plant growth and productivity when grown in Se-rich soils [11].

Lentil has a positive response to added low-dose Se fertilizer that varies by genotype. A field study demonstrated that the application of Se increases lentil grain yield, seed Se concentration, and antioxidant levels [12]. Specifically, lentil grain yield and antioxidant responses to added Se varied with genotype, with some cultivars (e.g., CDC Richlea, CDC Viceroy) increasing grain yield and antioxidant activity following Se treatment compared to untreated controls. Consequently, the lentil Se response appears to depend on where the genotype originated, i.e., in low vs. high Se soils [13]. Lentil Se biofortification is possible if based on genotype and location sourcing of diverse lentil germplasm resources including modern cultivated lentil and wild relatives [13–15]. Genetic mapping revealed that 352 accessions of cultivated lentil from South Asia and Canada have a narrow genetic diversity with respect to further genetic and breeding enhancement [16]. Therefore, further phenotyping studies with diverse wild accessions will provide greater opportunities to select the most potential breeding lines for Se biofortification efforts. The objectives of this study were to (1) determine the effect of low dose Se fertilizer on seedling biomass, antioxidant activity, and Se uptake of 26 cultivated lentil genotypes from low soil Se regions during germination; and (2) determine the seed Se concentration of 191 lentil wild accessions grown in low Se soils of Terbol, Lebanon.

## 2. Methods and Materials

### 2.1. Materials

Standards, chemicals, and high-purity solvents used for seed digestion, Se analysis, and antioxidant activity were purchased from VWR International, Sigma Aldrich Co. (St. Louis, MO, USA), and Alfa Assar—A Johnson Matthey Company (Ward Hill, MA, USA), and used without further purification. Water, distilled and deionized (ddH$_2$O) to a resistance of $\geq$18.2 M$\Omega$ (Milli-Q Water System, Millipore, Milford, MA, USA), was used for sample and reagent preparation.

### 2.2. Germination Study

Twenty-six cultivated lentil genotypes from the genus Lens subsp. *culinaris* were selected (Table 1). These genotypes were close relatives of high Se uptake lentil cultivars in current production [13,15] and had also been selected for a future genome-wide study based on contrasting physiological response to low dose Se fertilizer. Original lentil seeds were obtained from the USDA-ARS Grain Legume Genetics and Physiology Research Unit, Washington State University, WA, USA, and were multiplied using single plants at Washington State University [17]. Ten surface sterilized seeds from each lentil genotype were placed on sterile petri dishes with sterile absorbent paper, and germinated for five days in clean dark wooden boxes at 22 °C. The treatment design was a complete randomized design with two Se treatments [0 (control; no Se), and 30 kg of Se/ha (6ppm)] with three replicates for each lentil accession, and the entire experiment was repeated twice. Selenium treatment was applied as 4 mL of potassium selenate solution at day 0, 3 and 5 to provide a total dose of 30 kg of Se/ha per petri dish. The control treatment received a similar volume (12 mL for three days) of nano pure water at the same time points. A total of 156 petri dishes were randomly distributed in five similar wooden boxes located in a controlled environment for five days, and then germinating lentil seedlings were transferred to an automated light canopy with day/night temperatures of 22/20 °C, photosynthetically active radiation levels of 200–300 μmol/m$^2$/s using a 16-h photoperiod, and 50–60% relative humidity to complete the germination cycle for another two days. On the seventh day, lentils seedlings from each petri dishes were collected, seedling biomass determined (seedling weight), and moisture content of fresh sub-samples measured (via drying at 105 °C for 2 h). The remaining samples were then immediately stored at −40 °C until analysis. Seed Se, antioxidant activity, and biomass data are reported on a dry weight basis (10% moisture).

**Table 1.** *Lens culinaris* subsp. *Culinaris* genotypes used in the germination study.

| Origin/Source | n | Genotype (Plant Name) |
|---|---|---|
| Brazil | 3 | PI 518732 (CNPH 84–122)<br>PI 518733 (CNPH 84–123)<br>PI 518734 (CNPH 84–125) |
| Canada | 1 | PI 471917 (Eston) |
| France | 3 | PI 486128 (Dupuy)<br>PI 490288 (Anicia)<br>PI 490289 (Mariette) |
| Germany | 1 | PI 320937 (ILL505) |
| Spain | 4 | PI 533688 (870523-13)<br>PI 533690 (Pardina)<br>PI 533691 (Lenteja Verdina)<br>PI 533693 (Verdina) |
| USA | 14 | PI 477921 (Red Chief), PI 486127 (unknown), PI 508090 (Brewer)<br>W6 27754 (Parent of 1048-8R), W6 27758 (Parent of CDC Robin)<br>W6 27759 (Parent of Eston), W6 27760 (Parent of Giza-9)<br>W6 27762 (Parent of ILL 4605), W6 27763 (Parent of ILL 5588)<br>W6 27766 (Parent of ILL 7537), W6 27767 (Parent of ILL 8006 BM4)<br>W6 27780 (Parent of Milestone), W6 27781 (Parent of Pardina),<br>W6 27782 (Parent of Pennell) |
| Total | 26 | |

### 2.3. ICARDA Study

The wild lentil accessions considered in this study originated from 12 different countries and included genus Lens subsp. *culinaris* and five different sub-species (Table 2). A total of 191 lentil wild accessions were grown at the ICARDA field location in Terbol, Lebanon in 2014 using a single row seed per accession. Annual mean precipitation and temperature for 2014 were 247 mm and 8 °C, respectively. Terbol agricultural soils are clay loam with a slightly acidic pH (7.74), 3.24% organic matter, and 103 μg/kg of plant available Se [18]. The Selenium World Atlas indicates Middle Eastern soils are generally low in Se; however, exact soil Se levels were not reported [19]. According to available records, ICARDA fields have been not treated with Se fertilizer, and soil Se data for this field were not available due to logistical issues with importing soils to the USA. At physiological maturity, plants were hand harvested, thoroughly hand cleaned, and then 50 seeds from each of the 191 lentil wild accessions directly shipped to the Pulse Quality and Nutrition Laboratory, Clemson University, SC. Lentil seeds were hand ground using a mortar and pestle and passed through a 0.25 mm sieve prior to measurement of total seed Se concentration. Seed Se data are reported on a dry weight basis (10% moisture).

**Table 2.** Origin, species/subspecies and accession numbers of 191 lentil wild accessions used in the ICARDA study.

| Origin | Species/Subspecies | n | Accession Number |
|---|---|---|---|
| Armenia | *L. culinaris* subsp *orientalis* | 1 | 126939 |
| Cyprus | *L. culinaris* subsp *orientalis* | 2 | 72849, 72595 |
| Czech Republic | *L. culinaris* subsp *unknown* | 1 | 136657 |
| Iran | *L. culinaris* subsp *unknown* | 2 | 72593, 72594 |
| Jordan | *L. culinaris* subsp *orientalis* | 5 | 72847, 72848, 72858, 72864, 72865 |
| Lebanon | *L. culinaris* subsp *odemensis* | 1 | 110846 |
|  | *L. culinaris* subsp *unknown* | 2 | 72925, 110824 |
| Poland | *L. culinaris* subsp *orientalis* | 1 | 72600 |
|  | *L. culinaris* subsp *unknown* | 5 | 72597, 72598, 136652, 136653, 136658 |
| Syria | *L. culinaris* subsp *odemensis* | 15 | 72640, 72648, 72690, 72697, 72703, 72704 72706, 72758, 72759, 72760, 72893, 107449 119390, 126219, 126220 |
|  | *L. culinaris* subsp *orientalis* | 49 | 72534, 72638, 72639, 72645, 72647, 72680 72685, 72688, 72689, 72691, 72699, 72715 72719, 72720, 72750, 72751, 72754, 72761 72765, 72767, 72777, 72778, 72824, 72825 72853, 72854, 72866, 72868, 72869, 72872 72878, 72880, 72881, 72882, 72883, 72884 72887, 72888, 72890, 72892, 107448, 116048 116049, 116052, 126223, 135399, 135443 136777, 139285 |
|  | *L. culinaris* subsp *tomentosus* | 4 | 72686, 72820, 72845, 136814 |
|  | *L. culinaris* subsp *unknown* | 32 | 72643, 72644, 72646, 72666, 72668, 72669 72672, 72675, 72676, 72692, 72693, 72721 72769, 72770, 72818, 72852, 72870, 72871 72873, 72876, 72877, 107447, 110594, 110803 116043, 116045, 116047, 126221, 126222 135385, 135410, 135415 |
| Tajikistan | *L. culinaris* subsp *odemensis* | 1 | 72899 |
|  | *L. culinaris* subsp *orientalis* | 5 | 72904, 72905, 72907, 136679, 140379 |
| Turkmenia | *L. culinaris* subsp *orientalis* | 1 | 72901 |
| Turkey | *L. culinaris* subsp *odemensis* | 2 | 72562, 136673 |
|  | *L. culinaris* subsp *orientalis* | 33 | 72527, 72529, 72530, 72602, 72604, 72606 72608, 72610, 72611, 72612, 72613, 72616 72617, 72618, 72619, 72620, 72621, 72626 72627, 72628, 72629, 72632, 72726, 72746 72748, 72816, 72830, 114416, 116008, 116010 116029, 136677, 72623 |
|  | *L. culinaris* subsp *tomentosus* | 1 | 72625 |
|  | *L. culinaris* subsp *unknown* | 21 | 72724, 72742, 72743, 72744, 72800, 72801 72804, 72805, 72831, 72835, 72836, 72855 116015, 116027, 116028, 116034, 136662 136665, 136666, 136669, 136670 |
| Uzbekistan | *L. culinaris* subsp *odemensis* | 1 | 72900 |
|  | *L. culinaris* subsp *orientalis* | 5 | 72895, 72896, 72897, 72908, 72909 |
|  | *L. culinaris* subsp *unknown* | 1 | 72592 |
|  | Total | 191 |  |

## 2.4. Se Analysis

Seed Se concentration was determined using inductively coupled plasma optical emission spectrophotometry (ICP-OES; ICP-6500 Duo, Thermo Fisher Scientific, PA, USA) after nitric acid-hydrogen peroxide digestion [13]. Finely ground seed samples (500 mg) were digested in nitric acid (70% HNO$_3$) at 90 °C for 1 h. Samples were then further digested with hydrogen peroxide (30%) before being diluted to 10 mL with nanopure water. Total Se measurements using the ICO-OES method were validated using National Institute of Standards and Technology (NIST) standard reference material 1573a (apple leaves; [Se] = 0.054 ± 0.003 mg kg$^{-1}$). A homogenized laboratory reference material (CDC Redberry: [Se] = 400 ± 100 mg kg$^{-1}$) was also used periodically for quality control. A calibration curve for Se concentration was produced using serial dilutions from 1 to 40 mg L$^{-1}$. The limit of detection for this method was 10 ppt.

### 2.5. Antioxidant Activity

Antioxidant activity of fresh lentil seedlings was measured using the dipheny-picrylhydrazyl (DPPH) free radical scavenging method [12,20]. The DPPH stock solution (1 mM) was prepared by dissolving 19.7 mg DPPH (2, 2-diphenyl-1-picrylhydrazyl) in 50 mL of methanol. A 0.1 mM working solution was obtained by diluting the stock solution (10 mL) with methanol (90 mL). Seedlings were finely ground in liquid nitrogen using a mortar and pestle. One gram of ground seedling was mixed with 3 mL of water, vortexed for 30 s, and then centrifuged at 2000 rpm for 20 min. The supernatant was separated using a 10-mL syringe. Fifty μL of this sample extract were mixed with 3 mL of newly prepared working solution to prepare the sample solutions. A negative control was prepared by mixing 3 mL of 0.1 mM DPPH working solution with 50 μL of distilled water, and a blank control was prepared by mixing 3 mL of methanol with 50 μL of sample extract. All prepared sample solutions were kept in a dark chamber for 30 min at room temperature. After incubation, solutions were centrifuged at 2500 rpm for 10 min. Absorbance was measured at 518 nm and the following formula used to calculate the antioxidant activity (inhibition %):

$$\text{Inhibition } \% = \left( \frac{A_{\text{Negative}} - A_{\text{Sample}}}{A_{\text{Negative}}} \right) \times 100.$$

### 2.6. Statistical Analysis

For the germination study, the experiment used a completely randomized design with 26 lentil accessions, three replicates for each accession, and two Se rates ($n$ = 156). Data from replicates were combined and data error variances tested for homogeneity. For combined analysis, a mixed model analysis of variance was performed using the PROC GLM procedure of SAS version 9.4 (SAS Institute, Cary, NC, USA) [21], with genotypes and Se rates as the class variables and replicates as a random factor. Means were separated by Fisher's protected least significant difference (LSD) at $p < 0.05$. For lentil wild accessions, the PROC GCHART procedure of SAS version 9.4 was used for frequency distribution of seed Se concentration for 191 accessions.

## 3. Results

### 3.1. Germination Study

Combined statistical analysis of variance showed that lentil genotype and Se treatment significantly ($p < 0.05$) affected seedling biomass, antioxidant activity, and seedling Se concentration (Table 3). The interaction term, genotype × Se treatment, was also significant for all variables. In most cases, Se application increased seedling biomass, seedling antioxidant activity, and seedling Se concentration; however, the magnitude of this effect varied with lentil genotype (Figures 1 and 2). Positive significant correlations were observed between Se treatment and seedling biomass, antioxidant activity, and Se uptake, i.e., Se treatment significantly increased lentil seedling growth, antioxidant activity, and Se nutrition (Table 3).

Se treatment significantly increased seedling biomass in nine lentil genotypes vs. the control (Figure 1). Similarly, relative biomass increased; specifically, PI320937, PI533690, PI518732, W627767, W627754, and PI533693 demonstrated biomass increases of more than 50% compared to their controls (Figure 1). Among these genotypes, PI533690 and PI533693 demonstrated a >100% biomass increase compared to controls. As expected, Se treatment (30 kg of Se/ha) significantly increased seedling Se concentration for all genotypes; PI320937 showed the highest Se uptake (6.2 $\mu$g g$^{-1}$) and W627780 (1.1 $\mu$g g$^{-1}$) the lowest (Figure 2). Antioxidant activity for the control treatments was below the detection limit (data not shown). More than 20% antioxidant activity was observed in seven genotypes: PI320937, PI486128, PI508090, PI490289, PI471917, PI486127, and PI477921 (Figure 2). In general, antioxidant activity increased with Se application but responses varied across genotypes.

**Table 3.** Combined analysis of variance for 26 lentil genotypes during germination with response to Se fertilizer.

| Source | df | Mean Squares | | |
|---|---|---|---|---|
| | | Biomass | Antioxidant | Se Uptake |
| Genotype | 25 | * | * | * |
| Se treatment | 1 | * | * | * |
| Replication | 2 | NS | NS | NS |
| Genotype × Se treatment | 25 | * | * | * |
| Error | 102 | 0.5 | 3.1 | 0.1 |
| Person correlation coefficient (n = 156) | | | | |
| Biomass | | 1.00 | **0.41** * | **0.20** * |
| Antioxidant | | **0.41** * | 1.00 | **0.54** * |
| Se uptake | | **0.20** * | **0.54** * | 1.00 |

df, degree of freedom; * $p < 0.05$; NS, not significant. Values indicate Pearson correlation coefficients ($r$). Bold values indicate significant correlation. ** $p < 0.05$.

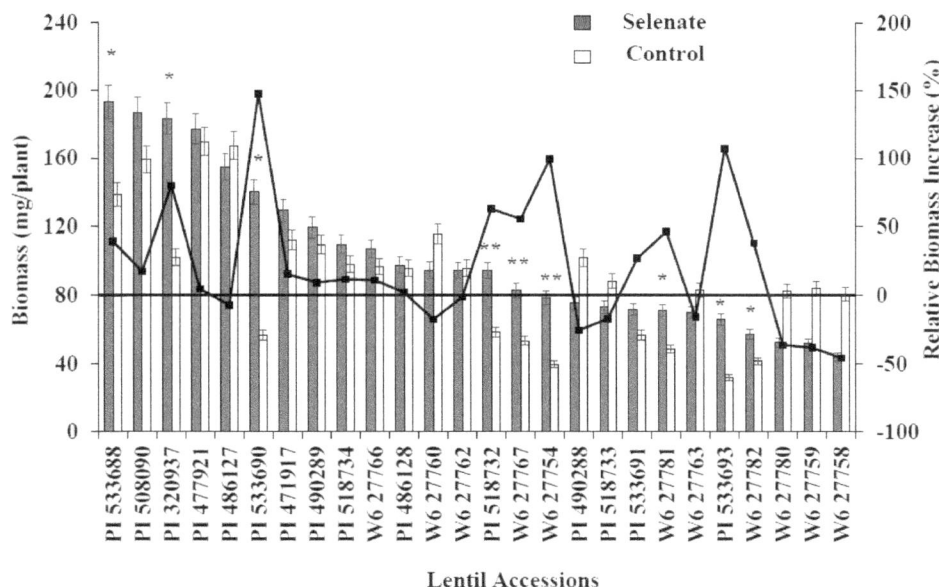

**Figure 1.** Variation in biomass and relative biomass change for 26 lentil genotypes during germination with added Se fertilizer. *, ** = Significant at $p < 0.05$ and $p < 0.1$, respectively. Relative biomass was calculated based on control biomass weight.

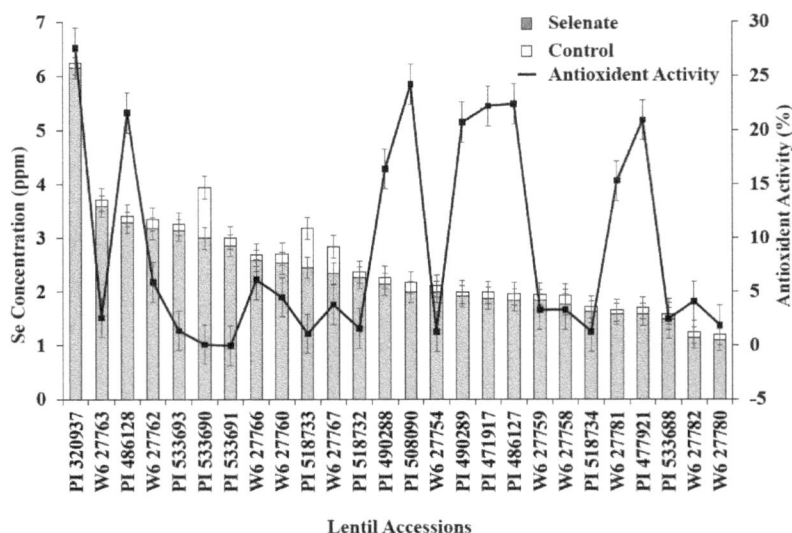

**Figure 2.** Variation of seedling Se concentration and antioxidant activity for 26 lentil genotypes during germination with added Se fertilizer. Control data are not shown as % inhibition and Se uptake levels were below the detection limit.

## 3.2. ICARDA Study

Wild lentil accessions originated from 12 different countries: 100 accessions were from Syria, 57 from Turkey, and the remaining 34 from lentil growing regions including the Middle East, Central Asia, and Europe (Table 4). Across the entire population, seed Se concentration ranged from 0 to 2.45 $\mu g\ g^{-1}$, with a mean of $0.41 \pm 0.21\ \mu g\ g^{-1}$. Seed Se concentrations of wild accessions originating from Syria ranged from 0 to 2.45 $\mu g\ g^{-1}$ with a mean value of 0.38 $\mu g\ g^{-1}$ and from Turkey ranged from 0 to 2.36 $\mu g\ g^{-1}$ with a mean value of 0.54 $\mu g\ g^{-1}$ (Table 4). The frequency distribution analysis indicated seed Se concentration for most accessions fell into the 0.25–0.75 $\mu g\ g^{-1}$ range (Figure 3). Most wild accessions (144 of 191) would provide an adequate amount of Se (notable % of recommended daily allowance) from a 50-g serving of lentil (Figure 4). For example, 54 accessions would provide at least 20% of the recommended daily allowance of Se and 16 accessions would provide more than 100%. Following 16 lines would provide more than 100% of the recommended daily allowance (RDA): 72868, 116047, 116048, 116052, 119390, 135385, 72527, 72617, 72619, 72628, 72726, 116027, 116028, 116029, 116034, and 114416. Above wild accessions with 1–2 $\mu g$ Se $g^{-1}$ can be used in lentil breeding programs as a source of high Se content for developing high Se lentil cultivars.

**Table 4.** Seed Se concentration of 191 lentil wild accessions grown in Lebanon.

| Country of Origin | No of Samples | Seed Se Concentration ($\mu g\ g^{-1}$) | |
|---|---|---|---|
| | | Range | Mean |
| Armenia | 1 | 0.59 | 0.59 |
| Cyprus | 2 | 0.20–0.51 | 0.36 |
| Czech Republic | 1 | 0.91 | 0.91 |
| Iran | 2 | 0.30–0.90 | 0.60 |
| Jordan | 5 | 0.18–0.81 | 0.34 |
| Lebanon | 3 | 0.02–0.49 | 0.23 |
| Poland | 6 | 0.01–0.66 | 0.32 |
| Syria | 100 | 0.00–2.45 | 0.38 |
| Tajikistan | 6 | 0.00–0.53 | 0.16 |
| Turkmenia | 1 | 0.31 | 0.31 |
| Turkey | 57 | 0.0–2.36 | 0.54 |
| Uzbekistan | 7 | 0.0–0.76 | 0.18 |
| Mean ± SD | | | 0.41 ± 0.22 |
| Total | 191 | | |

SD, Standard deviation ($n$ = 191).

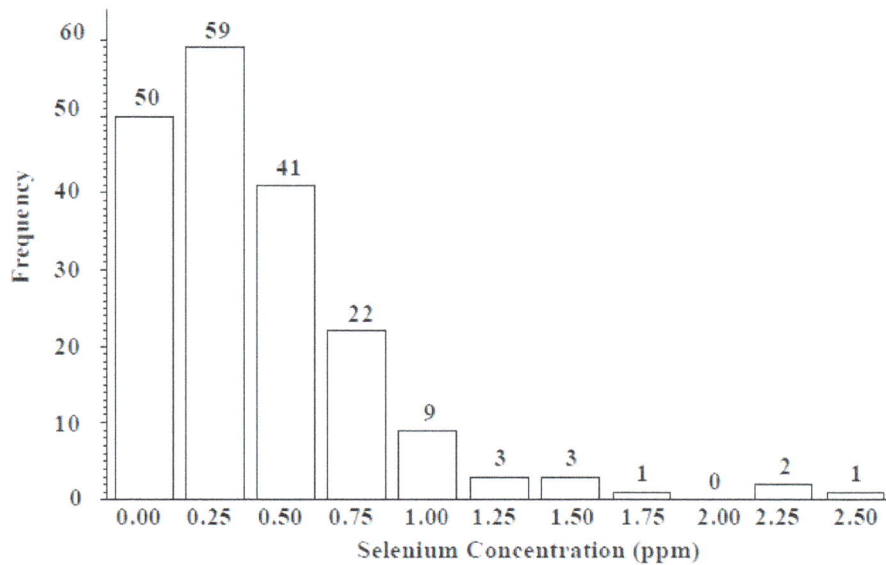

**Figure 3.** Frequency distribution of seed Se concentration of ICARDA lentil wild accessions grown in Lebanon.

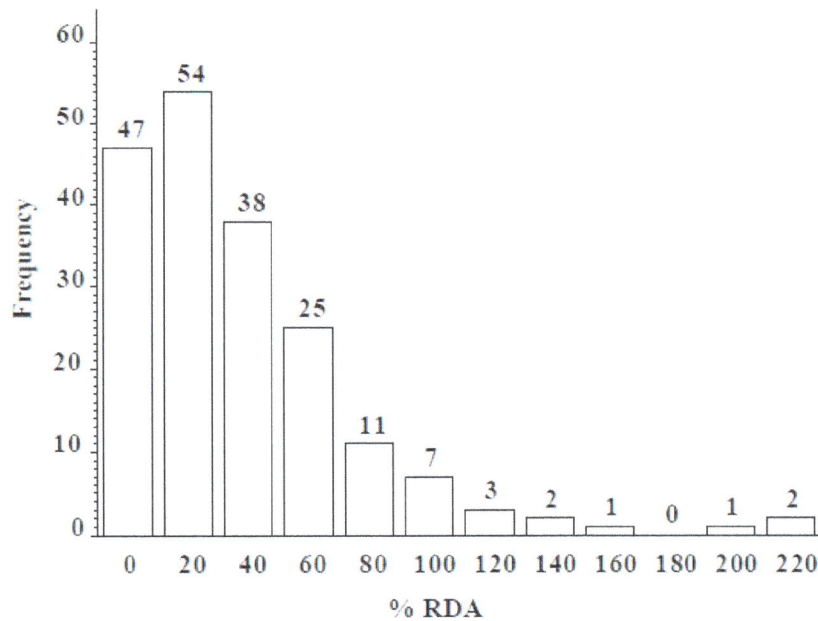

**Figure 4.** Frequency distribution of percent recommended daily allowance (%RDA: 55 µg/day for an adult) of Se from a 50-g serving of lentil seed from wild accessions grown in Lebanon.

## 4. Discussion

Selenium is an essential element for the general well-being of humans; however, Se is not an essential element for higher plant survival. Recent literature demonstrates that Se fertilization of food crops increases crop yield, antioxidant protection, drought tolerance, and ultimate Se nutritional quality [12,13,22–24]. Literature on the role of Se fertilization during germination on lentil seedling performance has not been reported, however. Results from the current study confirm that the application of Se during lentil germination significantly increases seedling biomass growth, antioxidant activity, plant health, and Se uptake, but these growth responses depend on the genotype. Selenium fertilization effects during the seedling stage may be a genetically driven function in lentil.

Seedling biomass responses to added Se varied with lentil genotype. Lentil accessions PI533690 and PI533693 showed >100% relative biomass increase while PI320937 and W6-27754 showed >50% relative biomass increase vs. their respective controls (Figure 1). PI320937 showed the greatest response in terms of increased antioxidant activity and seedling Se uptake (Figure 2). These findings challenge the current thinking that Se is not essential for higher plants, and provide a strong rationale to apply Se for improved plant growth and productivity.

The origin of lentil accessions can be used to explain Se growth responses during germination. Lentil is one of the oldest domesticated pulse crops, originating in the Mediterranean region in the Bronze Age where most soils are low in Se [1]. Our previous studies clearly support the fact that current lentil cultivars in production (e.g., Eston, ILL505, CDC Robin) that originated from low-Se soil countries show the greatest responses in terms of grain yield, seed Se concentration, and speciation to Se fertilization [12,15]. Se fertilization is therefore required to maximize the yield potential of these varieties when grown in low-Se soils. Similarly, data from the current study support the notion that adding low dose Se fertilizer at germination may also activate an initial Se growth response at the seedling stage, leading to healthier plants, increased grain yield, and improved nutritional quality.

Success in lentil Se biofortification efforts is a function of broad genetic diversity, heritability, and selection abilities. The Se uptake capability of lentil, especially when grown in naturally low-Se soils, could be an important trait and therefore discovering useful genes or alleles for Se uptake is of interest. This paper is the first to our knowledge to report the natural diversity of seed Se levels for the largest wild lentil accession collection. Seed Se concentration varied with *Lens* species and country of origin. Seed Se concentration ranged from 0 to 2.45 $\mu$g g$^{-1}$ for accessions originating from Syria, 0 to 2.35 $\mu$g g$^{-1}$ for accessions from Turkey, and 0 to 0.91 $\mu$g g$^{-1}$ for the other accessions (Table 4). Moreover, frequency distribution analysis clearly indicated approximately 59 wild accessions with 0.25 $\mu$g g$^{-1}$ (low seed Se), 41 wild accessions with 0.5 $\mu$g g$^{-1}$ (moderate seed Se), 22 accessions with 0.75 $\mu$g g$^{-1}$ (high seed Se), and nine accessions with 1.00 $\mu$g g$^{-1}$ (very high seed Se) of Se in their seed (Figure 3). Results from studies using molecular markers for lentil breeding efforts might have been limited due to the low genetic diversity among cultivated lentil species [14,16]. However, future genetic Se biofortification is possible by properly identifying genes or alleles responsible for Se uptake by using a larger and more diverse germplasm collection that includes wild accessions.

Lentil appears to be naturally biofortified with Se when grown in high-Se soils or when fertilized with Se when grown in Se-deficient soils; the latter approach would serve to increase both dietary Se and global lentil production. To address Se deficiency throughout the world, the biofortification of staple food crops with bioavailable Se is required. Several recent studies show Se fertilization may have beneficial effects on staple food crops [15,24,25]. Further, selenomethionine as the dominant Se-species in most food crops suggests that Se is likely to have a good biological availability [13,15,26]. Allaway et al. (1966) showed that Se fertilization to Se-deficient Oregon, USA soils increased the Se concentration of alfalfa from 0.01–0.04 mg kg$^{-1}$ to 2.6–2.7 mg kg$^{-1}$. Since then, the efficacy of Se-fertilization has been well documented around the world [1,25,27]. The efficacy of Se fertilization has been shown for many crops including barley, red clover, perennial ryegrass, and wheat [22–24,28].

Lentil Se biofortification can be achieved through conventional plant breeding and Se fertilization. Current research suggests that lentil grown in Canada and USA is naturally Se enriched, and fulfills daily dietary Se requirements when consumed in modest amounts (50–100 g d$^{-1}$) [13,25]. Recent biofortification efforts for other metals such as iron and zinc have only focused on plant breeding and biotechnology efforts due to their homeostatic nature. In contrast, Se uptake and biotransformation are controlled by plant physiological conditions and soil Se availability, although the exact metabolic role of Se in higher plants is yet to be determined. Incorporating diverse lentil accessions into a proper lentil mapping population may locate the genes or alleles responsible for Se uptake in lentil.

## 5. Summary

Lentil is an important food legume in many parts of the world. Exported lentil is currently produced in regions where the soil is relatively rich in Se. However, most regions where lentil is produced and consumed have low soil Se and lack the means to improve seed Se and total yields. Results from this study support the notion that the application of Se at germination not only increases lentil seedling Se uptake but also significantly increases plant productivity in terms of biomass for some genotypes. Incorporating a diverse group of wild accessions into global lentil Se biofortification programs will be of immense benefit as wild accession germplasm has greater phenotypic variation and will thus support effective genetic mapping studies.

**Acknowledgments:** Funding support for this study is provided by Clemson University Seed Grant (CU Seed), Clemson University, SC, USA, and the International Center for Agricultural Research in the Dry Areas (ICARDA), Morocco.

**Author Contributions:** D.T.: project PI, experimental design, data analysis, data interpretation, and manuscript writing; A.A. and I.M.: conduct lab analysis; P.T.: experimental design, and manuscript review; C.J.C.: provide seed for germination study, manuscript review; S.K.: conducted ICARDA study, experimental design, and manuscript review.

**Conflicts of Interest:** The authors declare no conflict of interest.

## References

1.  Combs, G.F. Selenium in global food systems. *Br. J. Nutr.* **2001**, *85*, 517–547. [CrossRef] [PubMed]
2.  CDC. Micronutrient facts, International Micronutrient Malnutrition Prevention and Control (IMMPaCt). Available online: http://www.cdc.gov/immpact/micronutrients/ (accessed on 5 May 2016).
3.  Welch, R.M.; Graham, R.D. Agriculture: The real nexus for enhancing bioavailable micronutrients in food crops. *J. Trace Elem. Med. Biol.* **2005**, *18*, 299–307. [CrossRef] [PubMed]
4.  Thavarajah, D.; Thavarajah, P.; Sarker, A.; Materne, M.; Vandemark, G.; Shrestha, R.; Idrissi, O.; Hacikamiloglu, O.; Bucak, B.; Vandenberg, A. A global survey of effects of genotype and environment on selenium concentration in lentils (*Lens culinaris* L.): Implications for nutritional fortification strategies. *Food Chem.* **2011**, *125*, 72–76. [CrossRef]
5.  Sen Gupta, D.; Thavarajah, D.; Knutson, P.; Thavarajah, P.; McGee, R.J.; Coyne, C.J.; Kumar, S. Lentils (*Lens culinaris* L.), a Rich Source of Folates. *J. Agric. Food Chem.* **2013**, *61*, 7794–7799. [CrossRef] [PubMed]
6.  Johnson, C.R.; Thavarajah, D.; Combs, G.F.; Thavarajah, P. Lentil (*Lens culinaris* L.): A prebiotic-rich whole food legume. *Food Res. Int.* **2013**, *51*, 107–113. [CrossRef]
7.  Pilon-Smits, E.A.H.; Quinn, C.F. Selenium metabolism in plants. In *Cell Biology of Metals and Nutrients Plant Cell Monographs*; Hell, R., Mendel, R.R., Eds.; Springer: Berlin, Germany, 2010; pp. 225–241.
8.  Yokota, A.; Shigeoka, S.; Onishi, T.; Kitaoka, S. Selenium as inducer of glutathione peroxidase in low-$CO_2$-grown *Chlamydomonas reinhardtii*. *Plant Physiol.* **1988**, *86*, 649–651. [CrossRef] [PubMed]
9.  Lobanov, A.V.; Hatfield, D.L.; Gladyshev, V.N. Eukaryotic selenoproteins and selenoproteomes. *Biochim. Biophys. Acta* **2009**, *1790*, 1424–1428. [CrossRef] [PubMed]
10. Fajardo, D.; Schlautman, B.; Steffan, S.; Polashock, J.; Vorsa, N.; Zalapa, J. The American cranberry mitochondrial genome reveals the presence of selenocysteine (tRNA-Sec and SECIS) insertion machinery in land plants. *Gene* **2014**, *536*, 336–343. [CrossRef]
11. Zhu, Y.G.; Pilon-Smits, E.A.H.; Zhao, F.J.; Williams, P.N.; Meharg, A.A. Selenium in higher plants: Understanding mechanisms for biofortification and phytoremediation. *Trends Plant Sci.* **2009**, *14*, 436–442. [CrossRef] [PubMed]
12. Ekanayake, L.J.; Thavarajah, D.; Vial, E.; Schatz, B.; McGee, R.; Thavarajah, P. The effect of selenium fertilization on lentil (*Lens culinaris* Medikus) grain yield, seed selenium concentration, and antioxidant activity. *Field Crops Res.* **2015**, *177*, 9–14. [CrossRef]
13. Thavarajah, D.; Ruszkowski, J.; Vandenberg, A. High potential for selenium biofortification of lentils (*Lens culinaris* L.). *J. Agric. Food Chem.* **2008**, *56*, 10747–10753. [CrossRef] [PubMed]

14. Khazaei, H.; Caron, C.T.; Fedoruk, M.; Diapari, M.; Vandenberg, A.; Coyne, C.J.; McGee, R.; Bett, K.E. Genetic Diversity of Cultivated Lentil (*Lens culinaris* Medik.) and Its Relation to the World's Agro-ecological Zones. *Front. Plant Sci.* **2016**, *7*, 1093. [CrossRef]

15. Thavarajah, D.; Thavarajah, P.; Vial, E.; Gebhardt, M.; Lacher, C.; Kumar, S.; Combs, G.F. Will selenium increase lentil (*Lens culinaris* Medik) yield and seed quality? *Front. Plant Sci.* **2015**, *6*, 356. [CrossRef] [PubMed]

16. Ates, D.; Sever, T.; Aldemir, S.; Yagmur, B.; Temel, H.Y.; Kaya, H.B.; Alsaleh, A.; Kahraman, A.; Ozkan, H.; Vandenberg, A.; et al. Identification QTLs Controlling Genes for Se Uptake in Lentil Seeds. *PLoS ONE* **2016**, *11*, e0149210.

17. US National Plant Germplasm System. Available online: https://npgsweb.ars-grin.gov/gringlobal/search.aspx (accessed on 5 May 2016).

18. Wakim, R.; Bashour, I.; Nimah, M.; Sidahmed, M.; Toufeili, I. Selenium levels in Lebanese environment. *J. Geochem. Explor.* **2010**, *107*, 94–99. [CrossRef]

19. Oldfield, J.E. Selenium World Atlas. Selenium—Tellurium Development Association, Belgium. Available online: http://www.369.com.cn/En/Se%20Atlas%202002.pdf (accessed on 5 May 2016).

20. Apostolidis, E.; Kwon, Y.I.; Shetty, K. Inhibitory potential of herb, fruit, and fungal-enriched cheese against key enzymes linked to type 2 diabetes and hypertension. *Innov. Food Sci. Emerg. Technol.* **2007**, *8*, 46–54. [CrossRef]

21. SAS Institute. *User's Guide: Statistics SAS Institute (Version 9.4)*; SAS Institute: Cary, NC, USA, 2012.

22. Hartikainen, H.; Xue, T. The promotive effect of selenium on plant growth as triggered by ultraviolet irradiation. *J. Environ. Qual.* **1999**, *28*, 1372–1375. [CrossRef]

23. Pezzarossa, B.; Petruzzelli, G.; Petacco, F.; Malorgio, F.; Ferri, T. Absorption of selenium by Lactuca sativa as affected by carboxymethylcellulose. *Chemosphere* **2007**, *67*, 322–329. [CrossRef] [PubMed]

24. Lyons, G.H.; Genc, Y.; Soole, K.; Stangoulis, J.C.R.; Liu, F.; Graham, R.D. Selenium increases seed production in Brassica. *Plant Soil* **2009**, *318*, 73–80. [CrossRef]

25. Thavarajah, D.; Thavarajah, P.; Combs, G.F. Selenium in lentils (*Lens culinaris* L.) and theoretical fortification strategies. In *Handbook of Food Fortification and Health: From Concepts to Public Health Applications*; Preedy, V.R., Srirajaskanthan, R., Patel, V.B., Eds.; Springer: New York, NY, USA, 2013.

26. Thavarajah, D.; Vandenberg, A.; George, G.N.; Pickering, I.J. Chemical form of selenium in naturally selenium-rich lentils (*Lens culinaris* L.) from Saskatchewan. *J. Agric. Food Chem.* **2007**, *55*, 7337–7341. [CrossRef] [PubMed]

27. Allaway, W.H.; Moore, D.P.; Oldfield, J.E.; Muth, O.H. Movement of physiological levels of selenium from soils through plants to animals. *J. Nutr.* **1966**, *88*, 411–418. [PubMed]

28. Turakainen, M.; Hartikainen, H.; Seppänen, M.M. Effects of selenium treatments on potato (*Solanum tuberosum* L.) growth and concentrations of soluble sugars and starch. *J. Agric. Food Chem.* **2004**, *52*, 5378–5382. [CrossRef] [PubMed]

# Influence of Nitrogen Availability on Growth of Two Transgenic Birch Species Carrying the Pine GS1a Gene

**Vadim G. Lebedev \*, Nina P. Kovalenko and Konstantin A. Shestibratov**

Branch of Shemyakin and Ovchinnikov Institute of Bioorganic Chemistry of the Russian Academy of Sciences, Science avenue 6, Pushchino, Moscow Region 142290, Russia; nina-kovalenko4@rambler.ru (N.P.K.); schestibratov.k@yandex.ru (K.A.S.)

\* Correspondence: vglebedev@mail.ru

Academic Editors: Vagner A. Benedito and Milan S. Stankovic

**Abstract:** An alternative way to increase plant productivity through the use of nitrogen fertilizers is to improve the efficiency of nitrogen utilization via genetic engineering. The effects of overexpression of pine glutamine synthetase (GS) gene and nitrogen availability on growth and leaf pigment levels of two *Betula* species were studied. Untransformed and transgenic plants of downy birch (*B. pubescens*) and silver birch (*B. pendula*) were grown under open-air conditions at three nitrogen regimes (0, 1, or 10 mM) for one growing season. The transfer of the GS1a gene led to a significant increase in the height of only two transgenic lines of nine *B. pubescens*, but three of five *B. pendula* transgenic lines were higher than the controls. In general, nitrogen supply reduced the positive effect of the GS gene on the growth of transgenic birch plants. No differences in leaf pigment levels between control and transgenic plants were found. Nitrogen fertilization increased leaf chlorophyll content in untransformed plants but its effect on most of the transgenic lines was insignificant. The results suggest that birch plants carrying the GS gene use nitrogen more efficiently, especially when growing in nitrogen deficient soil. Transgenic lines were less responsive to nitrogen supply in comparison to wild-type plants.

**Keywords:** *Betula*; chlorophyll; glutamine synthetase; nitrogen fertilization; transgenic birch

## 1. Introduction

Nitrogen plays a key role in the growth and development of plants as it is a component of amino acids, chlorophyll, nucleic acids, and growth regulators. This element is the main limiting factor for plant productivity in temperate and boreal forest ecosystems [1]. Forest plantations are also mainly established on poor soils, where nitrogen availability is limited. For this reason, nitrogen fertilizers are used intensively for increasing yield and reducing rotation age in forest plantations [2]. However, synthetic nitrogen fertilizers are expensive. At the same time, 50%–75% of the nitrogen applied to fields are not assimilated by plants and are lost [3], thus contaminating soil, water, and air. An alternative way to increase plant productivity is to improve the efficiency of nitrogen utilization, in particular, by means of genetic engineering.

For this purpose, the incorporation of additional glutamine synthetase (GS) genes is used most frequently. This is due to the fact that GS is a main enzyme for nitrogen assimilation in plants [4], as products of GS/GOGAT cycle (glutamate and glutamine) are precursors for organic compounds in plants [5]. GS genes were transferred into numerous herbaceous plants. However, up to now their effect has been only studied for a single woody plant. A hybrid *Populus tremula* × *P. alba* with a gene for thecytosolic form of pine GS has demonstrated an increased productivity under both greenhouse [6] and field conditions [7]. This gene was used because, the cytosolic form of the enzyme from gymnosperms is active in photosynthetic tissues, in contrast to angiosperms [8]. Results of studies [6,7] suggest that accelerated growth of transgenic trees of the *Populus* hybrid is not only explained by the primary nitrogen assimilation, but also by the reassimilation of ammonium, which is produced during various metabolic processes [7].

Most of the studies on both genetic transformation and influence of nitrogen availability are carried out in species and hybrids of *Populus* [9,10], which are used worldwide on short-rotation plantations. Birch is the most important broad-leaved tree species in the Northern and Eastern Europe forestry [11]. In order to increase productivity of forest plantations the gene of cytosolic GS from *Pinus sylvestris* was transferred into the genome of *Betula* [12]. Transgenic plants were tested for growth rate enhancement in the greenhouse [13] and resistance to phosphinothricin treatment [14]. As an extension of the evaluation of these plants, in the present work we estimated the effect of the pine GS1a gene expression on the growth and leaf pigment levels of two birch species (*Betula pubescens* and *B. pendula*) grown under three different nitrogen levels in outdoor conditions.

## 2. Results

### 2.1. RT-PCR Analysis

For analysis of the pine GS1a gene expression RT-PCR was performed with total RNA samples extracted from birch leaf tissue. The GS1a transcript was accumulated in all of the analyzed transgenic plants, including nine downy birch-based lines, and five silver birch-based lines. There was no transcription of GS1a gene in the untransformed plants of all genotypes. Results for some of the *B. pubescens* and *B. pendula* lines are shown in Figure 1a,b respectively.

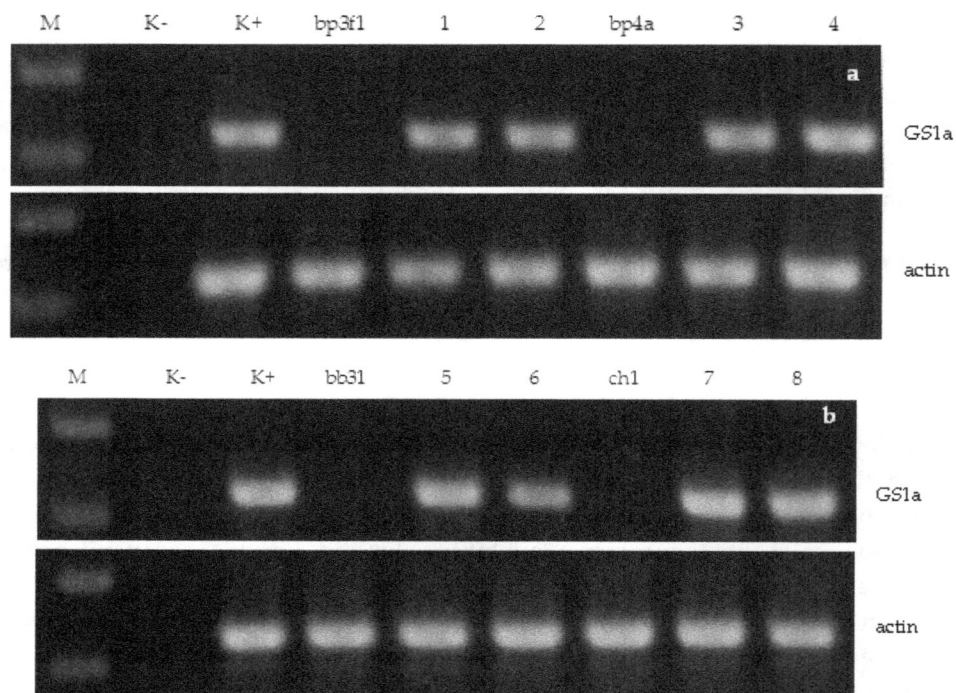

**Figure 1.** RT-PCR analysis of the GS1a gene expression in *B. pubescens* (**a**) and *B. pendula* (**b**) plants. The actin gene was used as an endogenous control. M—markers (750 and 500 bp); K—water; K+—pGS (for the GS1a gene) or non-transgenic birch plants (bp3f1 or bb31 for the actin gene); bp3f1, bp4a, bb31, ch1—wild-type plants; 1—F14GS8b; 2—F14GS9b; 3—P9GS18b; 4—P17GS1a, 5—B29GS1; 6—B29GS4; 7—N18GS8a; N18GS8b.

## 2.2. Growth Response of Birch Plants to Nitrogen Availability

Nitrogen supply affected height of birch plants (Table 1). By the end of the vegetation period, in all the control genotypes there were no significant differences when cultivated without nitrogen or in the presence of 1 mM nitrogen, whereas at 10 mM nitrogen plants were 31%–73% higher, depending on genotype. Regardless of the nitrogen availability, *B. pendula* plant heights were approximately 1.5-fold lower as compared to *B. pubescens* plants.

**Table 1.** Effect of nitrogen availability on growth (cm) of nontransgenic birch plants.

| Species | Genotype | Nitrogen, mM | | |
|---|---|---|---|---|
| | | 0 | 1 | 10 |
| [1] *B. pubescens* | bp3f1 | 67.1 ± 2.3 b [2] | 64.0 ± 1.7 b | 87.8 ± 3.2 a |
| | bp4a | 52.3 ± 2.8 b | 54.0 ± 3.9 b | 90.3 ± 2.6 a |
| *B. pendula* | bb31 | 40.9 ± 3.8 b | 44.2 ± 2.2 b | 59.5 ± 6.1 a |
| | ch1 | 39.4 ± 2.3 b | 41.7 ± 2.5 b | 51.4 ± 2.2 a |

[1] Data indicate mean ± SE; [2] Different letters in a line indicate significance of differences according to the Duncan test at $p < 0.05$.

In *B. pubescens* species, transformation with the GS1a gene led to a significant increase in heights only in the F14GS8b line (0 and 1 mM nitrogen) and P17GS1a (0 mM nitrogen) (Figure 2a). Plants of other transgenic lines under study, obtained on the basis of both bp3f1 and bp4a genotypes, did not grow as tall as the untransformed plants. Furthermore, the growth difference increased along with the rise of nitrogen availability. On the contrary, all plants of the *B. pendula* transgenic lines were higher than the control ones. However, this effect was significant only for three of five lines, all of which were based on the bb31 genotype (up to 63% in B29GS4 at 0 mM nitrogen) (Figure 2b).

Nitrogen supply had the opposite effect on growth of transgenic plants of *B. pendula* genotypes: the growth difference in bb31-based lines was reduced compared to the control, whereas such difference was increased in ch1-based lines.

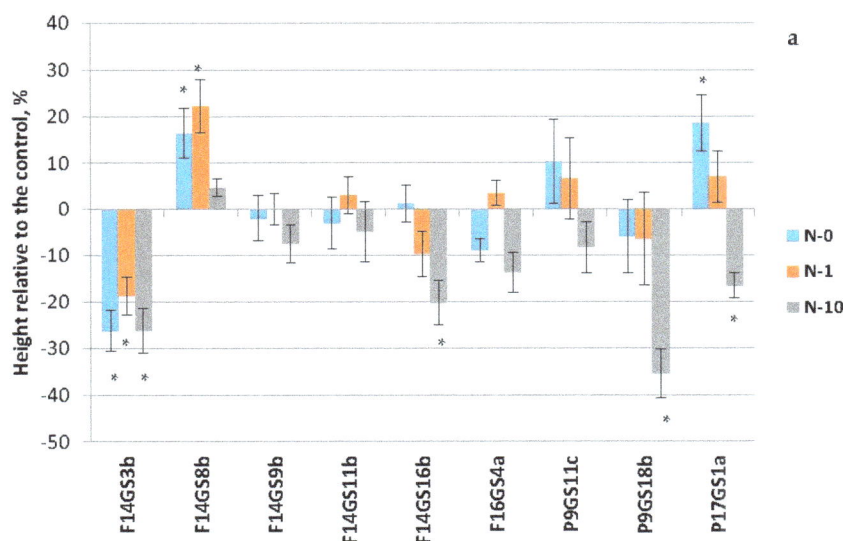

**Figure 2.** *Cont.*

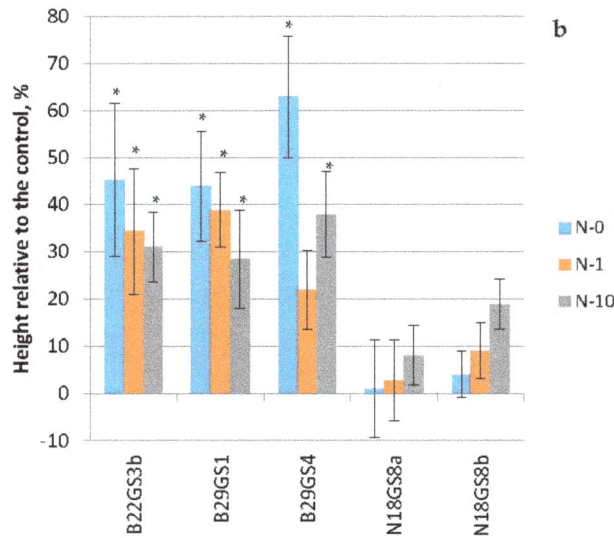

**Figure 2.** Effect of glutamine synthetase (GS) gene expression and nitrogen fertilization on the growth of transgenic *B. pubescens* (**a**) and *B. pendula* (**b**) plants. Asterisks over single bars indicates that the mean value of transgenic line were significantly higher than that of control plants at 0, 1, or 10 mM nitrogen, respectively, when analyzed by one-way ANOVA (* $p < 0.05$).

## 2.3. Pigment Levels in Leaves of Birch Plants

Chlorophyll content in nontransgenic plants did not differ among the genotypes of the same species, whereas in *B. pendula* plants there was more chlorophyll as compared to *B. pubescens* (Figure 3). Differences decreased along with an increase in nitrogen availability: $p = 0.0073$ in nitrogen-free variant, $p = 0.0139$ at 1 mM, and $p = 0.0494$ at 10 mM nitrogen. Significant differences in the chlorophyll b content were only observed at 1 mM nitrogen, which was higher in the genotype bb31 as compared to others. Carotenoid content varied from 0.285 to 0.355 µg/mg fresh weight and did not depend on the plant species and nitrogen availability (data not shown). Transfer of the GS gene did not affect the pigment levels in leaves—not a single transgenic line was different from the corresponding control.

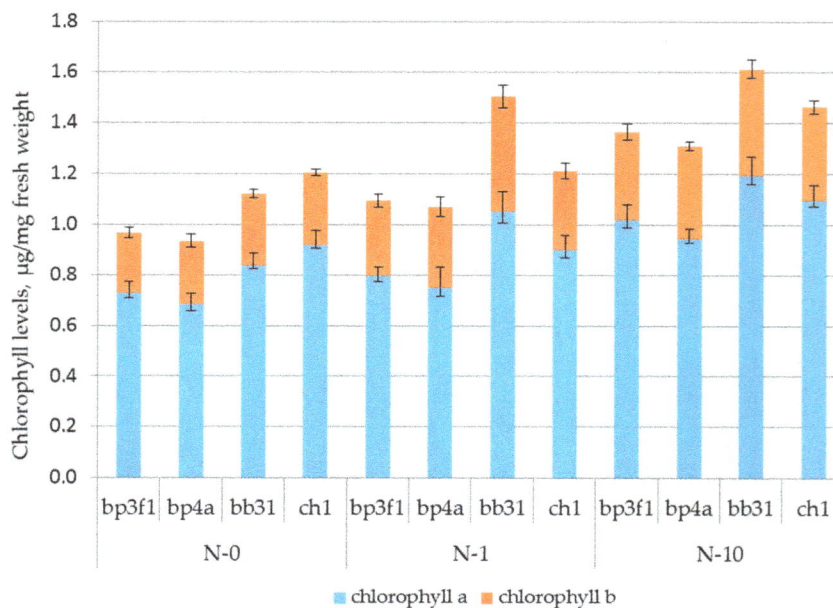

**Figure 3.** Chlorophyll levels in nontransgenic birch plants. Data bars represent mean ± SE.

Increased nitrogen availability led to an increase in pigment levels in birch leaves. However, this effect was dependent on the plant transgenic status (Table 2). In the response to nitrogen supply, levels of chlorophylls a and b increased in three out of four birch genotypes (except for ch1). However, in most of the transgenic lines of these genotypes nitrogen fertilization did not affect the chlorophyll content. ch1 genotype behaved differently: nitrogen supply almost did not affect the pigment levels in the control, whereas pigment levels were changed in transgenic plants.

**Table 2.** Statistical relevance of the effect of nitrogen availability on pigment levels in birch leaves.

| Species | Genotype | Chlorophyll a | Chlorophyll b | Carotenoids | Chlorophylls a + b |
|---|---|---|---|---|---|
| *B. pubescens* | bp3f1 (control) | *** [1] | * | ns [2] | ** |
| | F14GS3b | ns | ns | ns | ns |
| | F14GS8b | ns | ns | ns | ns |
| | F14GS9b | * | ns | ns | ns |
| | F14GS11b | ns | * | ns | ns |
| | F14GS16b | * | ns | ns | ns |
| | F16GS4a | ** | ** | ns | ** |
| | bp4a (control) | ** | * | ns | ** |
| | P9GS11c | * | * | ns | * |
| | P9GS18b | ns | ns | ns | ns |
| | P17GS1a | ns | ns | ns | ns |
| *B. pendula* | bb31 (control) | ** | ** | * | ** |
| | B22GS3b | ns | ns | ns | ns |
| | B29GS1 | ns | ns | ns | ns |
| | B29GS4 | * | * | * | * |
| | ch1 (control) | ns | ns | ns | * |
| | N18GS8a | * | ns | ns | * |
| | N18GS8b | ** | ns | ** | ** |

[1] Asterisks indicate significant differences between 0 and 10 mM nitrogen, when analyzed by one-way ANOVA (* $p < 0.05$; ** $p < 0.01$; *** $p < 0.001$); [2] ns = not significant.

## 3. Discussion

Birches are widely present in temperate forests, but they are relatively poorly studied compared to *Populus* species. In Europe, two commercially important birch species occur naturally: silver birch (*Betula pendula* Roth) and downy birch (*Betula pubescens* Ehrh.) [11]. However, to our knowledge, these two species were never compared for their reaction to nitrogen availability. The areas of these species overlap, but *B. pubescens* grows in moist sites and it is more resistant to the cold and occurs further to the north than *B. pendula*, which prefers warmer and drier sites [15].

Our studies have demonstrated that both species responded similarly to nitrogen supply: fertilization with solution containing 1 mM nitrogen was shown to marginally increase plant height, however, 10 mM nitrogen led to more significant increases in growth rates, with −37%–67% and 23%–35% increases for *B. pubescens* and *B. pendula*, respectively. Silver birch genotypes, in addition to displaying a less significant increase in growth rate, also demonstrated lower absolute height in comparison to downy birch (Table 1). Results of known studies of the nitrogen effect on birch growth have not been explicit. Use of fertilizers is not a common practice in the management of birch stands and several attempts made under field conditions in Finland have shown only a weak growth response to fertilization [11]. Studies under greenhouse conditions have shown that nitrogen fertilization had significantly increased the biomass of both *B. pubescens* [16] and *B. pendula* [17], however, a two-year experiment under open-field conditions showed that nitrogen supply significantly increased height of *B. pendula* plants only during the second year, but that the effect during the first year was insignificant [18]. As such, perhaps the effect of fertilization is more prominent under greenhouse conditions.

Significant differences between the plants of the two birch species transformed with a gene encoding the pine cytosolic GS were demonstrated. Whereas all the transgenic silver birch plants demonstrated an increased height in comparison with the nontransgenic control at all nitrogen regimes, only plants of certain downy birch lines were higher than control plants, and only in experiments with low nitrogen availability (0 or 1 mM (Figure 1)). In the work of Gallardo et al. [6], all the 22 lines of hybrid poplar carrying the GS gene were higher than the control following six months of growth in a greenhouse without fertilization. We observed a clear tendency for the reduction of plant heights in comparison with the control along with the increase in the nitrogen availability for all the transgenic lines except for the ones obtained on the basis of the *B. pendula* ch1 genotype. Apparently, additional copies of the GS gene provided enhanced nitrogen recycling in plants during conditions where there was insufficient uptake from the soil. However, under conditions of nitrogen abundance they became an obstacle. Man et al. [19] previously noted that hybrid poplars carrying GS were able to increase their height by 81% at 0.3 mM nitrate, whereas at 10 mM the heights only increased by 35% as compared to the control. In a study by Fuentes et al. [20], tobacco with the GS1 gene grew better than the control only in the absence of nitrogen, and the authors proposed that increased GS activity promoted re-assimilation of photorespiratory ammonium and recycling of other nitrogen-containing compounds. This is an advantage for plantation forestry, as plantations are commonly established on infertile soils. The response of transgenic birch plants to nitrogen availability was genotype-dependent: ch1 genotype based lines differed from other genotypes. It was already shown that genotypes of *Populus tremula* and *P. tremula* × *P. tremuloides* are very different in their reaction to nitrogen fertilization under open-air conditions [21].

Chlorophyll levels in leaves may be used as an indicator of the plant nitrogen status [22]. However, to the best of our knowledge, studies of correlations between chlorophyll levels and nitrogen availability have never been carried out in birch. Overexpression of the GS gene did not change the pigment levels in the leaves of the transgenic plants. Hybrid poplar plants with the GS gene also did not differ in the chlorophyll content from the control at 10 mM nitrogen, whereas at 50 mM levels of chlorophylls a and b significantly increased in the most upper leaves [23]. The chlorophyll content was significantly higher in silver birch plants by approximately 20% (Figure 3) regardless of the nitrogen availability. These data are partially consistent with the results of Li et al. [24], where it was shown that when grown on nutrient-deficient soils *Populus popularis* contained less photosynthetic pigments in its leaves as compared to *P. alba* × *P. glandulosa*, which usually grows on relatively fertile soils. However, these differences were only significant in the treatment without nitrogen, whereas nitrogen fertilization made the difference insignificant.

Our studies have demonstrated that untransformed plants had significantly increased chlorophyll levels (chlorophyll a, b, and their sum) in response to nitrogen supply (only total chlorophyll for *B. pendula* ch1 genotype) (Table 2). In similar studies, nitrogen fertilization increased chlorophyll levels in *Populus balsamifera* ssp. *trichocarpa* × *deltoides* [2], *P. simonii* [25], and *P. deltoides* [26], but did not change pigment levels in *P. alba* × *P. glandulosa* [24]. The differences between bb31 and ch1 genotypes of silver birch that we have shown are rather typical for different species. As in the case of the growth, most of the transgenic lines demonstrated less pronounced responses to nitrogen supply: in contrast to the control plants, differences between chlorophyll levels were insignificant.

Using nitrogen fertilizers to improve productivity is not an optimal solution: first, major parts of them are not assimilated by plants and thus pollute the environment; second, their abundance may disturb normal metabolism, for instance, nitrogen fertilization of *B. pubescens*, in addition to increasing biomass, reduced the concentration of most of the phenolic compounds which may result in reduced resistance against herbivores and pathogens [16]. It was also shown that along with increased inputs of nitrogen, the height and diameter increment of *Betula pendula* decreased [27]. Our data have shown that transgenic birch plants carrying the GS gene have increased nitrogen use efficiency under nitrogen deficient soil conditions. Additionally, we observed significant differences between transgenic downy birch and silver birch in response to nitrogen supply. Such trees with improved efficiency of nitrogen utilization are likely to be especially valuable for the establishment of forest plantations, for which poor soils are usually allocated, in order to avoid competition with agricultural crops.

## 4. Materials and Methods

### 4.1. Plant Material and Growth Conditions

The following four genotypes of two birch species were used in the study: *Betula pubescens* Ehrh. (bp3f1, bp4a), *Betula pendula* Roth (bb31, ch1), and transgenic lines obtained on their basis that contain the pine glutamine synthetase gene. Transgenic plants of *B. pubescens* were obtained via *Agrobacterium*-mediated transformation of birch leaf explants by pGS vector carrying the GS1a gene of *Pinus sylvestris* under the control of CaMV 35S promoter, and their status was confirmed by PCR [12]. Gene transfer and molecular analysis of *B. pendula* plants were carried out following the same procedure. Plants were micropropagated, transferred in a greenhouse at the beginning of April, planted into 1 liter pots with peat:perlite (3:1) at the end of May, and transferred outdoors. In total, 4 control genotypes and 14 transgenic lines (20–30 plants in each group) were planted. During nine weeks (from mid-June to mid-August) the plants were fertilized daily with solutions of macro- and micronutrients containing 0, 1, or 10 mM nitrogen in a calcium nitrate form (100 mL per plant). At the end of July, expression via RT-PCR and pigment levels were analyzed in the leaves from the middle part of the plants. Plant heights were measured once every two weeks during vegetation period.

### 4.2. RT-PCR Analysis

Total RNA was extracted from leaves of birch plants as described by Chang et al. [28]. RT-PCR reaction was carried out as previously described [14]. Information about (i) cDNA nucleotide sequences, (ii) location of primers, and (iii) length of amplicons are presented in Figure S1 (for the GS1a gene from *P.sylvestris*) and Figure S2 (for the Actin gene from *Populus tomentosa*).

### 4.3. Leaf Pigment Analyses

Chlorophyll and carotenoid contents were analyzed using the methods of Wellburn [29]: after pigment extraction with 80% acetone, optical density was measured at 663, 646, and 470 nm wavelengths (Shimadzu UV-1800, Kyoto, Japan).

### 4.4. Statistical Analysis

Statistical analysis was carried out using Statistica 10 software (StatSoft, Tulsa, OK, USA).

**Acknowledgments:** This work was supported by the Ministry of Education and Science of the Russian Federation (Project No.14.616.21.0013 from 17 September 2014, unique identifier RFMEFI61614X0013).

**Author Contributions:** Vadim Lebedev conceived and designed the experiments; Vadim Lebedev and Nina Kovalenko performed the experiments; Vadim Lebedev analyzed the data; Vadim Lebedev and Konstantin Shestibratov wrote the paper.

**Conflicts of Interest:** The authors declare no conflict of interest.

## References

1. Vitousek, P.M.; Aber, J.D.; Howarth, R.H.; Likens, G.E.; Matson, P.A.; Schindler, D.W.; Schlesinger, W.H.; Tilman, D.G. Human alteration of the global nitrogen cycle: Source and consequences. *Ecol. Appl.* **1997**, *7*, 737–750. [CrossRef]
2. Cooke, J.E.K.; Martin, T.A.; Davis, J.M. Short-term physiological and developmental responses to nitrogen availability in hybrid poplar. *New Phytol.* **2005**, *167*, 41–52. [CrossRef] [PubMed]

3.  Hirel, B.; Tétu, T.; Lea, P.J.; Dubois, F. Improving nitrogen use efficiency in crops for sustainable agriculture. *Sustainability* **2011**, *3*, 1452–1485. [CrossRef]

4.  Cai, H.; Zhou, Y.; Xiao, J.; Li, X.; Zhang, Q.; Lian, X. Overexpressed glutamine synthetase gene modifies nitrogen metabolism and abiotic stress responses in rice. *Plant Cell Rep.* **2009**, *28*, 527–537. [CrossRef] [PubMed]

5.  Ireland, R.J.; Lea, P.J. The enzymes of glutamine; glutamate; asparagines; and aspartate metabolism. In *Plant Amino Acids. Biochemistry and Biotechnology*; Singh, B.K., Ed.; Marcel Dekker: New York, NY, USA, 1999; pp. 49–109.

6.  Gallardo, F.; Fu, J.M.; Canton, F.R.; Garcia-Gutierrez, A.; Canovas, F.M.; Kirby, E.G. Expression of a conifer glutamine synthetase gene in transgenic poplar. *Planta* **1999**, *210*, 19–26. [CrossRef] [PubMed]

7.  Jing, Z.P.; Gallardo, F.; Pascual, M.B.; Sampalo, R.; Romero, J.; de Navarra, A.T.; Canovas, F.M. Improved growth in a field trial of transgenic hybrid poplar overexpressing glutamine synthetase. *New Phytol.* **2004**, *164*, 137–145. [CrossRef]

8.  Canovas, F.M.; Canton, F.R.; Gallardo, F.; García-Gutiérrez, A.; de Vicente, A. Accumulation of glutamine synthetase during early development of maritime pine (*Pinus pinaster*) seedlings. *Planta* **1991**, *185*, 372–378. [CrossRef] [PubMed]

9.  Ye, X.; Busov, V.; Zhao, N.; Meilan, R.; McDonnell, L.M.; Coleman, H.D.; Mansfield, S.D.; Chen, F.; Li, Y.; Cheng, Z.-M. Transgenic *Populus* trees for forest products, bioenergy, and functional genomics. *Crit. Rev. Plant Sci.* **2011**, *30*, 415–434. [CrossRef]

10. Dash, M.; Yordanov, Y.S.; Georgieva, T.; Kumari, S.; Wei, H.; Busov, V. A network of genes associated with poplar root development in response to low nitrogen. *Plant Signal. Behav.* **2016**, *11*, e1214792. [CrossRef] [PubMed]

11. Hynynen, J.; Vihera-Aarnio, A.; Velling, P.; Niemisto, P.; Brunner, A.; Hein, S. Silviculture of birch (*Betula pendula* Roth and *Betula pubescens* Ehrh.) in Northern Europe. *Forestry* **2010**, *83*, 103–119. [CrossRef]

12. Lebedev, V.G.; Schestibratov, K.A.; Shadrina, T.E.; Bulatova, I.V.; Abramochkin, D.G.; Miroshnikov, A.I. Cotransformation of aspen and birch with three T-DNA regions from two different replicons in one *Agrobacterium tumefaciens* strain. *Russ. J. Genetika* **2010**, *46*, 1282–1289. [CrossRef]

13. Shestibratov, K.; Lebedev, V.; Podrezov, A.; Salmova, M. Transgenic aspen and birch trees for Russian plantation forests. *BMC Proc.* **2011**, *5*. [CrossRef]

14. Lebedev, V.; Faskhiev, V.; Shestibratov, K. Lack of correlation between ammonium accumulation and survival of transgenic birch plants with pine cytosolic glutamine synthetase gene after "Basta" herbicide treatment. *J. Bot.* **2015**, *2015*, 1–6. [CrossRef]

15. Hytteborn, H.; Maslov, A.A.; Nazimova, D.I.; Rysin, L.P. Boreal forests of Eurasia. In *Coniferous Forests: Ecosystems of the World 6*; Andersson, F., Ed.; Elsevier: Amsterdam, The Netherlands, 2005; pp. 23–100.

16. Lappalainen, J.H.; Martel, J.; Lempa, K.; Wilsey, B.; Ossipov, V.V. Effects of resource availability on carbon allocation and developmental instability in cloned birch seedlings. *Int. J. Plant Sci.* **2000**, *161*, 119–125. [CrossRef] [PubMed]

17. Esmeijer-Liu, A.J.; Aerts, R.; Kürschner, W.M.; Bobbink, R.; Lotter, A.F.; Verhoeven, J.T.A. Nitrogen enrichment lowers *Betula pendula* green and yellow leaf stoichiometry irrespective of effects of elevated carbon dioxide. *Plant Soil* **2009**, *316*, 311–322. [CrossRef]

18. Pääkkonen, E.; Holopainen, T. Influence of nitrogen supply on the response of clones of birch (*Betula pendula* Roth.) to ozone. *New Phytol.* **1995**, *129*, 595–603. [CrossRef]

19. Man, H.M.; Boriel, R.; El-Khatib, R.; Kirby, E.G. Characterization of transgenic poplar with ectopic expression of pine cytosolic glutamine synthetase under conditions of varying nitrogen availability. *New Phytol.* **2005**, *167*, 31–39. [CrossRef] [PubMed]

20. Fuentes, S.I.; Allen, D.J.; Ortiz-Lopez, A.; Hernandez, G. Over-expression of cytosolic glutamine synthetase increases photosynthesis and growth at low nitrogen concentrations. *J. Exp. Bot.* **2001**, *52*, 1071–1081. [CrossRef] [PubMed]

21. Haikio, E.; Freiwald, V.; Silfver, T.; Beuker, E.; Holopainen, T.; Oksanen, E. Impacts of elevated ozone and nitrogen on growth and photosynthesis of European aspen (*Populus tremula*) and hybrid aspen (*P. tremula* x *Populus tremuloides*) clones. *Can. J. For. Res.* **2007**, *37*, 2326–2336. [CrossRef]

22.  Muñoz-Huerta, R.F.; Guevara-Gonzalez, R.G.; Contreras-Medina, L.M.; Torres-Pacheco, I.; Prado-Olivarez, J.; Ocampo-Velazquez, R.V. A review of methods for sensing the nitrogen status in plants: Advantages; disadvantages and recent advances. *Sensors* **2013**, *13*, 10823–10843. [CrossRef] [PubMed]

23.  Castro-Rodríguez, V.; García-Gutiérrez, A.; Canales, J.; Cañas, R.A.; Kirby, E.G.; Avila, C.; Cánovas, F.M. Poplar trees for phytoremediation of high levels of nitrate and applications in bioenergy. *Plant Biotechnol. J.* **2016**, *14*, 299–312. [CrossRef] [PubMed]

24.  Li, H.; Li, M.; Luo, J.; Cao, X.; Qu, L.; Ga, Y.; Jiang, X.; Liu, T.; Bai, H.; Janz, D.; et al. N-fertilization has different effects on the growth; carbon and nitrogen physiology; and wood properties of slow- and fast-growing *Populus* species. *J. Exp. Bot.* **2012**, *63*, 6173–6185. [CrossRef] [PubMed]

25.  Luo, J.; Zhou, J.; Li, H.; Shi, W.; Polle, A.; Lu, M.; Sun, X.; Luo, Z.-B. Global poplar root and leaf transcriptomes reveal links between growth and stress responses under nitrogen starvation and excess. *Tree Physiol.* **2015**, *35*, 1283–1302. [CrossRef] [PubMed]

26.  Li, J.; Dong, T.; Guo, Q.; Zhao, H. *Populus deltoides* females are more selective in nitrogen assimilation than males under different nitrogen forms supply. *Trees* **2015**, *29*, 143–159. [CrossRef]

27.  Kula, E.; Pešlová, A.; Martinek, P. Effects of nitrogen on growth properties and phenology of silver birch (*Betula pendula* Roth). *J. For. Sci.* **2012**, *58*, 391–399.

28.  Chang, S.; Puryear, J.; Cairney, J.A. A simple and efficient method for isolating RNA from pine trees. *Plant Mol. Biol. Rep.* **1993**, *11*, 113–116. [CrossRef]

29.  Wellburn, A.R. The spectral determination of chlorophyll *a* and *b*, as well as total carotenoids, using various solvents with spectrophotometers of different resolution. *J. Plant Physiol.* **1994**, *144*, 307–313. [CrossRef]

# Effect of Drought on Herbivore-Induced Plant Gene Expression: Population Comparison for Range Limit Inferences

**Gunbharpur Singh Gill** [1], **Riston Haugen** [1], **Steven L. Matzner** [2], **Abdelali Barakat** [3] and **David H. Siemens** [1,*]

1   Integrative Genomics Program, Black Hills State University, Spearfish, SD 57789, USA; Gunbharpur.Gill@yellowjackets.bhsu.edu (G.S.G.); Riston.Haugen@yellowjackets.bhsu.edu (R.H.)
2   Biology Department, Augustana University, Sioux Falls, SD 57197, USA; steven.matzner@augie.edu
3   Biology Department, University of South Dakota, Vermillion, SD 57069, USA; Abdelali.Barakat@usd.edu
*   Correspondence: David.Siemens@bhsu.edu

Academic Editor: Debora Gasperini

**Abstract:** Low elevation "trailing edge" range margin populations typically face increases in both abiotic and biotic stressors that may contribute to range limit development. We hypothesize that selection may act on ABA and JA signaling pathways for more stable expression needed for range expansion, but that antagonistic crosstalk prevents their simultaneous co-option. To test this hypothesis, we compared high and low elevation populations of *Boechera stricta* that have diverged with respect to constitutive levels of glucosinolate defenses and root:shoot ratios; neither population has high levels of both traits. If constraints imposed by antagonistic signaling underlie this divergence, one would predict that high constitutive levels of traits would coincide with lower plasticity. To test this prediction, we compared the genetically diverged populations in a double challenge drought-herbivory growth chamber experiment. Although a glucosinolate defense response to the generalist insect herbivore *Spodoptera exigua* was attenuated under drought conditions, the plastic defense response did not differ significantly between populations. Similarly, although several potential drought tolerance traits were measured, only stomatal aperture behavior, as measured by carbon isotope ratios, was less plastic as predicted in the high elevation population. However, RNAseq results on a small subset of plants indicated differential expression of relevant genes between populations as predicted. We suggest that the ambiguity in our results stems from a weaker link between the pathways and the functional traits compared to transcripts.

**Keywords:** drought; herbivory; glucosinolates; carbon isotope ratio; gene expression

## 1. Introduction

The study of factors and processes affecting species range limits has a long history in ecology and evolution starting with Darwin, but there has been a recent resurgence of interest in range limits in part to understand some of the consequences of climate change [1–4]. Because most transplant experiments show poorer performance across range boundaries (Sexton *et al.* 2009 [2] for review), many range margin populations must face stressful environments that they are not adapted to. Therefore, understanding what prevents this adaptation may be key to understanding the development of range limits. Since there is often sufficient genetic variation for traits that matter within range margin populations, if there are also no barriers to dispersal, possible constraints on the process of adaptation include swamping gene flow from elsewhere in the range and tradeoffs. But because many range margin populations are also geographically and genetically isolated, it is thought that the study of range limit development should often focus on molecular, physiological or developmental tradeoffs [5]. What kind of tradeoffs might be constraining low elevation trailing edge populations?

At low latitudinal or altitudinal "trailing edge" range limits, populations are thought to more commonly face both abiotic and biotic stressors compared to high altitudes or latitudes at leading edges where abiotic stressors predominate [6]. Although there would be some exceptions to this pattern depending on latitude, altitude and local climate conditions, many cases should comply. For example, transplant experiments with the upland mustard species *Boechera stricta*, a close relative of *Arabidopsis*, resulted in lower survivorship across low elevation range boundaries that coincided with increased abiotic and biotic stressors such as decreased water availability and increased herbivory by generalist insect herbivores [7]. Presumably, increased drought tolerance and chemical defense levels would allow upland species like *B. stricta* to expand low elevation range boundaries. However, negative genetic correlations, *i.e.*, evolutionary tradeoffs, between glucosinolate (GS) toxin defense allocation and abiotic stress tolerances associated with low elevation range limits have been observed in *B. stricta* [7–10].

One hypothesis for the evolutionary tradeoff involves natural selection acting on antagonistic plastic response pathways [7,11]. The process of adaptation often proceeds by modifying existing structures and pathways. Within ranges, stress response signal transduction pathways help plants to survive temporary challenges from abiotic and biotic stressors [12]. Just across range boundaries, some of these same stressors increase in frequency; therefore, one would predict that adaptation to stressful environments across range limits would involve the up-regulation of stress response pathways such that the pathways and the traits that they regulate were expressed more frequently or stably. However, evolutionary models predict that a problem may arise when antagonistic response pathways are co-opted simultaneously for evolutionary change [13]. For example, from work on *Arabidopsis* it is well known that stress response pathways, such as Abscisic acid (ABA) signaling for coping with temporary challenges of abiotic stressors (e.g., drought), and Jasmonic acid (JA) signaling for coping with bouts of biotic stressors (e.g., herbivores) may negatively interfere with one another ([14–16] for reviews). Thus, the simultaneous co-option of these antagonistic pathways for low elevation range expansion where organisms face both increased abiotic and biotic stressors may be problematic because of the crosstalk.

The problem is with negative pleiotropic and epistatic effects that may constrain evolution. Multiple signaling pathways often form networks involving regulatory genes—transcription factors (TFs)—that may interact to produce multiple positive and negative integrative effects. If natural selection acts on genetic variation in either the coding regions or the *cis* regulatory regions of TFs involved in the networks, multiple pathways may be affected. Although some of the effects may be adaptive, many may be mal-adaptive. For example, the flowering time signaling network in *Arabidopsis* consists of many positive and negative interactions among photoperiod, circadian clock, vernalization, autonomous and Gibberellic acid pathways. Epistatic interactions between the major flowering time network genes FRI and FLC, were one of the contributing factors in the maintenance of genetic variation in *Arabidopsis* flowering time [17]. Thus, by preventing fixation of alleles, the epistasis represents an evolutionary constraint. FRI and FLC are major TFs in the flowering time signaling network that allow large behavioral shifts involving many genes, but these major effects might impede evolution.

It has also been shown that high and low elevation populations of *B. stricta* have diverged with respect to abiotic stress tolerance and glucosinolate defense levels as one would predict based on the above hypothesis; neither population had high values of both kinds of traits [11]. In a common garden growth chamber experiment, the two populations showed genetic divergence for glucosinolate content ($F_{18, 701} = 7.101$, $p < 0.001$) and stress tolerance traits such as root:shoot ratio ($F_{5, 251} = 3.576$, $p = 0.004$). The high elevation population showed higher inherent root:shoot ratios, while the lower elevation population was higher in glucosinolate levels. Further, in experimental crosses between the populations, the two kinds of traits would not segregate independently of one another in the F2 generation (Siemens *et al.*, unpublished data [18]: $F_{1,599} = 65.987$, $p < 0.001$). Together, these results indicate a genetic tradeoff that probably involves negative pleiotropic or epistatic interactions.

Here, we compared the genetically diverged high and low elevation populations of *B. stricta* in a double challenge drought-herbivory growth chamber experiment to test the following predictions based on the central hypothesis that the evolutionary tradeoff derives from antagonistic plastic responses: (1) Drought stress inhibits herbivore-induced defense responses; (2) Induced abiotic stress responses are attenuated in the high elevation population that already shows high constitutive levels of the tolerance traits; (3) Likewise, herbivore-induced defense responses are attenuated in the low elevation population that already shows high basal levels of the defensive traits. In other words, predictions 2 and 3 state that high constitutive levels of traits would coincide with lower plasticity. Prediction 1 is just an expectation to determine whether antagonistic signaling exists.

## 2. Results

### 2.1. Flat Weights

Drought treated flats were watered progressively with less amounts and less frequently than control treated flats over a 4-week period as documented by flat weights just after watering (Figure 1a). Between watering, the flat weights of the control watered flats fluctuated, but were always higher than those of the drought treated flats. By contrast, the flat weights of the drought treated plants remained low between watering treatments and steadily declined (Figure 1b). Eventually, on day 51 post-planting, 24 days after the drought treatments began, the flat weights of drought treated plants fell just below 6.5 kg. At this time, two plants in the drought treatment group had curled rosette leaves indicating early signs of wilting (Figure S1). Flat weights were then monitored more frequently and the flats were watered just enough (eventually to 7.5 kg, similar to flat weights on day 42 post-planting) to allow wilted plants to recover to previous non-wilted stress levels and to survive without further wilting, but still stressed, through the 2-day herbivore feeding treatments, which began four days later on day 55.

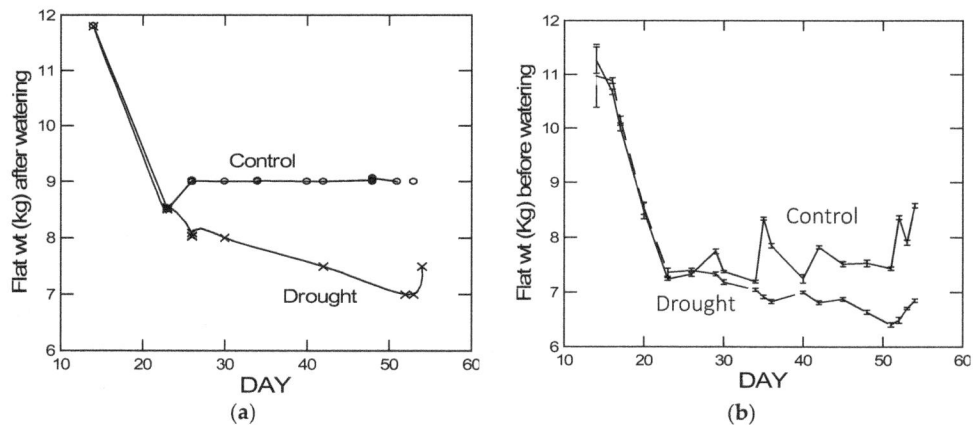

**Figure 1.** Flat weights (**a**) just after and (**b**) before or between watering for the control and drought treatments. All flats in each treatment were watered to the same weight, so there are no error bars for the after watering flat weights. For the before watering flat weights, error bars are ± 1SE across four flats for each watering treatment.

### 2.2. Betacyannin Color Score

By day 36, as water deficiency treatments progressed, the lower epidermis of drought treated plants began to turn a more violet red color (Betacyannin antioxidant stress response: $F_{1,6} = 15.1877$, $p = 0.008$). This visual Betacyannin indicator of stress was the same for plants of both populations (no drought-by-population interaction: $F_{1,235} = 0.135$, $p = 0.713$) (Table 1, Figure S2a).

**Table 1.** ANOVA for violet red "Betacyannin" leaf color score. Census was taken on 30 December and 14 January and then the analysis was conducted on the cumulative score. $r^2 = 26.8\%$.

| Source | df | Mean Squares | $F$-Ratio | $p$-Value |
|---|---|---|---|---|
| Population | 1 | 0.064 | 0.069 | 0.793 |
| Drought | 1, 6 | 57.012 | 15.187 | 0.000 |
| Drought × pop | 1 | 0.125 | 0.135 | 0.713 |
| Flat (Drought) | 6 | 3.754 | 4.059 | 0.001 |
| Error | 235 | 0.925 | | |

### 2.3. Plant Growth Response

There was an eventual difference in the size of shoots between drought treatments (repeated measures within subjects ANOVA: time-by-drought interaction—$F_{4, 24} = 6.281$, $p = 0.001$) (Figure S2b). Although the populations differed in shoot growth (repeated measures between subjects ANOVA: $F_{4, 940} = 6.946$, $p < 0.001$), the effect of drought did not differ between the populations (no drought-by-population interaction: $F_{4, 940} = 0.396$, $p = 0.811$) (Table 2, Figure S2b). By day 42, shoots of drought treated plants were 15.9% smaller than control watered plants, but shoots of plants from the high elevation Big Horn Mountain population were still 19.0% larger than the Black Hills population. Shoot growth rates remained positive, even towards the end of the drought period (e.g., between days 43 and 51) for both the high elevation ($0.61 \pm 0.080$ mm/day) and low elevation ($0.44 + 0.079$) populations.

**Table 2.** Repeated measures analysis of shoot size recorded five times during the experiment.

| Between Subjects | | | | |
|---|---|---|---|---|
| Source | df | Mean Squares | $F$-Ratio | $p$-Value |
| Population | 1 | 1453.863 | 31.175 | 0.000 |
| Drought | 1, 6 | 3394.266 | 7.079 | 0.037 |
| Drought × pop | 1 | 1.850 | 0.040 | 0.842 |
| Flat (Drought) | 6 | 479.496 | 10.282 | 0.000 |
| Error | 235 | 46.636 | | |

| Within Subjects | | | | |
|---|---|---|---|---|
| Source | df | Mean Squares | $F$-Ratio | $p$-Value |
| Day | 4 | 20,292.390 | 1401.337 | 0.000 |
| Day × Population | 4 | 100.583 | 6.946 | 0.000 |
| Day × Drought | 4, 24 | 788.871 | 6.281 | 0.001 |
| Day × Drought × Pop | 4 | 5.738 | 0.396 | 0.811 |
| Day × Flat (Drought) | 24 | 113.572 | 7.843 | 0.000 |
| Error | 940 | 14.481 | | |

### 2.4. Population Divergence

As previously reported [11], high and low elevation populations of *B. stricta* showed genetic divergence with respect to total glucosinolate production and root:shoot ratio; neither population had high values of both kinds of traits (Figure 2). The high elevation population was higher in inherent dry mass root:shoot ratios across drought treatments (Population effect: $F_{1, 166} = 27.522$, $p < 0.001$). However, the 41.4% higher ratio under control watering of the high elevation population decreased to 25.8% under drought conditions (drought-by-population interaction: $F_{1, 166} = 5.742$, $p = 0.018$) (Table 3). The difference occurred because of a greater decline in the high elevation population. By contrast, the high elevation population had 20.8% lower inherent total glucosinolate levels ($F_{1, 162} = 18.138$, $p < 0.001$), a difference that was not affected by the environmental treatments of drought and herbivory (no treatment-by-population interaction: $F_{3, 162} = 1.728$, $p = 0.163$) (Table 4).

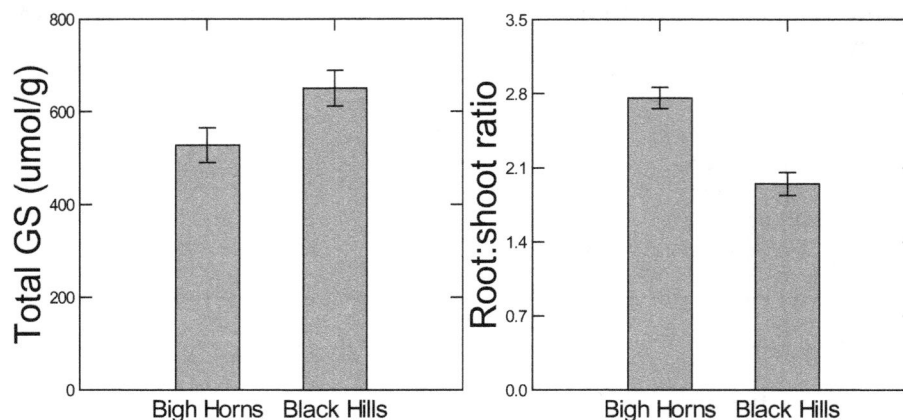

**Figure 2.** Genetic divergence between high and low elevation populations in basal dry mass root:shoot ratio and total glucosinolate concentration. Values are least squares means. Statistical analyses in Tables 3 and 4. Error bars are $\pm 1SE$, total sample size $n = 175$.

**Table 3.** ANOVA of dry root:shoot mass ratio. $r^2 = 49.2\%$.

| Source | df | Mean Squares | F-Ratio | p-Value |
|---|---|---|---|---|
| Population | 1 | 45.266 | 37.725 | 0.001 |
| Drought | 1, 6 | 13.293 | 27.522 | 0.000 |
| Drought × pop | 1 | 2.773 | 5.742 | 0.018 |
| Flat (Drought) | 6 | 1.200 | 2.484 | 0.025 |
| Seedling size * | 1 | 4.950 | 10.249 | 0.002 |
| Error | 166 | 0.483 | | |

\* width of seedling across cotyledons.

**Table 4.** ANOVA of total glucosinolate concentration.

| Source | df | Mean Square Error | F-Ratio | p-Value |
|---|---|---|---|---|
| Drought & Herbivory | 3 | 314,545.772 | 7.271 | 0.000 |
| Population | 1 | 784,711.045 | 18.138 | 0.000 |
| (D & H) × Pop | 3 | 74,748.963 | 1.728 | 0.163 |
| Flat (D & H) | 4 | 131,583.387 | 3.041 | 0.019 |
| Seedling size | 1 | 141,881.984 | 3.280 | 0.072 |
| Dry leaf wt | 1 | 1,312,877.577 | 30.346 | 0.000 |
| Error | 162 | 43,263.071 | | |

## 2.5. Principal Component Analysis of Drought Tolerance Traits

Several other trait measures in addition to growth, Betacyannin color score and root:shoot ratio were made to help assess evolutionary and ecological tolerance responses to drought. In addition, we measured glucosinolate production in several ways besides total glucosinolate content. Thus, we used multivariate principle component analysis to further assess the differences between populations in responses to drought and then herbivory.

Principal component analysis of 10 traits that may contribute to drought tolerance resulted in four significant PCs, each explaining at least 10% of the total variance (Table 5A). However, only for PC1 and PC4 were there effects of the environmental treatments or population line (Table 5B). PC1 was mainly positively correlated with carbon isotope ratio and negatively correlated with root:shoot ratio and LMA, while PC4 was mainly correlated with trichome density and stomata size (see component loadings, Table 5A). Although the significant effects on these PCs were attributable to the drought treatments and population line, but there was no interaction between these factors (Table 5B, Figure S3). An exception was for the separate analysis of the carbon isotope ratio.

**Table 5.** Principal component analysis of traits that may contribute to drought tolerance. (**A**) Component loadings of PCs; and (**B**) F-ratios from ANOVA on the effects of all combinations of drought and herbivore treatments (Drought & herbivory), and of population lines (Line). Relatively high loadings are indicated in bold.

(A)

| Traits | PC1 | PC2 | PC3 | PC4 |
|---|---|---|---|---|
| Carbon isotope ratio | **0.721** | −0.428 | −0.023 | 0.164 |
| Betacyannin color score | **0.532** | 0.106 | **−0.490** | −0.051 |
| Root:shoot ratio | **−0.694** | 0.115 | −0.228 | −0.102 |
| LMA | **−0.668** | −0.051 | 0.083 | 0.375 |
| Trichome size | 0.281 | 0.286 | **0.539** | 0.367 |
| Trichome density | 0.145 | **0.513** | 0.305 | **0.508** |
| Stomata density | 0.137 | **0.577** | 0.260 | −0.439 |
| Number of rosette leaves | **−0.476** | **0.483** | −0.335 | 0.101 |
| Stomata length | 0.220 | 0.172 | **−0.689** | **0.456** |
| Growth rate | −0.395 | **−0.649** | 0.165 | 0.204 |
| Variance explained | 22.8% | 15.7% | 13.6% | 10.2% |

(B)

| Source | df | PC1 | PC2 | PC3 | PC4 | D13C |
|---|---|---|---|---|---|---|
| Drought & Herbivory | 3, 4 | 17.823 ** | 0.333 | 0.009 | 0.847 | 9.284 * |
| Pop Line | 3 | 17.276 *** | 0.825 | 1.263 | 13.177 *** | 30.023 *** |
| (D & H) × Pop Line | 9 | 0.858 | 1.266 | 0.824 | 0.892 | 3.114 ** |
| Flat (D & H) | 4 | 4.641 ** | 5.258 *** | 2.308 | 3.225 * | 7.421 *** |
| Seedling size | 1 | 2.049 | 8.170 ** | 6.926 ** | 1.101 | 5.353 * |
| Error | 96 | | | | | |
| $r^2$ | | 80% | 34.9% | 23.7% | 44.1% | 81.5% |

In a separate analysis of the carbon isotope ratio (D13C, Table 5B), a main component of PC1, there was a significant effect of herbivory, but only in the drought treated plants (Figure 3). The effect of herbivory was to decrease D13C in the drought treated plants, an effect that was more dramatic in the Black Hills population (Table 5B, environmental treatments-by-population line interaction: $F_{9, 95} = 3.114$, $p = 0.003$).

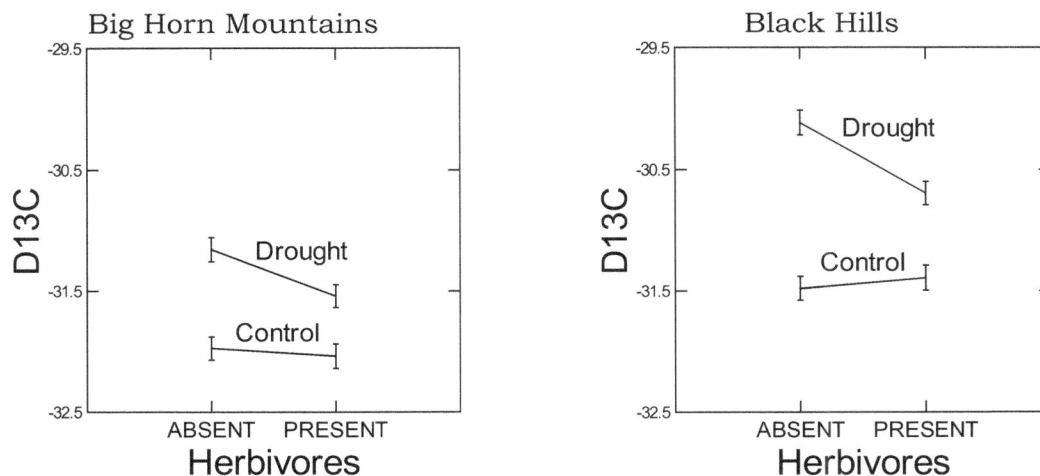

**Figure 3.** Effects of herbivory and drought on the carbon isotope ratio. See Table 5B for statistical analysis. The effect of herbivory was only significant in the drought treated plants (LSD multiple comparisons *post hoc* test: $p$'s < 0.05). Error bars are ±1 SE, total sample size $n = 116$.

## 2.6. Principal Component Analysis of Glucosinolate Measures

Principal component analysis (PCA) of GS resulted in two significant PCs (Table 6A). The main components of PC1 included the GS ratios and the MET-GS 6-mthylsulfinylhexyl, whereas the main components of PC2 included the BCGS 2-hydroxyl-1-methylethyl and 1-methylethyl. Although there were significant differences among the population lines in all measures of GS production, the responses to herbivory and drought (Figure 4) did not depend on population line (no treatment-by-population line interaction, Table 6B).

**Table 6.** Principal component analysis of glucosinolates. (**A**) Component loadings of GS PCs and (**B**) $F$-ratios of ANOVAs on the GS-PCs and their important component loadings. Relatively high loadings are indicated in bold.

(A)

| GS Measure | GS-PC1 | GS-PC2 |
|---|---|---|
| Total GS | 0.653 | 0.655 |
| 2-hydroxyl-1-methylethyl GS (GS1) | 0.502 | **0.811** |
| 1-methylethyl GS (GS2) | 0.528 | **0.748** |
| 6-mthylsulfinylhexyl GS (GS3) | **0.996** | 0.040 |
| BCGS/METGS (GS1 + GS2)/GS3 | **−0.861** | 0.489 |
| GS1/GS3 | **−0.861** | 0.477 |
| GS2/GS3 | **−0.750** | 0.582 |
| Variance explained | 57.0% | 35.0% |

(B)

| Source | df | GS-PC1 | GS-PC2 | GS-Ratio | BCGS | METGS |
|---|---|---|---|---|---|---|
| Drought & Herbivory | 3 | 6.606 *** | 1.606 | 6.788 *** | 6.309 *** | 5.789 *** |
| Pop line | 3 | 5.662 *** | 45.930 *** | 17.321 *** | 14.882 *** | 4.993 ** |
| (D & H) × Pop line | 9 | 0.437 | 1.410 | 0.459 | 1.075 | 0.409 |
| Flat (D & H) | 4 | 2.081 | 0.744 | 1.792 | 2.492 * | 1.810 |
| Dry leaf wt | 1 | 13.178 *** | 127.507 *** | 1.086 | 33.259 *** | 21.848 ** |
| Seedling size | 1 | 0.257 | 8.338 ** | 1.571 | 4.097 * | 0.332 |
| Error | 154 | | | | | |
| $r^2$ | | 33.6 | 68.9% | 37.2% | 50.8% | 35.2% |

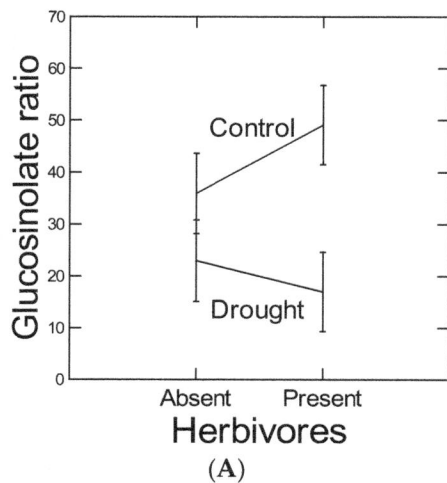

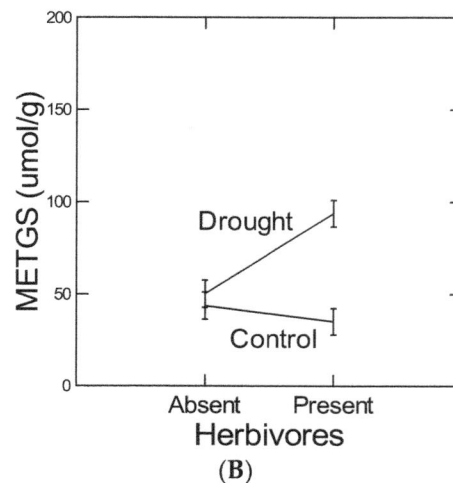

**Figure 4.** *Cont.*

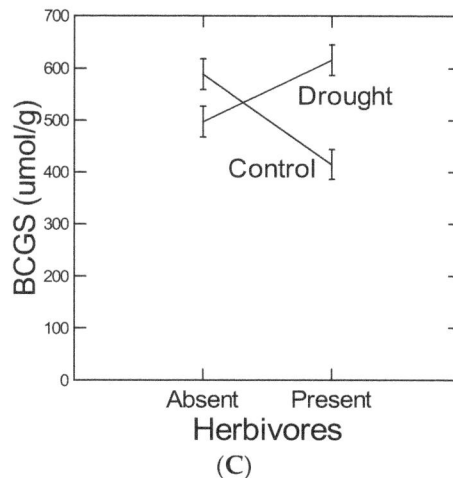

(C)

**Figure 4.** Effect of drought on the herbivore-induced responses of (**A**) ratio of branch chain (BCGS) to Methionine-derived straight-chain glucosinolates (METGS), (**B**) METGS, and (**C**) BCGS. Statistical analysis in Table 6. Error bars are $\pm$1SE, total sample size $n = 175$.

## 2.7. Correlation between Carbon Isotope Ratio and Glucosinolate Production

There was a negative correlation between carbon isotope ratio and glucosinolate concentration that depended on environmental treatments and population (Table 7). In the Black Hills population, drought, but especially herbivory, was associated with the negative correlation. This was evident for BCGS (Figure S4A) or METGS (Figure S4B). In the high elevation Big Horn Mountain population, the correlation was negative across treatments, but the pattern was complicated by variation among the population lines. We did not have the statistical power to also include the lines in the statistical analysis. In contrast to the negative correlations involving GS concentrations, the correlation was positive for GS ratio (Figure S4C).

**Table 7.** $F$-ratios from ANCOVAs on carbon isotope ratios. The analysis was conducted separately for different combinations of glucosinolates (BCGS—branch chain GS, METGS—methionine-derived GS, GSRATIO—ratio of BCGS to METGS) and population (BIHM—Big Horn Mountains, BLHI—Black Hills).

| | df | BCGS | | METGS | | GSRATIO | |
|---|---|---|---|---|---|---|---|
| | | BIHM | BLHI | BIHM | BLHI | BIHM | BLHI |
| GS | 1 | 10.754 ** | 33.509 *** | 32.322 *** | 30.752 *** | 26.655 *** | 25.120 *** |
| Drought | 1, 1 | 144,624 ** | 9.666 | 16.385 | 57.407 | 1.625 | 9.805 |
| Herbivore | 1, 1 | 3.502 | 0.480 | 2.198 | 10.304 | 1.879 | 31.006 |
| Drought × GS | 1 | 0.000 | 4.564 * | 1.520 | 1.564 | 7.880 ** | 9.915 ** |
| Herbivore × GS | 1 | 0.130 | 7.761 ** | 0.091 | 0.356 | 1.124 | 0.311 |
| Drought × Herbivore | 1, 1 | 0.077 | 3.746 | 0.105 | 1.713 | 0.697 | 15.705 |
| Drought × Herbivore × GS | 1 | 1.967 | 0.375 | 0.458 | 9.766 *** | 4.227 * | 0.525 |
| Error | 82 | | | | | | |
| $r^2$ | | 34.3% | 75.1% | 50.7% | 75.9% | 45.6% | 71.5% |

$* p \leqslant 0.05; ** p \leqslant 0.01; *** p \leqslant 0.001.$

## 2.8. Gene Expression

Drought and herbivory treatments were not additive for the number of genes differentially expressed. This was evident from the relatively high number of unique genes expressed in the double challenge drought/herbivore treatment combination (Figure 5). There were 290 (82.4%) unique genes up-regulated and 133 (79.2%) unique genes down-regulated in the double-challenge condition. Because of the high overlap between herbivory and the double challenge treatment, we assume that the response to herbivory alone was not blocked by drought. By contrast, because of the low overlap between drought and the double challenge combination, we assume that herbivory could have blocked some of the genes expressed in the drought alone treatment. To better understand the potential function of the unique genes expressed in the double challenge combination, the comparison was also made with single-challenge controls. Relatively more unique genes were expressed in the double-challenge treatment combination when the single-challenge control was drought (Figure S5), suggesting that the drought treatment could have enhanced the response to herbivory.

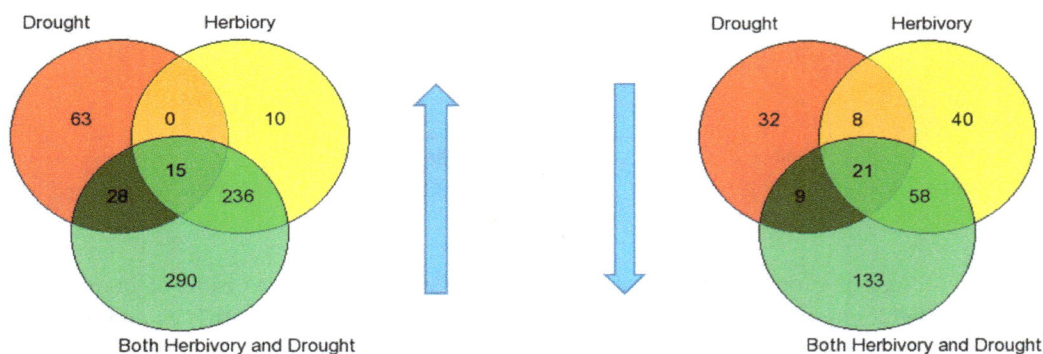

**Figure 5.** Venn diagrams for number of significantly up- and down-regulated genes (direction of blue arrows indicate up- or down-regulation). Red circle is Drought *vs.* control (no stress), Yellow circle = Herbivory *vs.* control (no stress), Green circle is double challenge Drought + Herbivory *vs.* control (no stress). Number of biological replicates, *n*, in comparisons: red circle, drought *vs.* control (*n* = 8); yellow circle, herbivory *vs.* control (*n* = 6), green circle, double challenge drought + herbivory *vs.* control (*n* = 8).

A functional analysis (gene ontology enrichment analysis) of the unique up-regulated genes in the double-challenge combination of drought and herbivory included 262 *Arabidopsis* analogs. Of these, 40.0% or 105 genes were involved in responses to abiotic and biotic stressors. But there was a greater proportion of defense (39.0%) and wound-related genes (28.5%) compared to water deprivation (18.1%). A similar functional analysis was conducted on 119 uniquely down-regulated *Arabidopsis* analogs. Of these, 25.2% or 30 genes were involved in biotic and abiotic (water deprivation) stress responses; none in defense and wounding, and 33.3% involved in water deprivation. Thus, the prolonged drought treatment up-regulated unique defense genes in response to herbivory, but the herbivory down-regulated unique water deprivation genes in drought-treated plants.

When the number of genes differentially regulated was examined by defense- and drought tolerance-related gene categories, some other notable patterns were evident (Figure 6). For example, although genes for JA signaling (e.g., LOX, JAZ—see heat map, Figure S6a for other examples) were up-regulated in response to *S. exigua* larvae feeding, defense response genes (e.g., CYPs, WRKYs—see Figure S6b for other examples) were down-regulated (Figure 6—light blue bar). However, under conditions of drought, herbivory up-regulated both JA signaling and defense related genes (Figure 6—darker green bar). Thus, drought apparently blocked the inhibitory effects of the herbivore to regulate the plant's defense response. Of particular note is the dependence of the CYP79 herbivore-induced response on the presence of drought. The *B. stricta* homolog of CYP79 is involved in the GS-ratio response [19].

Signaling pathways that may mediate the apparent inhibitory effect of the herbivore and release of this inhibition by drought were identified in the hierarchical clustering analysis (Figure S6a). These candidate genes (transcription factors—TFs) were up-regulated in response to herbivory, but down-regulated when herbivores fed on drought stressed plants. Or, in the case of candidates involved in the ability of drought to block the inhibitory effects of the herbivore, activated only in the double challenge situation. These candidate TFs that were up-regulated by the herbivore, but down-regulated in the herbivore-drought combination, and their respective signaling pathway (in parentheses), included MYB95 (ETH), MYB13 (SA, ABA), AIB (ABA), and GRP-5 (ABA). However, all of the pathways had some TFs that were only up-regulated in the double challenge, such as MYC2 (JA), that could be involved in the blocking of the herbivore inhibition of the defense response. TFs such as MYC2 have been previously implicated in the crosstalk between responses to biotic and abiotic stressors [14]. MAPK signaling was also implicated in a similar analysis using gProfiler. MAPK signaling may be involved in the regulation of coordinated signaling responses under abiotic and biotic stress [20].

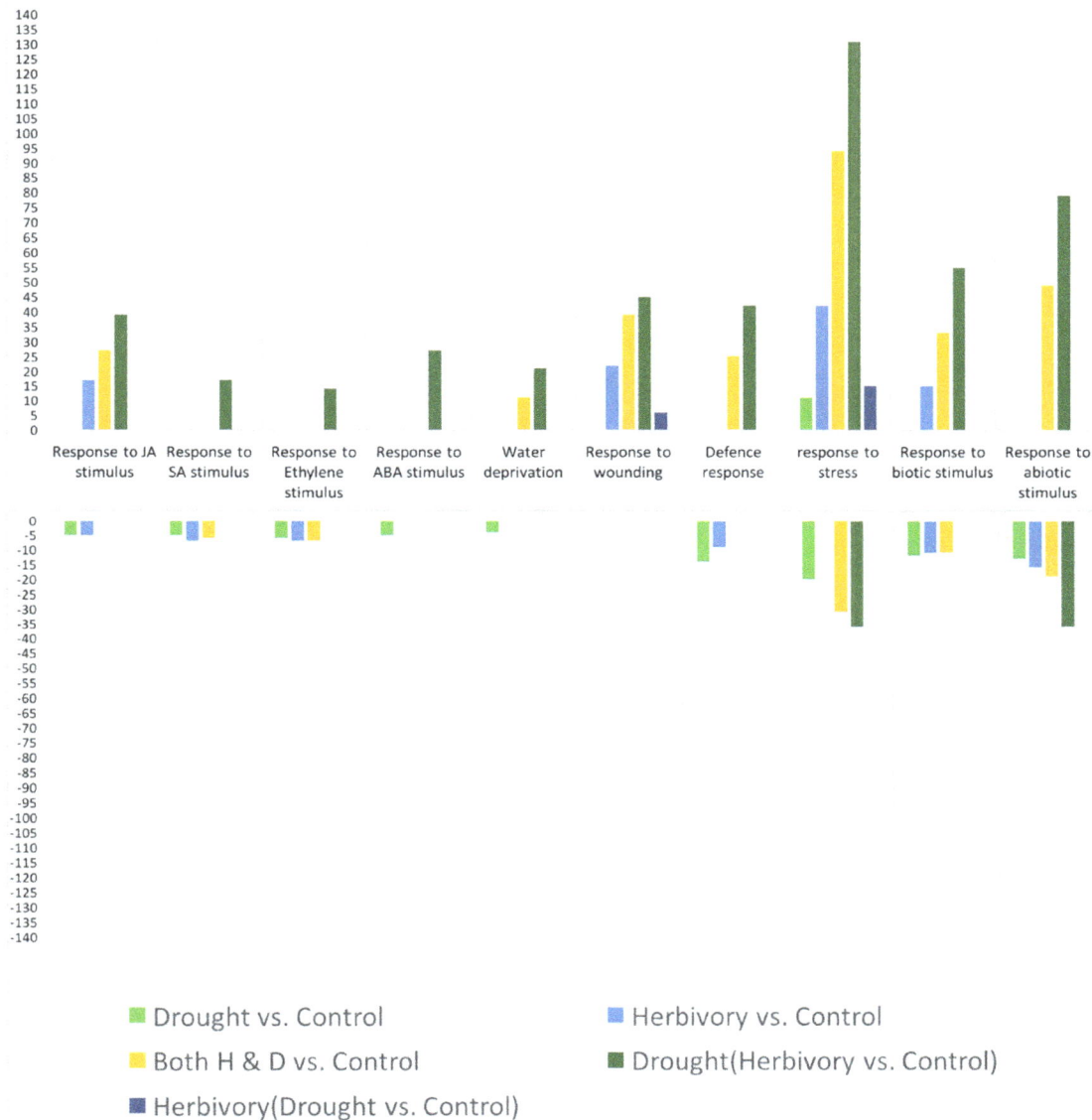

**Figure 6.** Number of significantly up- and down-regulated genes by defense and drought-tolerance functional categories. Herbivory (Drought *vs.* Control) means that both drought and control-watered treatments were fed upon by herbivores. Similarly, Drought (Herbivory *vs.* Control) means that plants in the presence and absence of herbivores were under drought stress. Number of biological replicates for the comparisons ranged between 6 and 10.

Finally, a comparison between high and low elevation populations for the number of relevant differentially expressed genes indicated that the low elevation population was more responsive to drought, as predicted, while the high elevation population was more responsive to herbivory, as also predicted (Figure 7). While this was clear for up-regulated genes, it was, however, less clear for down-regulated genes.

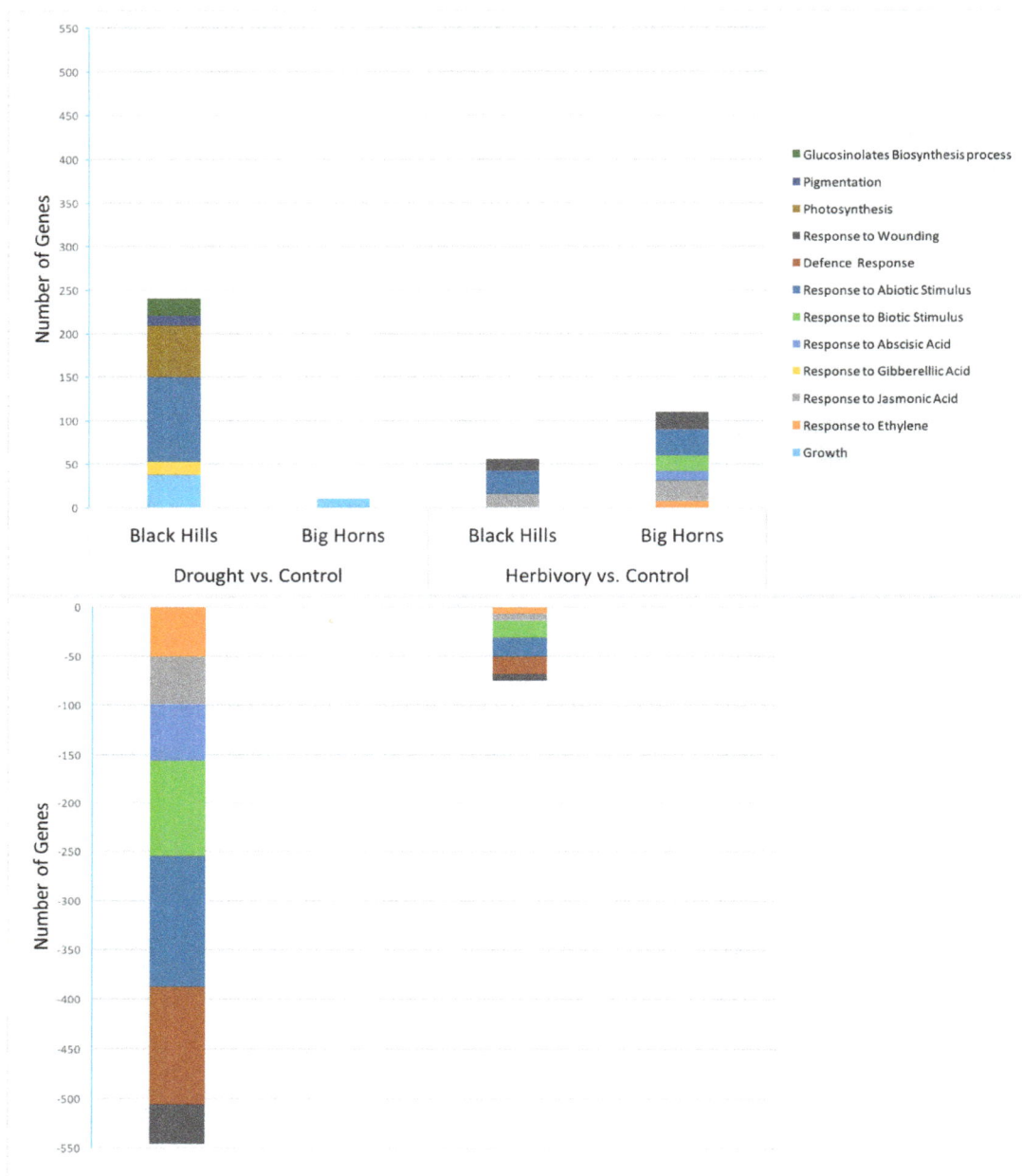

**Figure 7.** Number of differentially expressed genes of relevant biological processes between populations. The number of biological replicates for the Black Hills population comparisons was $n = 4$ and for the Big Horn population, $n = 5$.

## 3. Discussion

Many factors and processes, alone or in combination, may contribute to species range limits development by preventing adaptation to stressful environments. Some major factors include lack of genetic variation in range margin populations, barriers to dispersal, swamping gene flow from elsewhere within the range and various kinds of tradeoffs [2]. Of these, relatively little is known about possible molecular, physiological or developmental tradeoffs [5].

Here, we tested predictions from a hypothesis explaining the existence of an apparent genetic tradeoff between defense allocation and abiotic stress tolerance. This tradeoff may contribute to low elevation range limit development. The hypothesis states that antagonistic plastic response pathways may inhibit their simultaneous co-option for more stable expression that is needed for range expansion. If a plastic response pathway in a signaling network is co-opted in evolution for more stable expression, then one might predict that other pathways in the network would also be affected [13]. To test this prediction, we compared two genetically diverged populations for their plastic responses to drought and herbivory. For example, the high elevation population from the Big Horn Mountains has diverged with respect to higher constitutive levels of some abiotic stress tolerant traits, such as root:shoot ratio (Figure 2). Since the root:shoot ratio is in part regulated by ABA signaling [21], we tested the prediction that ABA-regulated traits, such as stomatal aperture, would be less plastic in the high elevation population. We also expected there to be less gene expression in response to drought in the high elevation population. Similarly, we expected the low elevation population to be less plastic in defensive traits. In general, we predicted that relatively high constitutive levels of the functional traits would coincide with lower plasticity.

Although our central hypothesis involved evolutionary constraints caused by potentially antagonistic signaling pathways, we did not make direct measurement of the pathway hormones, making the link between the signaling and the ecological and evolutionary responses speculative. However the predictions that we tested did not necessitate direct measurement of the hormones that trigger the candidate pathways. Instead, we relied on existing literature that makes a link between the candidate signaling pathways, gene expression and the functional traits. The candidate signaling pathways that were invoked for the evolutionary tradeoff between chemical defense allocation and stress tolerance in *B. stricta* were the jasmonic acid/ethylene (JA/ET) and the abscisic acid (ABA) pathways, respectively. These pathways are generally antagonistic to one another in *Arabidopsis*, probably because a stress response under dry conditions takes precedence over defense that may function primarily against pathogens under moist conditions [14–16,20] for reviews). It is also known that JA/ET signaling regulates aliphatic glucosinolate (GS) toxin-induced defense responses in *Arabidopsis* [22,23], induced resistance against generalist insect herbivores in another closely related species *Boechera divaricarpa* [24], and that GS are defensive against generalist insect herbivores as has also been found to be the case for *B. stricta* (e.g., [7,19]). We therefore assumed that JA/ET signaling regulates aliphatic GS defense responses to generalist insect herbivores in *B. stricta*. Likewise, ABA is a general stress response hormone, produced most notably in response to abiotic stressors, especially to drought or salinity, but also to other factors such as nutrient deficiency ([21,25] for reviews, [26]). For example, in response to limiting supplies of water or soil nitrate levels and subsequent increases in ABA concentration or sensitivity, stomata close, leaves grow more slowly, and root growth is maintained and characterized by lateral root proliferation. Whether the genetic divergence between high and low elevation populations in levels of aliphatic GS or the various other functional traits associated with abiotic stress tolerance that we measured represents genetic divergence in the joint JA/ET regulatory pathway or ABA signaling was tested here indirectly without direct measurement of the hormones.

As predicted, in the high elevation population, we observed a reduced plastic stomatal aperture response to drought as measured by the carbon isotope ratio (Figure 3). That is, for the carbon isotope ratio, the difference between drought and control treatments was smaller in the Big Horn Mountain population compared to the Black Hills population. However, less regulation on stomatal control is typical of alpine plants where water deficiency is usually not a problem [27]. The drought-induced change in the root:shoot ratio was actually greater in the high elevation population, which was not predicted. However, the change was a decrease in the root:shoot ratio, which may not be mediated by ABA signaling. Under natural circumstances, root:shoot ratios increase in response to drought, but gradual water deprivation in small pots when watered from above is not conducive to increases in root:shoot ratio. Yet there was also no population-by-drought treatment interaction in the ANOVA for any of the four significant PCs constructed from the 10 putative abiotic stress tolerance response variables (Table 5B). This interaction would be an indication of any differences in plasticity between the populations. It was not until we analyzed the gene expression data that there was relatively clear evidence of differential plasticity in response to drought between the populations. For the number of relevant up- and down-regulated genes, the high elevation Big Horn population was indeed less plastic to drought.

In a similar study, transcript profiles resulting from cDNA-AFLPs were compared between high and low elevation populations of *Boechera holboellii* in a dry-down growth chamber experiment [28]. The focus was on identifying candidate genes involved in local adaptation by noting population-specific expression patterns. Although the dry-down treatments resulted in differential expression patterns, only a couple of *Arabidopsis* homologs were identified as candidates. Probably because of the method used for transcript profiling, there were too few gene expression differences between populations that were reported to evaluate predictions of our hypothesis.

Similarly, because the low elevation Black Hills population was higher in constitutive levels of total glucosinolate content (Figure 2), we predicted lower plastic responses to herbivory from that population. However, we did not detect any population differences in herbivore-induced glucosinolate measures (no indication of any herbivory-by-population interaction in the ANOVAs, Tables 4 and 6). Many other studies have checked for negative genetic correlations between constitutive and herbivore-induced defense or resistance levels. This is because popular theory on the evolution of herbivore-induced defenses assumes resource allocation costs of constitutive defenses and cost-savings of inducible defenses ([29] and references therein). The results have been mixed, probably because several factors may influence constitutive and induced defense levels [30]. In response to herbivory, the low elevation Black Hills population was clearly less plastic in terms of the number of relevant up-regulated genes expressed, but this was not the case for the number of down-regulated genes.

Our predictions were based on the existence of antagonistic defense and abiotic stress tolerance plastic responses in *B. stricta*, as has been well documented in *Arabidopsis* [20]. Specifically, we checked whether a glucosinolate defense response to the generalist insect herbivore *S. exigua* was negatively affected under conditions of water deficiency. The assumption that we made was that under drought conditions, ABA signaling would interfere with JA/ET signaling to attenuate herbivore-induced defense responses [14]. We found that an herbivore-induced GS ratio (BCGS/METGS) response was attenuated under conditions of drought. The BCGS/METGS ratio has been shown to cause resistance to generalist insect herbivores in *B. stricta* [19,31] and therefore may be the most relevant GS response variable. That is, in these other studies, this GS ratio in *B. stricta* has been fine mapped, and candidate genes have been transformed and near isogenic lines created and tested with positive results. Interestingly, feeding by *S. exigua* caused GS concentrations to decrease, which was reversed under drought (Figure 4; Note, BCGS and total GS concentrations are highly correlated r = 0.98 because of the relative abundance of BCGS). Because increased GS concentrations render plants more susceptible to specialists like flea beetles (e.g., [32]), we suggest that the plants responded to feeding by increasing resistance to both generalists (by increasing GS ratio) and specialists (by decreasing total GS concentrations). If so, the drought treatments were antagonistic to herbivore-induced defense responses. To our knowledge, it is not known why the GS ratio is effective independent of total GS concentration.

In apparent contrast to the GS results, the gene expression results suggested that the *S. exigua* larvae may down-regulate the *B. stricta* induced defense response, but that this inhibition by the herbivore was blocked by the drought treatments. Of course, to verify this suggestion, direct hormone, proteome, metabolome and other experimental molecular and genetic analysis would help. Nonetheless, although an herbivore-induced JA response was evident, a defense response was notably absent (Figure 6). Patterns of gene expression further suggested that the putative herbivore inhibition may be mediated by other pathways such as ET, SA and ABA. That herbivores may inhibit defense responses downstream of JA signaling via an ethylene burst was reported previously for feeding by the specialist herbivore

*Manduca sexta* on wild tobacco *Nicotiana attenuata* [33]. Here, we show that apparent signaling-mediated manipulation of plant defense responses by herbivores may be dependent on abiotic conditions. The mechanism of drought to block the herbivore inhibition of defense apparently involved further signaling crosstalk. However, in apparent agreement with the GS results, was the dependence of the CYP79 herbivore-induced response on the presence of drought. The *B. stricta* homolog of CYP79 is involved in the GS-ratio response [19].

A negative correlation was also observed between smaller stomatal apertures, as indicated by less negative carbon isotope ratios, and GS levels (Figure S4). Interestingly, this pattern held for both BCGS and METGS, but the correlation was positive for GS ratio. In addition, these correlations were dependent on drought and herbivore treatments, the particulars of which differed between populations (Table 7) and probably lie within populations. For example, in the low elevation Black Hills population, the correlations were dependent on the presence of drought and especially herbivory. What caused these correlations? These correlations may be dependent on the area of leaf tissue consumed, which we did not record, but which presumably would have influenced water loss and stomatal behavior. Alternatively, these correlations may reflect interactions between hormone-mediated response pathways.

## 4. Experimental Section

### 4.1. Study System

*Boechera stricta* (Brassicaceae) is a monophyletic, predominantly self-fertilizing, diploid, genetically diverse perennial and close relative of *Arabidopsis thaliana* that ranges across western North America at higher altitudes [34–37]. Unlike *Arabidopsis* in North America, *B. stricta* and many other species of *Boechera* are native, occur in natural habitats, and because of longer life cycles, face, and presumably adapt to, more ecological stressors [38–40]. Here we focused on plastic responses of genetically diverged high and low elevation populations of *B. stricta* in a double challenge experiment; herbivore induced defense responses under prolonged laboratory experimental drought.

The high elevation population was located at 44°18′22″N, 107°18′33″W, elevation 2780 m in the Big Horn Mountains, Wyoming and the low elevation population 44°24′50″N, 103°56′18″W, elevation 1365 m, the Black Hills, South Dakota. These are geographically isolated and genetically divergent populations [9,35]. The populations are located at different ends of the altitudinal range of *B. strict*, typically 1700 to 3000 m [36], and thus the sites differ by several environmental factors.

### 4.2. Growth Chamber Experiment

A double-challenge growth chamber experiment was conducted to determine the effects of drought stress on plant defense responses to herbivores. In nature, plants experience slow increases in drought stress, but encounter shorter bouts with herbivores. Therefore, we slowly increased plant exposure to more severe water deficiency (drought) and then challenged the plants with a relatively short bout of attack by herbivores. Of particular interest was the comparison of high and low elevation populations of *B. stricta*. As detailed below, the 4-week long drought treatment consisted of progressively reduced amounts and frequencies of watering for a gradual soil dry-down that slowly brought the plants close to the wilting point. During the gradual drought treatment process, plants were frequently monitored non-destructively for stress by examining decreased growth rates and color changes indicative of antioxidant production. These assessments of plant status determined the amount and frequency of watering treatments, which was recorded by weighing planting flats. After a brief 3-day recovery period where plants were brought back to a slightly more mild level of stress, early instar larvae of the generalist herbivore *Spodoptera exigua* were allowed to feed on stressed and control plants over a 2-day period to induce a glucosinolate defense response. At the end of the experiment, plant tissue was analyzed for several additional drought-tolerance related traits: carbon isotope ratio (water use efficiency—WUE), leaf trichome and stomata size and density, root:shoot mass ratio, leaf mass area (LMA) and the number of leaves per rosette. Glucosinolate levels were measured to assess defense response. On a subset of plants, whole-genome gene expression was also examined to identify signaling pathways involved.

### 4.2.1. Experimental Design

The design of the growth chamber experiment was split-plot [41]; drought and herbivory treatments varied among planting flats, while population varied within flats. There were 256 plants total in the experiment distributed among eight flats, each containing 32 pots. We used two uniparental lines per population, and we randomized eight replicates of each line within each flat (2 pops/flat × 2 lines/pop × 8 plants/line = 32 plants/flat). Seeds were planted in 0.2-L pots filled with a soil mix of 2/3 *Premier ProMix BX* and 1/3 sand. Pots also contained 45.78 mg of 7:40:6 NPK MagAmp time release fertilizer. Plants were grown in a BioChambers growth room with a 16/8 h D/N photoperiod and 23/21 °C D/N temperatures. Light intensity was 360 µmoles/m$^2$/s from a combination of 1220 mm T5HO fluorescent and halogen lamps.

Watering and herbivory varied among flats for practical reasons. Drought treated flats were watered less and less often (see below), which was monitored by weighing whole flats with a postal scale rather than weighing pots individually. Similarly, although individual plants were caged (see below), to reduce effects of any caterpillars moving between plants within a flat, flats either had no caterpillars or had caterpillars on all plants. However, because plants were in separate pots with their own soil and water, and because caterpillars were caged on individual plants, individual plants were otherwise treated independently of one another.

### 4.2.2. Drought Treatments

Watering treatments began on day 26 after planting, and herbivore induction treatments on day 54. When watering treatments began, all flats were also watered with 0.7 g/500 mL 20:20:20 NPK + micronutrient Peter's fertilizer. For the control and drought watering treatments, the experiment was divided into two groups of four flats each. There was no difference in average rosette diameters among the two groups of flats ($F_{1,6}$ = 0.058, $p$ = 0.817) when watering treatments began.

During the period prior to herbivore feeding, drought treated plants were gradually stressed such that growth rates decreased relative to controls, but remained positive. While control flats were watered to 9 kg every four days, the drought treated flats were watered less and less often in a progressively decreasing manner. The amount of water was monitored by flat weights (see Results). The same amount of water was distributed among pots within flats by moving with the watering can across flats at a constant rate.

### 4.2.3. Plant Growth Rates

In the process of decreased turgor and subsequent ABA synthesis caused by water deficiencies, growth decreases before photosynthesis and subsequent wilting [42]; therefore, rosette size was used to help monitor effects of the watering treatments and to determine the amount and frequency of watering treatments. Thus, watering treatments were administered such that drought treated plants grew slower than did controls. Rosette size was measured using digital calipers beginning as soon as true leaves appeared and continued every 9 to 10 days during the experiment.

### 4.2.4. Leaf Betacyannin Production

Additionally, leaf color change as an indicator of antioxidant production, and any signs of wilting (drooping rosette leaves) were also used to monitor drought treatments. Under drought stress, *B. stricta* leaves turn a red violet color because of Betacyannin production. The color change is mainly on the lower epidermis of the leaves. Betacyannins are pigments that are produced in response to abiotic stress such as drought and function as ROS scavengers [43]. The Betacyannin response of an individual plant was recorded visually on a subjective scale (0, 1, 2, 3) corresponding to the extent of the red violet coloration of the leaves.

### 4.2.5. Herbivore Induction Treatments

Larvae of the generalist insect herbivore *Spedoptera exigua* (Beet armyworm) were used to induce a chemical defense response in *B. stricta* plants. In particular, we assumed that aliphatic glucosinolate levels would change in response to *S. exigua* feeding as can occur in *Arabidopsis* [44]. Eggs from Benzon Research were hatched onto cabbage in Styrofoam cups and allowed to feed before being placed on plants. Four early instar caterpillars were place on each plant in two of the flats in each watering treatment. Individual plants were caged in clear plastic cylinders (*i.e.*, clear plastic cups with the bottom cut out and wedged into the soil around the plants) covered with netting. Plants in herbivore-control plants were also caged to control for any cage effects. After two days of feeding, the experiment was terminated for plant tissue harvest.

### 4.2.6. Water Use Efficiency

Water use efficiency is a ratio of $CO_2$ uptake to water loss, but the carbon isotope ratio ($\delta^{13}C$) is also used to estimate WUE in $C_3$ plants ([45] and references therein, [46]). This is because the $^{13}C/^{12}C$ ratio can be modeled as a function of the ratio of intercellular to atmospheric partial pressure of $CO_2$ ($C_i/C_a$), which is also supported empirically, and $C_i/C_a$ is empirically correlated with WUE in C3 plants. Values of $\delta^{13}C$ are usually negative, and less negative values indicate greater WUE. For carbon isotope discrimination, whole basal rosette shoots were freeze-dried, ground to <0.5 mm and analyzed on a Thermo Delta V isotope ratio mass spectrometer (IRMS) interfaced to a NC2500 elemental analyzer at the Cornell Isotope Laboratory (COIL). Values were expressed as per mL (‰) $^{13}C$ values.

### 4.2.7. Leaf Trichomes and Stomata

Trichome and stomata size and densities are known to affect plant water relations in montane plants [27]. Cell size and densities were recorded from freeze-dried leaves using a compound light microscope and fingernail polish leaf peels of the lower epidermis. We used fully expanded leaves near the center of rosettes. One leaf from each plant was used, but two measures, one on either side of the leaf midrib, were averaged for each leaf. Stomata cell size was measured from a length measurement at $400\times$ magnification, and stellate trichome cell size was calculated from length and width measures at $100\times$. Cell lengths and widths were measured using a calibrated ocular micrometer. The number of stomata cells were counted in the entire field of view at $400\times$, which was an area of 0.196 $mm^2$, while trichome counts were conducted at $100\times$, and area of 0.314 $mm^2$. The areas allowed us to convert the counts to densities.

### 4.2.8. Root:Shoot Ratio and Number of Leaves

At the end of the experiment, fresh and freeze-dried weights were obtained for whole shoots (basal rosettes) and roots. Roots were floated in water and rinsed to remove sand and soil materials before weighing and freeze drying. Shoots and roots were flash frozen in liquid nitrogen before freeze-drying. The number of leaves per rosette were also counted from freeze-dried shoots.

### 4.2.9. Glucosinolates

Glucosinolates were extracted in methanol, isolated on Sephadex ion-exchange columns, and measured on a HPLC [47,48] as summarized elsewhere [10]. Briefly, weighed, freeze-dried basal rosette leaves were extracted in 1.2 mL methanol, separated on a 0.6-mL DEAE A-25 Sephadex column, and eluted after 12 h incubation with sulfatase (Sigma-Aldrich, St. Louis, MO, USA). A Lichroshpere (RP-C18, endcapped) 250 × 4-mm analytical column was used on the HPLC, and chromatograms generated at 229 nm were analyzed.

4.2.10. Statistical Analysis

We used SYSTAT version 13.0 for all statistical analyses of defense and drought tolerance phenotypes. To analyze the split-plot statistical model, we used mixed-model (*i.e.*, the presence of fixed and random effects) ANCOVA with Type III sums of squares. For example, we used the model

$$\text{Response} = C + \text{Drought} + \text{Flat (Drought)} + \text{Pop} + (\text{Drought} \times \text{pop}) + \text{size}$$

in the analyses of the effects of drought treatments and population. Response represents the response variable (e.g., carbon isotope ratio), C = constant, Flat (Drought) represents unmeasured random variation among flats not accounted for by the whole-flat treatment factors (e.g., drought), Pop is for population (*i.e.*, high or low elevation populations) and size represents seedling size, which was included to control for any initial developmental differences. In this case, drought and population were fixed effects and seedling size was a random covariate effect. We also conducted repeated measures ANCOVA when using rosette diameters and as the response variables. We computed $F$-ratios from appropriate mean-square errors for this split-plot design (Zar 1996, [49]). Preliminary inspection of the data for the interaction between population or population line and any unmeasured whole-flat factors (population-by-Flat (Drought)) determined that this interaction was not important; therefore this interaction was not included in the analyses (Montgomery 1997, [50]). Eliminating this interaction simplified the analyses and did not affect the main results. For example, in the above model, the $F$-ratio for Drought was calculated using the Flat (Drought) mean square error in the denominator, otherwise for the other factors, F-ratios were calculated over the residual mean square error.

Principal components were constructed to reduce the dimensionality of drought-tolerance-related variables, of which there were $n = 10$, and of glucosinolate measures, of which there were $n = 7$. Essentially, SYSTAT uses the rule of thumb that PCs are significant if they explain at least $1/n$ of the variance [51]. Significant PCs were used as response variables in the split-plot ANCOVA modes (see previous paragraph).

4.2.11. Gene Expression

Shoots of 13 plants representing all four treatment combinations (presence and absence of drought and herbivory) and the two populations were used for RNAseq analysis. There were at least 2 to 3 replicates per population in each herbivory treatment (presence, absence) under drought conditions, but there were just 1 to 2 replicates per population in the herbivory treatments under control watering conditions. This unbalanced design occurred because of funding limitations. The replicates within populations were from the same line; line 63 from the Big Horn Mountains and line 48 from the Black Hills.

QIAGEN RNeasy Mini Kit was used for RNA extraction. In brief, plants samples were crushed/powdered in liquid nitrogen and 33 mg of the powder was extracted RLT buffer. After precipitation, RNA was purified using DNase digestion and washes. RNA quality was checked using a Qubit Fluorometer and a denaturing gel.

Extracted RNA samples were sent to DHM (David H. Murdock) Research Institute for Illumina sequencing. RNA libraries were constructed for each sample and each uniquely tagged with a molecular barcode or "index". Those libraries were quantified using Real Time PCR. Two pools were generated from these libraries and sequenced via a 100 base pairs single read sequencing run on the Illumina HIseq2500 platform.

Data analyses were performed by utilizing tools in Discovery Environment, iplant collaborative (https://de.iplantcollaborative.org/de/). Trimmed quality reads obtained from BHM Research Institute were mapped to *Boechera stricta* reference genome utilizing Tophat2 tool. Max intron length used in Tophat2 was 5000, while the default setting was used for other parameters. *Boechera stricta* gene annotation file (Gff3 format) was used; Tophat2 aligns the sequence to the transcriptome first, then only unmapped files are aligned to the reference genome. Cufflinks2 was used to build the transcripts by using each Tophat2 mapping file, which was further used in Cuffmerge2 to merge the newly identified transcripts with already predicted transcripts from *Boechera stricta* genome. The merged file with reference ids (gtf format) was used as annotation in Cuffdiff2 along with individual mapping files (bam files by Tophat2) from every sample to examine the differential expression of genes due to comparison of treatment with control. From the output of differentially expressed genes provided by Cuffdiff2, only the genes with Log2 fold change $\leqslant -2$ or $\geqslant 2$ and false discover rate of 0.05 was considered for further analysis.

AGRIGO Singular Enrichment Analysis tool was used for GO enrichment analysis of differentially expressed genes in individual treatments. For examining the number of genes involved in various biological processes related to drought and herbivory in different treatments, we used gprofiler (http://biit.cs.ut.ee/gprofiler/gcocoa.cgi). Venn diagrams were made by utilizing "Genevenn tool" (http://genevenn.sourceforge.net) for examining overlapping of genes in different treatments. Hierarchical clustering of genes related to drought and herbivory was performed using Cluster 3.0 software by using complete linkage function. Output files produced by Cluster 3.0 were used in Java TreeView to visualize Hierarchical clustering maps.

RNAseq data from this study have been submitted to the NCBI Gene Expression Omnibus (GEO; http://www.ncbi.nlm.nih.gov/geo/) under accession number GSE78101'.

## 5. Conclusions

Although it is now relatively well-known that plant responses to simultaneous challenges of biotic and abiotic stressors are not additive and involve signaling crosstalk [20], much less is known about the evolutionary implications of these interactions. Here, we tested a hypothesis for range limit development involving these interactions. The hypothesis states that antagonistic crosstalk between signaling pathways may be an evolutionary constraint, preventing adaptation to stressful environments across the range boundary. The constraint occurs if the crosstalk inhibits the simultaneous co-option of the antagonistic response pathways. For example, evolution of transcription factors involved in the crosstalk may have negative pleiotropic effects. We tested the hypothesis by comparing populations that have diverged with respect to defense and drought tolerance traits as one would predict if natural selection acted on antagonistic signaling pathways; neither population was high in both kinds of traits—defense and stress tolerance. The populations that had diverged with respect to traits regulated by the abiotic or biotic stress response pathways were in some cases less inducible, as we predicted based on the central hypothesis; however the results were ambiguous, depending on the level of analysis. That is, while some support was found at the level of gene expression, relatively little support was found in the analysis of trait phenotypes. We therefore suggest that genetic assimilation ([52]) of signaling pathways and the evolutionary consequences of crosstalk may be better studied at the molecular level. Recent modeling efforts predict that range shift response to climate change may be more pessimistic when variation among populations in phenotypic plasticity is accounted for [53]. While some studies indicate that complex traits controlled by signaling networks may be inherently evolutionarily constrained [17], others do not [54]. Thus, more work is needed to simultaneously connect molecular, evolutionary and ecological contexts for plant responses to biotic and abiotic stressors.

**Acknowledgments:** We thank Ethan Thompson and Jesse Larson for assistance in all aspects of the experiment, from planting, measuring, and harvesting tissues. Support was also provided by NIH INBRE P20GM103443, NSF EPSCoR, and startup funds from the University of South Dakota to A.B.

**Author Contributions:** Gunbharpur Singh Gill contributed to all aspects of the study; Riston Haugen contributed to conceptual and experimental design, tissue harvest and extractions, and interpretation and presentation of gene expression results; Steven L. Matzner contributed to all aspects of carbon isotope ratio analysis; Abdelali Barakat contributed to conceptual and experimental design, RNA extraction; and David H. Siemens helped with all aspects of the study except analysis of RNAseq data. Gunbharpur Singh Gill and David H. Siemens wrote the paper, but all authors read and commented on the manuscript.

**Conflicts of Interest:** The authors declare no conflict of interest.

## References

1.  Parmesan, C.; Gaines, S.; Gonzalez, L.; Kaufman, D.; Kingsolver, J.; Townsend Peterson, A.; Sagarin, R. Empirical perspectives on species borders: From traditional biogeography to global change. *Oikos* **2005**, *108*, 58–75. [CrossRef]

2.  Sexton, J.P.; McIntyre, P.J.; Angert, A.L.; Rice, K.J. Evolution and ecology of species range limits. *Annu. Rev. Ecol. Evol. Syst.* **2009**, *40*, 415–436. [CrossRef]

3.  Gaston, K.J. Geographic range limits: Achieving synthesis. *Proc. R. Soc. B Biol. Sci.* **2009**, *276*, 1395–1406. [CrossRef] [PubMed]

4.  Wiens, J.J. The niche, biogeography and species interactions. *Philos. Trans. R. Soc. B Biol. Sci.* **2011**, *366*, 2336–2350. [CrossRef] [PubMed]

5.  Kawecki, T.J. Adaptation to marginal habitats. *Annu. Rev. Ecol. Evol. Syst.* **2008**, *39*, 321–342. [CrossRef]

6.  Ettinger, A.F.; Ford, K.R.; HilleRisLambers, J. Climate determines upper, but not lower, altitudinal range limits of pacific northwest conifers. *Ecology* **2011**, *92*, 1323–1331. [CrossRef] [PubMed]

7.  Siemens, D.H.; Haugen, R.; Matzner, S.; VanAsma, N. Plant chemical defense allocation constrains evolution of local range. *Mol. Ecol.* **2009**, *18*, 4974–4983. [CrossRef] [PubMed]

8.  Siemens, D.H.; Duvall-Jisha, J.; Jacobs, J.; Manthey, J.; Haugen, R.; Matzner, S. Water deficiency induces stress tolerance-chemical defense evolutionary tradeoff that may help explain restricted range in plants. *Oikos* **2012**, *121*, 790–780. [CrossRef]

9.  Siemens, D.H.; Haugen, R. Plant chemical defense allocation constrains evolution of tolerance to community change across a range boundary. *Ecol. Evol.* **2013**, *3*, 4339–4347. [CrossRef] [PubMed]

10. Alsdurf, J.D.; Ripley, T.J.; Matzner, S.L.; Siemens, D.H. Drought-induced trans-generational tradeoff between stress tolerance and defence: Consequences for range limits? *AoB Plants* **2013**, *5*. [CrossRef] [PubMed]

11. Gill, G.S.; Haugen, R.; Larson, J.; Olsen, J.; Siemens, D.H. Plant evolution in response to abiotic and biotic stressors at "rear-edge" range boundaries. In *Abiotic and Biotic Stress in Plants—Recent Advances and Future Perspectives*; Shanker, A.K., Shanker, C., Eds.; InTech Europe: Rijeka, Croatia, 2016.

12. Van Straalen, N.M.; Roelofs, D. *An Introduction to Ecological Genomics*, 2nd ed.; Oxford University Press: Oxford, UK, 2012; p. 376.

13. Des Marais, D.L.; Juenger, T.E. Pleiotropy, plasticity, and the evolution of plant abiotic stress tolerance. *Ann. N. Y. Acad. Sci.* **2010**, *1206*, 56–79. [CrossRef] [PubMed]

14. Fujita, M.; Fujita, Y.; Noutoshi, Y.; Takahashi, F.; Narusaka, Y.; Yamaguchi-Shinozaki, K.; Shinozaki, K. Crosstalk between abiotic and biotic stress responses: A current view from the points of convergence in the stress signaling networks. *Curr. Opin. Plant Biol.* **2006**, *9*, 436–442. [CrossRef] [PubMed]

15. Asselbergh, B.; De Vieesschauwer, D.; Hofte, M. Global switches and fine-tuning—ABA modulates plant pathogen defense. *Mol. Plant-Microbe Interact.* **2008**, *21*, 709–719. [CrossRef] [PubMed]

16. Ton, J.; Flors, V.; Mauch-Mani, B. The multifaceted role of aba in disease resistance. *Trends Plant Sci.* **2009**, *14*, 310–317. [CrossRef] [PubMed]

17. Korves, T.M.; Schmid, K.J.; Caicedo, A.L.; Mays, C.; Stinchcombe, J.R.; Purugganan, M.D.; Schmitt, J. Fitness effects associated with the major flowering time gene frigida in arabidopsis thaliana in the field. *Am. Nat.* **2007**, *169*, E141–E157. [CrossRef] [PubMed]

18. Siemens, D.H.; Olsen, J.; Haugen, R.; Gill, G. Black Hills State University, Spearfish, SD, USA, 2016, Unpublished work.

19. Prasad, K.V.; Song, B.H.; Olson-Manning, C.; Anderson, J.T.; Lee, C.R.; Schranz, M.E.; Windsor, A.J.; Clauss, M.J.; Manzaneda, A.J.; Naqvi, I.; *et al.* A gain-of-function polymorphism controlling complex traits and fitness in nature. *Science* **2012**, *337*, 1081–1084. [CrossRef] [PubMed]

20. Atkinson, N.J.; Urwin, P.E. The interaction of plant biotic and abiotic stresses: From genes to the field. *J. Exp. Bot.* **2012**, *63*, 3523–3543. [CrossRef] [PubMed]

21. Wilkinson, S.; Davies, W.J. Aba-based chemical signalling: The co-ordination of responses to stress in plants. *Plant. Cell. Environ.* **2002**, *25*, 195–210. [CrossRef] [PubMed]

22. Hirai, M.Y.; Sugiyama, K.; Sawada, Y.; Tohge, T.; Obayashi, T.; Suzuki, A.; Araki, R.; Sakurai, N.; Suzuki, H.; Aoki, K.; *et al.* Omics-based identification of arabidopsis myb transcription factors regulating aliphatic glucosinolate biosynthesis. *Proc. Natl. Acad. Sci. USA* **2007**, *104*, 6478–6483. [CrossRef] [PubMed]

23. Beekwilder, J.; van Leeuwen, W.; van Dam, N.M.; Bertossi, M.; Grandi, V.; Mizzi, L.; Soloviev, M.; Szabados, L.; Molthoff, J.W.; Schipper, B.; *et al.* The impact of the absence of aliphatic glucosinolates on insect herbivory in arabidopsis. *PLoS ONE* **2008**, *3*, e2068. [CrossRef] [PubMed]

24. Vogel, H.; Kroymann, J.; Mitchell-Olds, T. Different transcript patterns in response to specialist and generalist herbivores in the wild arabidopsis relative boechera divaricarpa. *PLoS ONE* **2007**, *2*, e1081. [CrossRef] [PubMed]

25. Hirayama, T.; Shinozaki, K. Research on plant abiotic stress responses in the post-genome era: Past, present and future. *Plant J.* **2010**, *61*, 1041–1052. [CrossRef] [PubMed]

26. Seki, M.; Umezawa, T.; Urano, K.; Shinozaki, K. Regulatory metabolic networks in drought stress responses. *Curr. Opin. Plant Biol.* **2007**, *10*, 296–302. [CrossRef] [PubMed]

27. Körner, C. *Alpine Plant Life*; Springer: Heidelberg, Germany, 1999; p. 338.

28. Knight, C.; Vogel, H.; Kroymann, J.; Shumate, A.; Witsenboer, H.; Mitchell-Olds, T. Expression profiling and local adaptation of boechera holboellii populations for water use efficiency across a naturally occurring water stress gradient. *Mol. Ecol.* **2006**, *15*, 1229–1237. [CrossRef] [PubMed]

29. Siemens, D.H.; Mitchell-Olds, T. Evolution of pest-induced defenses in brassica plants: Tests of theory. *Ecology* **1998**, *79*, 632–646. [CrossRef]

30. Rasmann, S.; Chassin, E.; Bilat, J.; Glauser, G.; Reymond, P. Trade-off between constitutive and inducible resistance against herbivores is only partially explained by gene expression and glucosinolate production. *J. Exp. Bot.* **2015**, *66*, 2527–2534. [CrossRef] [PubMed]

31. Schranz, M.E.; Manzaneda, A.J.; Windsor, A.J.; Clauss, M.J.; Mitchell-Olds, T. Ecological genomics of boechera stricta: Identification of a QTL controlling the allocation of methionine- *vs* branched-chain amino acid-derived glucosinolates and levels of insect herbivory. *Heredity* **2009**, *102*, 465–474. [CrossRef] [PubMed]

32. Siemens, D.H.; MitchellOlds, T. Glucosinolates and herbivory by specialists (coleoptera: Chrysomelidae, lepidoptera: Plutellidae): Consequences of concentration and induced resistance. *Environ. Entomol.* **1996**, *25*, 1344–1353. [CrossRef]

33. Kahl, J.; Siemens, D.H.; Aerts, R.J.; Gäbler, R.; Kühnemann, F.; Preston, C.A.; Baldwin, I.T. Herbivore-induced ethylene suppresses a direct defense but not a putative indirect defense against an adapted herbivore. *Planta* **2000**, *210*, 336–342. [CrossRef] [PubMed]

34. Song, B.H.; Windsor, A.J.; Schmid, K.J.; Ramos-Onsins, S.; Schranz, M.E.; Heidel, A.J.; Mitchell-Olds, T. Multilocus patterns of nucleotide diversity, population structure and linkage disequilibrium in boechera stricta, a wild relative of arabidopsis. *Genetics* **2009**, *181*, 1021–1033. [CrossRef] [PubMed]

35. Lee, C.R.; Mitchell-Olds, T. Quantifying effects of environmental and geographical factors on patterns of genetic differentiation. *Mol. Ecol.* **2011**, *20*, 4631–4642. [CrossRef] [PubMed]

36. Song, B.; Clauss, M.J.; Pepper, A.; Mitchell-Olds, T. Geographic patterns of microsatellite variation in *Boechera stricta*, a close relative of *Arabidopsis*. *Mol. Ecol.* **2006**, *15*, 357–369. [CrossRef] [PubMed]

37. Schranz, M.E.; Dobes, C.; Koch, M.A.; Mitchell-Olds, T. Sexual reproduction, hybridization, apomixis, and polyploidization in the genus boechera (brassicaceae). *Am. J. Bot.* **2005**, *92*, 1797–1810. [CrossRef] [PubMed]

38. Mitchell-Olds, T. *Arabidopsis thaliana* and its wild relatives: A model system for ecology and evolution. *Trends Ecol. Evol.* **2001**, *16*, 693–700. [CrossRef]

39. Rushworth, C.A.; Song, B.H.; Lee, C.R.; Mitchell-Olds, T. Boechera, a model system for ecological genomics. *Mol. Ecol.* **2011**, *20*, 4843–4857. [CrossRef] [PubMed]

40. Lovell, J. Boechera summit 2011. *Mol. Ecol.* **2011**, *20*, 4840–4842. [CrossRef] [PubMed]

41. Snedecor, G.W.; Cochran, W.B. *Statistical Methods*; Iowa State University: Ames, IA, USA, 1967.

42. Fitter, A.H.; Hay, R.K.M. *Environmental Physiology of Plants*, 3rd ed.; Academic press: New York, NY, USA, 2002; p. 367.

43. Casique-Arroyo, G.; Martínez-Gallardo, N.; González de la Vara, L.; Délano-Frier, J.P. Betacyanin biosynthetic genes and enzymes are differentially induced by (a)biotic stress in *Amaranthus hypochondriacus*. *PLoS ONE* **2014**, *9*, e99012.

44. Rohr, F.; Ulrichs, C.; Schreiner, M.; Zrenner, R.; Mewis, I. Responses of arabidopsis thaliana plant lines differing in hydroxylation of aliphatic glucosinolate side chains to feeding of a generalist and specialist caterpillar. *Plant Physiol. Biochem.* **2012**, *55*, 52–59. [CrossRef] [PubMed]

45.   McKay, J.K.; Richards, J.H.; Mitchell-Olds, T. Genetics of drought adaptation in arabidopsis thaliana: I. Pleiotropy contributes to genetic correlations among ecological traits. *Mol. Ecol.* **2003**, *12*, 1137–1151. [CrossRef] [PubMed]

46.   Farquhar, G.D.; Richards, R.A. Isotopic composition of plant carbon correlates with water-use efficiency of wheat genotypes. *Aust. J. Plant Physiol.* **1984**, *11*, 539–552. [CrossRef]

47.   Brown, P.; Tokuhisa, J.; Reichelt, M.; Gershenzon, J. Variation of glucosinolate accumulation among different organs and developmental stages of *Arabidopsis thaliana*. *Phytochemistry* **2003**, *62*, 471–481. [CrossRef]

48.   Prestera, T.; Fahey, J.W.; Holtzclaw, W.D.; Abeygunawardana, C.; Kachinski, J.L.; Talalay, P. Comprehensive chromatographic and spectroscopic methods for the separation and identification of intact glucosinolates. *Anal. Biochem.* **1996**, *239*, 168–179. [CrossRef] [PubMed]

49.   Zar, J.H. *Biostatistical Analysis*, 3rd ed.; Prentice Hall: Upper Saddle River, NJ, USA, 1996.

50.   Montgomery, D.C. *Design and Analysis of Experiments*, 4th ed.; John Wiley and Sons: New York, NY, USA, 1997.

51.   Afifi, A.A.; Clark, V. *Computer-Aided Multivariate Analysis*; Lifetime Learning: Belmont, CA, USA, 1984.

52.   Pigliucci, M.; Murren, C.J.; Schlichting, C.D. Phenotypic plasticity and evolution by genetic assimilation. *J. Exp. Biol* **2006**, *209*, 2362–2367. [CrossRef] [PubMed]

53.   Valladares, F.; Matesanz, S.; Guilhaumon, F.; Araújo, M.B.; Balaguer, L.; Benito-Garzón, M.; Cornwell, W.; Gianoli, E.; Kleunen, M.; Naya, D.E. The effects of phenotypic plasticity and local adaptation on forecasts of species range shifts under climate change. *Ecol. Lett.* **2014**, *17*, 1351–1364. [CrossRef] [PubMed]

54.   Thaler, J.S.; Humphrey, P.T.; Whiteman, N.K. Evolution of jasmonate and salicylate signal crosstalk. *Trends Plant Sci.* **2012**, *17*, 260–270. [CrossRef] [PubMed]

# Litter Accumulation and Nutrient Content of Roadside Plant Communities in Sichuan Basin, China

**Huiqin He [1] and Thomas Monaco [2],***

[1]    College of Resources and Environmental Engineering, Yibin University, Yibin 644000, China; huiqinhe211@sina.com

[2]    US Department of Agriculture, Agricultural Research Service, Forage and Range Research Laboratory, Utah State University, Logan, UT 84322-6300, USA

*    Correspondence: tom.monaco@ars.usda.gov

Academic Editor: Günter Hoch

**Abstract:** It is widely recognized that feedbacks exist between plant litter and plant community species composition, but this relationship is difficult to interpret over heterogeneous conditions typical of modified environments such as roadways. Given the need to expedite natural recovery of disturbed areas through restoration interventions, we characterized litter accumulation and nutrient content (i.e., organic carbon, total N, and P) and quantified their association with key plant species. Plant species cover and litter characteristics were sampled at 18 successional forest plant communities along major roadways in Sichuan Basin, western China. Variation in litter across communities was assessed with principal component analysis (PCA) and species with the highest correlation to PCA axes were determined with Pearson's $r$ coefficients. Plant communities with the longest time since road construction (i.e., 70 years) were distinctly different in litter total N and organic carbon compared to plant communities with a shorter disturbance history. We encountered 59 plant species across sampling plots, but only four rare species (i.e., frequency < 5) were strongly correlated with litter characteristics ($p < 0.01$); none of which were the most abundant where they occurred. These results highlight the importance of site-specific factors (i.e., geographic location, disturbance age) regulating plant litter across heavily disturbed landscapes and how litter characteristics and rare plant species are correlated.

**Keywords:** forest succession; plant community assembly; disturbance age; species colonization; principal component analysis

## 1. Introduction

Plant litter is an important component of ecosystem functioning and nutrient cycling [1]. Litter accumulation provides ground cover and reduces soil erosion and water runoff [2]. In addition, litter creates soil microenvironments that may preferentially support seed germination of select plant species [3–5] and provides substrates for soil nutrient and resource pools [6,7]. Assessing litter attributes can improve our understanding of vegetation dynamics following natural disturbances as well as expedite ongoing efforts to restore plant communities [8,9].

While plant species and their relative contributions to bulk, mixed-species litter is widely studied, feedbacks between bulk litter and species composition are less clear. For example, when chemical composition of the litter from each species as well as their respective contribution to bulk litter accumulation are known [10], relationships between litter and species composition can be deciphered across environmental and successional gradients [11,12]. However, when litter composition and accumulation values are not available and species contributions to bulk litter are uncertain due to complex multi-species mixtures [13,14], associating litter traits and plant species composition must resort to exploratory multivariate analyses and/or correlative measures to gain further insights [15–17]. When considering mixed-species litter, identifying which species are correlated with litter accumulation may provide insights in plant community development across successional gradients [18,19]. For example, correlations between bulk litter traits and individual species can reveal site-specific information regarding the influence of litter on microenvironmental conditions, soil stability, and the establishment of select species within a vegetation type [3,20,21], which could be used to expedite the selection of suitable species for revegetation of disturbed areas [22,23].

Bulk litter nutrient content can also strongly influence plant community development [24,25]. Consequently, complex, mixed-species communities may contain a broad range of litter nutrient contents depending on species composition [26,27]. For example, grass and herb dominated plant communities produce relatively fine litter, characterized by high nutrient content (i.e., organic carbon, nitrogen [28]) and fast decomposition rates [29]. In contrast, coarse litter with greater lignin and woody plant structures has relatively lower nutrient content and decomposition rates due to resistance to enzymatic decomposition [30]. Correspondence between litter nutrient content and the abundance of individual species often reflects the contribution of dominant species within plant communities [31], whereas association with subordinate or rare species is largely unknown [17,32]. If rare species are habitat specialists, it is possible that they may actually show higher correlations with site-specific litter traits than dominant, generalist species [33,34].

Drastic disturbances, such as when soil and vegetation are removed and soil substrates are redistributed, often create heterogeneous vegetation patterns characterized by variable plant species composition [35] and litter characteristics [19]. Such heterogeneity within an ecosystem provides a unique setting to assess variation in litter accumulation and nutrient content [18,36]. For example, road construction unavoidably creates extensive disturbances [37,38], which results in successional plant communities where litter accumulation and nutrient content may similarly vary [39]. Consequently, in this study we collected bulk, mixed-litter samples from 54 plots located within 18 heterogeneous recovering plant communities that were previously destroyed by road construction in Sichuan Basin, China. Our objectives were to assess the variability of litter accumulation and nutrient content among plant communities and identify which species have the greatest correlation with bulk litter traits. We hypothesized that if rare species are habitat specialists, it is possible that they may actually show higher correlations with site-specific litter traits than dominant, generalist species [33,34].

## 2. Results

### 2.1. Sampling Plot Variation in Canopy Cover of Primary Life Forms

A total of 59 species were encountered within the 54 sampling plots. Species richness was greatest for herbs (i.e., 22 species), followed by an equal number of graminoid and shrub species (15), and only 7 tree species. Cover of all life forms was variable among sampling plots (Figure 1a), illustrated by large ranges between the 25th and 75th percentiles, especially herbs. Among the 59 species, percentage cover (mean ± SE) was greatest for the large statured graminoid *Imperata cylindrica* (9.3 ± 2.7) and the herb *Artemisia argyi* (5.5 ± 1.7). Species also varied widely in the number of plots in which they occurred (i.e., frequency). Most species were actually rare based on the fact that 36 of the 59 species occurred in 5 or fewer sampling plots. By comparison, only 9 species occurred in 10 or more plots, and the two most common species (i.e., *Imperata cylindrica* and *Artemisia argyi*) also had the highest average cover. Similar to life form cover and species frequency, litter characteristics were highly variable across the sampling plots, illustrated by skewed data distributions and low correspondence between mean and median values (Figure 1b–e).

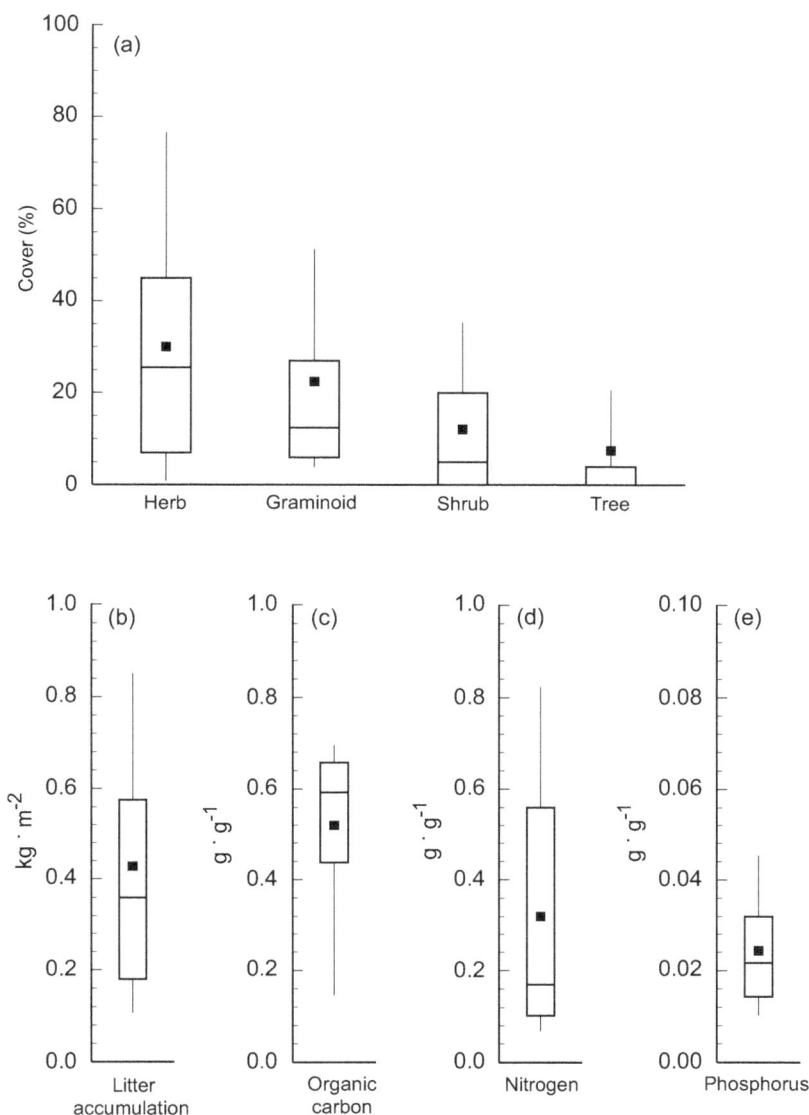

**Figure 1.** (**a**) Box-and-whisker plots showing percentage cover of plant life forms; (**b**) litter accumulation; (**c–e**) litter nutrient content for 54 sampling plots located along roadsides in Sichuan Basin, China. For each variable, the top, bottom, and middle lines of the box correspond to the 75th percentile, 25th percentile, and median, respectively; vertical lines extending from the bottom and top of the box correspond with the 10th and 90th percentile, respectively; and the black squares indicate the mean value. Note differences in y-axis scale.

## 2.2. Principal Component Analysis of Litter Characteristics

The PCA solution explained 94.2% of total variation among the 54 plots on three axes (Table 1). Axes were best defined by the following: axis 1, inverse relationship between litter N and organic carbon; axis 2, positive relationship between litter accumulation and P; and axis 3, inverse relationship between litter accumulation and P (Table 1). Ordination of PCA scores for axes 1 and 3 showed the most distinction among sampling plots. Axis 1 emphasized that litter N was higher and organic carbon was lower for sampling plots associated the longest period since road construction (i.e., 70 years; Figure 2). In addition, axis 3 featured high variability, both within and among sampling plots, for litter accumulation and P content; however, variation was relatively lower for plots associated with the shortest period since road construction, which tended to have higher organic carbon and P than the other roads.

**Table 1.** Results of principal component analysis (PCA) based on four litter variables for the 54 sampling plots located along roadsides in Sichuan Basin, China. The PCA solution identified three axes, whose eigenvalues explained 94.2% of the total variance among plots.

|                        | Axis 1  | Axis 2  | Axis 3  |
|------------------------|---------|---------|---------|
| Eigenvalues            | 1.8     | 1.2     | 0.8     |
| Total variance (%)     | 44.1    | 29.7    | 20.4    |
| Litter accumulation    | 0.028   | 0.775   | −0.630  |
| Litter organic carbon  | 0.938   | −0.650  | 0.041   |
| Litter total N         | −0.939  | 0.012   | 0.067   |
| Litter total P         | 0.065   | 0.764   | 0.642   |

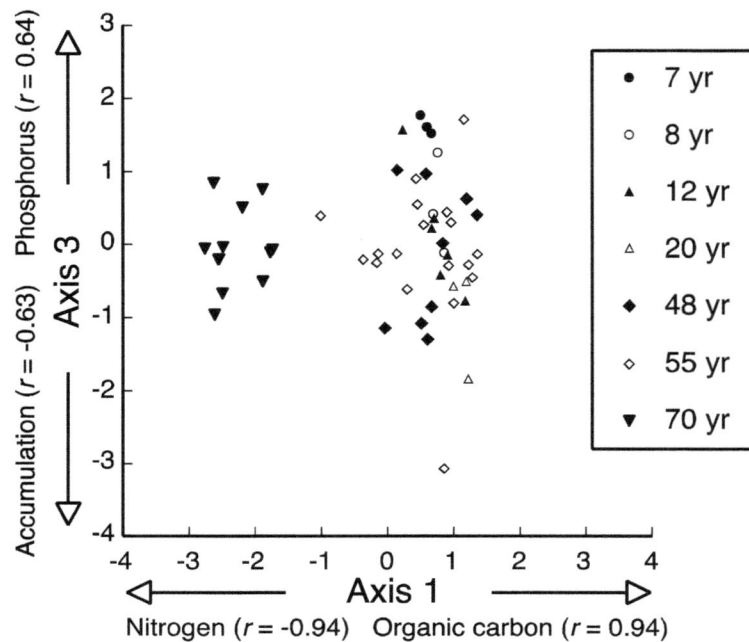

**Figure 2.** Ordination of 54 roadside sampling plots in Sichuan Basin, China, based on principal component analysis (PCA) of litter variables showing bi-plots for axes 1 and 2 (see Table 1). Plots were located adjacent to major roadways radiating from Chengdu City that varied in time since road construction (i.e., years shown with different symbols). Arrows indicate direction of Pearson's $r$ coefficient, which represents pair-wise correlation between PCA axes the litter variables.

## 2.3. Plant Species Correlation with PCA Axes

Of the 59 species encountered, only four were significantly correlated with PCA axes ($p < 0.01$; Table 2); however, no species were correlated with axis 2 (data not shown). One tree, *Dalbergia hupeana*, and two graminoids, *Arthraxon hispidus* and *Eragrostis ferruginea*, showed negative association with PCA axis 1, and thus, greater affinity for plant communities with litter composed of higher total N, but lower organic carbon. In contrast, only one species, the shrub *Lycium chinense*, was positively correlated with axis 3, where it tended to occur in plant communities with higher P content and lower litter accumulation. Interestingly, all three of the species that were highly correlated with axis 1 only occurred in plant communities associated with the longest period since road construction (i.e., 70 years. Alternatively, *Lycium chinense* only occurred in plant communities accompanying a recovery period of 55 years.

**Table 2.** Pearson's *r* coefficients showing correlations between species canopy cover and PCA axes 1 and 3 for 54 sampling plots located along roadways in Sichuan Basin, China. Bold text indicates significant correlations; $p < 0.01$. Frequency (i.e., number of times species was encountered) and mean percentage cover across the 54 plots are shown.

| | Axis 1 | Axis 3 | Frequency | Cover % |
|---|---|---|---|---|
| **Herbs** | | | | |
| *Anemone vitifolia* | 0.1207 | 0.0985 | 5 | 2.5 |
| *Artemisia argyi* | 0.1785 | 0.3175 | 35 | 8.5 |
| *Bidens pilosa* | −0.1387 | 0.0777 | 16 | 5.8 |
| *Boenninghausenia albiflora* | 0.0856 | 0.1851 | 3 | 24.7 |
| *Cayratia japonica* | 0.1271 | 0.1274 | 9 | 4.1 |
| *Clematis florida* | 0.0296 | 0.1864 | 3 | 2.7 |
| *Commelina communis* | 0.0106 | 0.1898 | 2 | 3.5 |
| *Dendranthema indicum* | 0.1225 | 0.0496 | 17 | 6.9 |
| *Dryopteris bissetiaha* | 0.0696 | 0.0040 | 10 | 6.9 |
| *Epimedium brevicornum* | −0.2596 | −0.0192 | 2 | 9.0 |
| *Erigeron acer* | 0.0280 | 0.1753 | 6 | 17.2 |
| *Gelsemium elegans* | 0.1109 | 0.0302 | 4 | 1.0 |
| *Humulus japonicus* | 0.1143 | 0.0100 | 7 | 2.7 |
| *Iris japonica* | −0.2758 | −0.0025 | 3 | 16.7 |
| *Paederia scandens* | 0.0564 | 0.1638 | 4 | 15.3 |
| *Spora lygodii* | −0.2032 | −0.0601 | 6 | 3.0 |
| *Strobilanthes cusia* | −0.0082 | −0.0142 | 6 | 2.7 |
| *Taraxacum officinale* | 0.1309 | 0.1963 | 4 | 2.5 |
| *Torilis japonica* | 0.1725 | −0.035 | 4 | 3.0 |
| *Trifolium repens* | 0.1435 | −0.1317 | 4 | 2.5 |
| *Vicia carcca* | 0.1194 | 0.2471 | 2 | 7.5 |
| *Youngia japonica* | 0.2022 | 0.0007 | 5 | 2.6 |
| **Graminoids** | | | | |
| *Arthraxon hispidus* | **−0.4535** | −0.0474 | 4 | 3.8 |
| *Carex rigescens* | 0.1604 | −0.0622 | 14 | 10.3 |
| *Cyperus microiria* | −0.1556 | −0.2874 | 16 | 9.5 |
| *Cynodon dactylon* | 0.0269 | 0.0664 | 2 | 15.0 |
| *Cymbopogon goeringii* | −0.0978 | −0.074 | 7 | 18.6 |
| *Digitaria sanguinalis* | 0.0281 | −0.0768 | 6 | 5.7 |
| *Eragrostis ferruginea* | **−0.4156** | −0.0924 | 5 | 5.6 |
| *Eriophorum vaginatum* | 0.1036 | 0.1219 | 11 | 17.5 |
| *Festuca arundinacea* | 0.1253 | −0.3227 | 2 | 26.0 |
| *Fimbristylis dichotoma* | −0.2301 | 0.1107 | 2 | 9.0 |
| *Imperata cylindrica* | 0.0196 | −0.2238 | 19 | 26.5 |
| *Miscanthus sinensis* | 0.1812 | −0.0570 | 14 | 20.9 |
| *Panicum brevifolium* | 0.1014 | 0.0091 | 3 | 20.3 |
| *Pogonatherum paniceum* | 0.1480 | 0.0755 | 8 | 1.9 |
| *Setaria viridis* | 0.0709 | 0.1270 | 8 | 3.5 |
| **Shrubs** | | | | |
| *Berberis julianae* | −0.2049 | −0.0133 | 2 | 8.5 |
| *Boehmeria nivea* | 0.0793 | −0.0142 | 3 | 7.7 |
| *Broussonetia papyrifera* | 0.1293 | −0.1376 | 2 | 9.5 |
| *Clerodendrum bungei* | 0.0958 | 0.1108 | 2 | 22.5 |
| *Coriaria sinica* | −0.0082 | 0.0061 | 7 | 14.7 |
| *Hedera nepalensis* | −0.0138 | −0.0243 | 5 | 5.6 |
| *Lespedeza bicolor* | −0.0929 | −0.2740 | 6 | 2.3 |
| *Lindera glauca* | 0.0062 | −0.1711 | 2 | 35.5 |
| *Lycium chinense* | 0.0783 | **−0.4635** | 2 | 11.0 |
| *Pueraria lobata* | 0.1002 | −0.0367 | 3 | 13.3 |
| *Rhus chinensis* | −0.3156 | −0.1573 | 3 | 3.7 |

**Table 2.** *Cont.*

|  | Axis 1 | Axis 3 | Frequency | Cover % |
|---|---|---|---|---|
| *Rubus corchorifolius* | −0.1512 | −0.0418 | 6 | 7.2 |
| *Solanum nigrum* | 0.1061 | 0.2683 | 2 | 21.5 |
| *Vitex negundo* | −0.1467 | 0.0362 | 6 | 16.2 |
| *Pyracantha fortuneana* | −0.1172 | 0.0299 | 6 | 13.3 |
| **Trees** | | | | |
| *Cupressus funebris* | −0.2208 | 0.0285 | 4 | 15.5 |
| *Dalbergia hupeana* | **−0.3564** | −0.0056 | 3 | 7.7 |
| *Diospyros kaki* | 0.0575 | −0.1534 | 2 | 55.0 |
| *Myrsine africana* | −0.0512 | −0.1608 | 8 | 8.4 |
| *Populus adenopoda* | 0.0119 | −0.2225 | 3 | 30.7 |
| *Ulmus pumila* | 0.0577 | 0.1363 | 2 | 11.5 |
| *Vaccinium bracteatum* | −0.2469 | −0.0175 | 2 | 15.0 |

## 3. Discussion

The scientific literature is replete with examples illustrating a strong association between species composition and litter attributes—especially nutrient content [40,41]. Because bulk, mixed-litter content is a reflection of the plant community as a whole, dominant species often contribute a disproportionate amount to litter [42,43], which can strongly dictate nutrient cycling as well as establish plant-soil-feedbacks that favor themselves or other species in the community [44,45]. In contrast, subordinate, rare species contribute little to bulk litter accumulation but may actually be more sensitive than dominant species to site-specific litter attributes [33]. Our results support this hypothesis because only non-dominant, rare species were correlated with litter characteristics. These results may be a reflection of our study design that considered bulk, mixed litter and its relationship with heterogeneous species composition within a natural system. Indeed, experiments using single- as well as mixed-species litter in controlled settings have greater precision to evaluate how litter quantity and quality impact individual species [31,46]. Our results are also likely a consequence of rare species having greater likelihood of being correlated with litter because most species across the sampled plant communities were rare and sparsely distributed relative to the few dominant species. Nonetheless, these subordinate, rare species were correlated with litter traits and illustrate high affinity for specific litter conditions. Our results are also specific to only a few litter variables, and we recognize that other nutrients, which were not considered, might have a significant effect on species abundance across our study locations.

The higher affinity we observed for a few subordinate species to high total litter N conditions is most apparent for one graminoid species (i.e., *Arthraxon hispidus*) and one tree species (i.e., *Dalbergia hupeana*). The graminoid is an annual grass known to flourish in ruderal environments [47,48] where soil N is often enriched [49], whereas *Dalbergia hupeana* is an important agroforestry tree that produces litter with higher N and P content relative to other co-occurring species [50]. In contrast, it is not clear why the other two species showed greater abundance on locations characterized by higher litter N content. Further experimentation is needed to understand how these site-specific conditions favor seedling establishment, growth, and persistence capacity of these rare species (i.e., [3,31,46]).

The exclusive occurrence of these four species at study locations with different periods since road construction emphasizes that site history also plays a role in species assemblages within these plant communities. Accordingly, given the site-specific nature of litter characteristics and species affinity, our results should only be viewed in the context of independent case studies for these disparate plant communities and locations. For example, longer post-disturbance periods (i.e., 55–70 years) may have contributed to unique soil development and plant community assemblages with fundamentally different litter characteristics that were particularly suitable for these rare species. Since none of the dominant species were correlated with litter patterns, it can only be speculated that the responses we observed indicate that rare species have unique requirements to initially colonize or persist over the long term in these successional forest plant communities [39]. Further research is needed to combine what is currently known about plant traits associated with colonization of roadside slopes [51] and the feedbacks between litter characteristics and the performance of individual plant species, which are typically very complex [44,46].

Greater association between litter and rare species relative to dominant species in our study stresses the importance of rare species to biodiversity of ecosystems [52]. Although most species are in fact rare in many plant communities, recent research illustrates that rare species actually contribute disproportionately to the structure and function of species assemblages [53]. In addition, rare species have been shown to disproportionately increase the range of functions provided within ecosystems, thus, their conservation value should be heightened to prevent the loss of important ecosystem services [54,55]. Roadside vegetation in the highly modified region of the Sichuan Basin thus serves as a refuge to support a high number of rare endemic species [56], where conservation is a high priority to support local plant diversity and protect numerous threatened species [57]. Our research illustrates a high degree of variation in litter characteristics among plant communities and offers a promising outlook for ongoing efforts to identify species best suited for existing roadside microhabitats (i.e., [58]).

In this study, we conclude that litter N and organic carbon explained most of the variation in litter among plant communities, and that none of the dominant species were correlated with litter characteristics. Instead, we found that four, rare and sparsely represented species were strongly associated with gradients in litter nutrient content and accumulation. High affinity of these species for specific conditions that only existed at specific locations underscores the importance of site disturbance history on litter characteristics. Consequently, our results can only offer case studies for these locations, yet they suggest that future research should pay more attention to (1) defining the contribution of both dominant and subordinate species to bulk, mixed litter pools and (2) exploring the possibility that rare species can potentially be more sensitive than dominant species to variation in litter characteristics.

## 4. Materials and Methods

### 4.1. Vegetation Sampling

The study was conducted in Sichuan Basin, China (26°03′–34°19′ N, 97°21′–108°31′ W), a region characterized by a subtropical moist climate, with a distinct dry season between June and October. Mean annual temperature and precipitation values (1994–2014) were 17.3 °C and 826 mm, respectively [59]. The primary soil types within the basin include soils developed from either basalt or sand-shale substrates with high amounts of clay minerals, especially Fe-Al oxides [60]. The original vegetation is locally known as broad-leaved forest, often dominated by large-statured trees (i.e., *Castanopsis fargesii* and *Cyclobalanopsis glanuca* [61]). We sought to study vegetation adjacent to major roads radiating from Chengdu City where the original forest vegetation was destroyed during road construction and the current vegetation reflects various stages of natural succession over a 7–70 year period. Given the high level of urban development in this region [62], and uncertain disturbance history, large study locations to support multiple within-site sampling plots were limited. Consequently, we acquired road construction records from the Sichuan Province highway department and visited potential study locations at a set distance of 250 km along major roadways from Chengdu City. We specifically sought study locations with natural vegetation that covered an area greater than 2 hectares contained a mix of various successional stages composed of herbaceous/graminoid, shrub and tree life forms, and where no subsequent disturbances have occurred since road construction. Our search yielded 18 suitable locations that met these criteria (Figure 3). Because our objective was to assess the widest possible variation in litter and successional stages within each location and across the region, we randomly selected a 20 m × 100 m (i.e., 2000 m$^2$) sampling area at each location and sampled vegetation and litter within 3, 5 m × 5 m randomly located plots, each of which were situated at least 5 m from the bottom of the slope and 10 m from other plots. This design resulted in a total of

54 sampling plots, which were subsequently analyzed as independent experimental units given the high level of within-location heterogeneity.

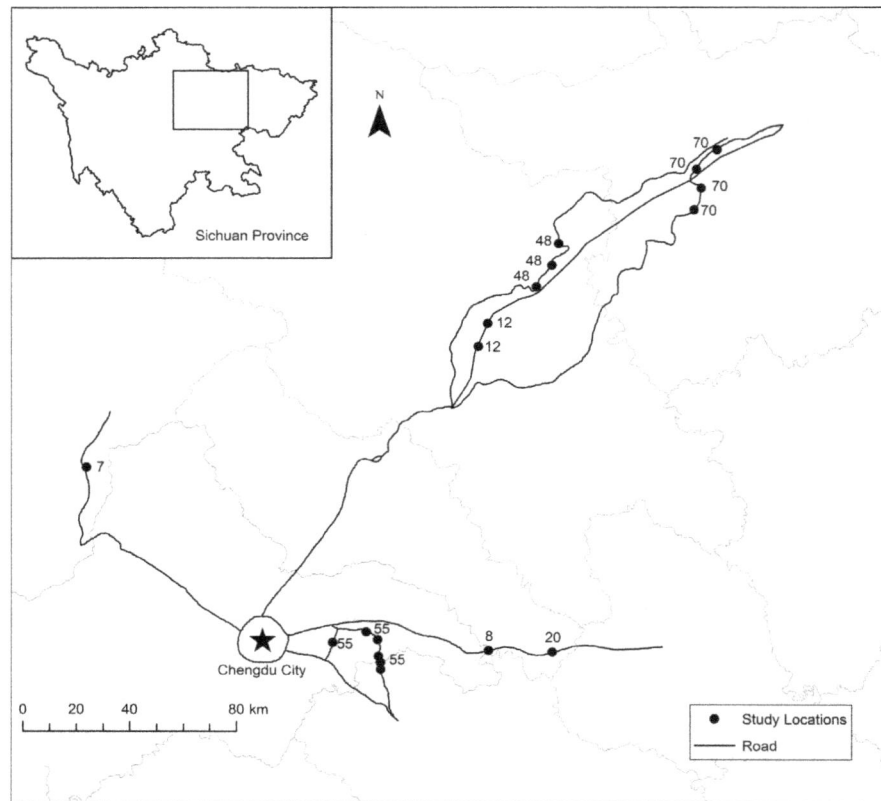

**Figure 3.** Map showing Sichuan Province in southwestern China (inset) and the 18 study locations within Sichuan Basin along major roadways near Chengdu City. Numbers indicate the period of time since road construction (i.e., years) for study locations. Gray lines represent county boundaries.

Vegetation was sampled in plots by visually estimating the percentage canopy cover of each species (i.e., nearest 1%). Estimates were made by one individual to avoid subjective differences among samplers. Canopy cover was defined as the proportion of ground area occupied by the aboveground parts of plants, i.e., the area covered by the vertical projection of plant canopies and tree crowns [63]. To aid in estimating cover of large trees, a cord was placed on the ground within plots to indicate canopy edges. Thus, total plot cover often exceeded 100% when canopies were multi-layered. Each species was classified by life form (i.e., herb, graminoid, shrub, and tree) and composite cover percentages were summed for each category.

*4.2. Litter Sampling*

Within each of the 54 plots, bulk, mixed-species litter was hand-collected from 3, 1 m × 1 m subplots, sealed in paper bags, and transported to a laboratory at Sichuan University in Chengdu City. Subsamples were combined and dried in a convective oven for 48 h at 65 °C to determine litter accumulation (kg dry mass m$^2$). Dried samples were thoroughly mixed and 50 g subsamples were milled through a 2 mm diameter screen. Litter organic C was analyzed according to the $K_2Cr_2O_7$-$H_2SO_4$ oxidation method with external heating [64]. Samples were also analyzed for total nitrogen (litter N) and total phosphorus (litter P) using the Kjeldahl method and digestion in $NaOH/H_2SO_4$, respectively, using a Technicon Autoanalyzer (Seal Analytical Inc., Mequon, WI, USA). All chemical analyses were analyzed in triplicate.

*4.3. Statistical Analyses*

Data for the 54 sampling plots were analyzed with principal component analysis (PCA) to assess variation based on the four litter characteristics. In brief, this multivariate statistical procedure identifies the relative influence of each litter characteristic on PCA axes (i.e., largest absolute eigenvector values) by placing sampling plots within an orthogonal coordinate system based on the multivariate relationship among the four litter characteristics [65]. In addition, the strength of pair-wise associations between PCA axes and percentage cover of plant species were evaluated with correlation coefficients (Pearson's *r*) and null hypothesis tests (i.e., the true correlation coefficient is equal to zero; $\alpha < 0.01$). All analyses were performed with Jump (JMP) version 13.0.0.

**Acknowledgments:** This research was supported by the Science and Technology Department of Sichuan Province (2012SZ0176).

**Author Contributions:** Huiqin He conceived and designed the study, oversaw data collection, and conducted all litter analyses; Huiqin He and Thomas Monaco analyzed the data and co-wrote the paper. Both authors have read and approved the final manuscript.

**Conflicts of Interest:** The authors declare no conflict of interest. The founding sponsors had no role in the design of the study; in the collection, analyses, or interpretation of data; in the writing of the manuscript, and in the decision to publish the results.

## References

1.  Powers, J.S.; Montgomery, R.A.; Adair, E.C.; Brearley, F.Q.; DeWalt, S.J.; Castanho, C.T.; Chave, J.; Deinert, E.; Ganzhorn, J.U.; Gilbert, M.E.; et al. Decomposition in tropical forests: A pan-tropical study of the effects of litter type, litter placement and mesofaunal exclusion across a precipitation gradient. *J. Ecol.* **2009**, *97*, 801–811. [CrossRef]

2.  Li, D.; Peng, S.; Chen, B. The effects of leaf litter evenness on decomposition depend on which plant functional group is dominant. *Plant Soil* **2013**, *365*, 255–266. [CrossRef]

3.  Loydi, A.; Eckstein, R.L.; Otte, A.; Donath, T.W. Effects of litter on seedling establishment in natural and semi-natural grasslands: A meta-analysis. *J. Ecol.* **2013**, *101*, 454–464. [CrossRef]

4.  Santos, S.L.D.; Valio, I.F.M. Litter accumulation and its effect on seedling recruitment in a southeast brazilian tropical forest. *Braz. J. Bot.* **2002**, *25*, 89–92. [CrossRef]

5.  Villalobos-Vega, R.; Goldstein, G.; Haridasan, M.; Franco, A.C.; Miralles-Wilhelm, F.; Scholz, F.G.; Bucci, S.J. Leaf litter manipulations alter soil physicochemical properties and tree growth in a neotropical savanna. *Plant Soil* **2011**, *346*, 385. [CrossRef]

6.  Tiegs, S.D.; Akinwole, P.O.; Gessner, M.O. Litter decomposition across multiple spatial scales in stream networks. *Oecologia* **2009**, *161*, 343–351. [CrossRef] [PubMed]

7.  Engber, E.A.; Varner, J.M.; Arguello, L.A.; Suihara, N.G. The effects of conifer encroachment and overstory structure on fuels and fire in an oak woodland landscape. *Fire Ecol.* **2012**, *7*, 32–50. [CrossRef]

8.  Wood, T.E.; Lawrence, D.; Clark, D.A.; Chazdon, R.L. Rain forest nutrient cycling and productivity in response to large-scale litter manipulation. *Ecology* **2009**, *90*, 109–121. [CrossRef] [PubMed]

9.  Fahey, T.J.; Yavitt, J.B.; Sherman, R.E.; Groffman, P.M.; Fisk, M.C.; Maerz, J.C. Transport of carbon and nitrogen between litter and soil organic matter in a northern hardwood forest. *Ecosystems* **2011**, *14*, 326–340. [CrossRef]

10. Jacob, M.; Viedenz, K.; Polle, A.; Thomas, F.M. Leaf litter decomposition in temperate deciduous forest stands with a decreasing fraction of beech (*Fagus sylvatica*). *Oecologia* **2010**, *164*, 1083–1094. [CrossRef] [PubMed]

11. Sariyildiz, T.; Küçük, M. Influence of slope position, stand type and rhododendron (*Rhododendron ponticum*) on litter decomposition rates of oriental beech (*Fagus orientalis* Lipsky.) and spruce [*Picea orientalis* (L.) link]. *Eur. J. For. Res.* **2009**, *128*, 351–360. [CrossRef]

12. Rowe, E.C.; Healey, J.R.; Edwards-Jones, G.; Hills, J.; Howells, M.; Jones, D.L. Fertilizer application during primary succession changes the structure of plant and herbivore communities. *Biol. Conserv.* **2006**, *131*, 510–522. [CrossRef]

13. Hui, D.; Jackson, R.B. Assessing interactive responses in litter decomposition in mixed species litter. *Plant Soil* **2009**, *314*, 263–271. [CrossRef]

14.  Berger, T.W.; Berger, P. Does mixing of beech (*Fagus sylvatica*) and spruce (*Picea abies*) litter hasten decomposition? *Plant Soil* **2014**, *377*, 217–234. [CrossRef] [PubMed]

15.  Schwilk, D.W.; Caprio, A.C. Scaling from leaf traits to fire behaviour: Community composition predicts fire severity in a temperate forest. *J. Ecol.* **2011**, *99*, 970–980. [CrossRef]

16.  Hale, B.; Robertson, P. Plant community and litter composition in temperate deciduous woodlots along two field gradients of soil ni, cu and co concentrations. *Environ. Pollut.* **2016**, *212*, 41–47. [CrossRef] [PubMed]

17.  Camill, P.; McKone, M.J.; Sturges, S.T.; Severud, W.J.; Ellis, E.; Limmer, J.; Martin, C.B.; Navratil, R.T.; Purdie, A.J.; Sandel, B.S.; et al. Community- and ecosystem-leve changes in species-rich tallgrass prairie. *Ecol. Appl.* **2004**, *14*, 1690–1694. [CrossRef]

18.  Wu, D.; Li, T.; Wan, S. Time and litter species composition affect litter-mixing effects on decomposition rates. *Plant Soil* **2013**, *371*, 355–366. [CrossRef]

19.  Loydi, A.; Lohse, K.; Otte, A.; Donath, T.W.; Eckstein, R.L. Distribution and effects of tree leaf litter on vegetation composition and biomass in a forest–grassland ecotone. *J. Plant Ecol.* **2014**, *7*, 264–275. [CrossRef]

20.  Facelli, J.M.; Carson, W.P. Heterogeneity of plant litter accumulation in successional communities. *Bull. Torrey Bot. Soc.* **1991**, *118*, 62–66. [CrossRef]

21.  Bonet, A. Secondary succession of semi-arid mediterranean old-fields in south-eastern spain: Insights for conservation and restoration of degraded lands. *J. Arid Environ.* **2004**, *56*, 213–233. [CrossRef]

22.  Walck, J.L.; Baskin, J.M.; Baskin, C.C. Ecology of the endangered species *Solidago shortii*. Vi. Effects of habitat type, leaf litter, and soil type on seed germination. *J. Torrey Bot. Soc.* **1999**, *126*, 117–123. [CrossRef]

23.  Becerra, P.I.; Celis-Diez, J.L.; Bustamante, R.O. Effects of leaf litter and precipitation on germination and seedling survival of the endangered tree *Beilschmiedia miersii*. *J. Veg. Sci.* **2004**, *7*, 253–257. [CrossRef]

24.  Solly, E.F.; Schöning, I.; Boch, S.; Kandeler, E.; Marhan, S.; Michalzik, B.; Müller, J.; Zscheischler, J.; Trumbore, S.E.; Schrumpf, M. Factors controlling decomposition rates of fine root litter in temperate forests and grasslands. *Plant Soil* **2014**, *382*, 203–218. [CrossRef]

25.  Liao, J.H.; Wang, H.H.; Tsai, C.C.; Hseu, Z.Y. Litter production, decomposition and nutrient return of uplifted coral reef tropical forest. *For. Ecol. Manag.* **2006**, *235*, 174–185. [CrossRef]

26.  Ma, Y.; Filley, T.R.; Szlavecz, K.; McCormick, M.K. Controls on wood and leaf litter incorporation into soil fractions in forests at different successional stages. *Soil Biol. Biochem.* **2014**, *69*, 212–222. [CrossRef]

27.  Kazakou, E.; Violle, C.; Roumet, C.; Pintor, C.; Gimenez, O.; Garnier, E. Litter quality and decomposability of species from a mediterranean succession depend on leaf traits but not on nitrogen supply. *Ann. Bot.* **2009**, *104*, 1151–1161. [CrossRef] [PubMed]

28.  McLaren, J.R.; Turkington, R. Plant identity influences decomposition through more than one mechanism. *PLoS ONE* **2011**, *6*, e23702. [CrossRef] [PubMed]

29.  Scherer-Lorenzen, M. Functional diversity affects decomposition processes in experimental grasslands. *Funct. Ecol.* **2008**, *22*, 547–555. [CrossRef]

30.  Laiho, R.; Prescott, C.E. Decay and nutrient dynamics of coarse woody debris in northern coniferous forests: A synthesis. *Can. J. For. Res.* **2004**, *34*, 763–777. [CrossRef]

31.  Quested, H.; Eriksson, O. Litter species composition influences the performance of seedlings of grassland herbs. *Funct. Ecol.* **2006**, *20*, 522–532. [CrossRef]

32.  Yilmaz, H.; Yilmaz, O.Y.; Akyuz, Y.F. Determining the factors affecting the distribution of *muscari latifolium*, an endemic plant of turkey, and a mapping species distribution model. *Ecol. Evol.* **2017**, *7*, 1112–1124. [CrossRef] [PubMed]

33.  Walker, B.; Kinzig, A.; Landgridge, J. Plant attribute diversity, resilience, and ecosystem function: The nature and significance of dominant and minor species. *Ecosystems* **1999**, *2*, 95–113. [CrossRef]

34.  Jain, M.; Flynn, D.F.; Prager, C.M.; Hart, G.M.; Devan, C.M.; Ahrestani, F.S.; Palmer, M.I.; Bunker, D.E.; Knops, J.M.; Jouseau, C.F.; et al. The importance of rare species: A trait-based assessment of rare species contributions to functional diversity and possible ecosystem function in tall-grass prairies. *Ecol. Evol.* **2014**, *4*, 104–112. [CrossRef] [PubMed]

35.  Jobidon, R.; Cyr, G.; Thiffault, N. Plant species diversity and composition along an experimental gradient of northern hardwood abundance in *picea mariana* plantations. *For. Ecol. Manag.* **2004**, *198*, 209–221. [CrossRef]

36.  Nakamura, T. Ecological relationships between seedling emergence and litter cover in the earliest stage of plant succession on sandy soil. *Ecol. Res.* **1996**, *11*, 105–110. [CrossRef]

37. Spooner, P.G.; Lunt, I.D.; Briggs, S.V.; Freudenberger, D. Effects of soil disturbance from roadworks on roadside shrubs in a fragmented agricultural landscape. *Biol. Conserv.* **2004**, *117*, 393–406. [CrossRef]

38. Agherkakli, B.; Najafi, A.; Sadeghi, S.H. Ground based operation effects on soil disturbance by steel tracked skidder in a steep slope of forest. *J. For. Sci.* **2010**, *56*, 278–284.

39. He, H.; Li, S.; Sun, H.; Yang, T. Environmental factors of road slope stability in mountain area using principal component analysis and hierarchy cluster. *Environ. Earth Sci.* **2011**, *62*, 55–59. [CrossRef]

40. Bansal, S.; Sheley, R.L.; Blank, B.; Vasquez, E.A. Plant litter effects on soil nutrient availability and vegetation dynamics: Changes that occur when annual grasses invade shrub-steppe communities. *Plant Ecol.* **2014**, *215*, 367–378. [CrossRef]

41. Meier, C.L.; Bowman, W.D. Links between plant litter chemistry, species diversity, and below-ground ecosystem function. *Proc. Natl. Acad. Sci. USA* **2008**, *105*, 19780–19785. [CrossRef] [PubMed]

42. Facelli, J.M.; Pickett, S.T.A. Plant litter—Its dynamics and effects on plant community structure. *Bot. Rev.* **1991**, *57*, 1–32. [CrossRef]

43. Grime, J.P. Benefits of plant diversity to ecosystems: Immediate, filter and founder effects. *J. Ecol.* **1998**, *86*, 902–910. [CrossRef]

44. Eppinga, M.B.; Kaproth, M.A.; Collins, A.R.; Molofsky, J. Litter feedbacks, evolutionary change and exotic plant invasion. *J. Ecol.* **2011**, *99*, 503–514. [CrossRef]

45. Ehrenfeld, J.G.; Ravit, B.; Elgersma, K. Feedback in the plant-soil system. *Ann. Rev. Environ. Res.* **2005**, *30*, 75–115. [CrossRef]

46. Xiong, S.J.; Nilsson, C. The effects of plant litter on vegetation: A meta-analysis. *J. Ecol.* **1999**, *87*, 984–994. [CrossRef]

47. Dick, D.A.; Gilliam, F.S. Spatial heterogeneity and dependence of soils and herbaceous plant communities in adjacent seasonal wetland and pasture sites. *Wetlands* **2007**, *27*, 951–963. [CrossRef]

48. Tang, F.K.; Cui, M.; Lu, Q.; Liu, Y.G.; Guo, H.Y.; Zhou, J.X. Effects of vegetation restoration on the aggregate stability and distribution of aggregate-associated organic carbon in a typical karst gorge region. *Solid Earth* **2016**, *7*, 141–151. [CrossRef]

49. James, J.J. Leaf nitrogen productivity as a mechanism driving the success of invasive annual grasses under low and high nitrogen supply. *J. Arid Environ.* **2008**, *72*, 1775–1784. [CrossRef]

50. Hossain, M.; Siddique, M.R.H.; Rahman, M.S.S.; Hossain, M.Z.; Hassan, M.M. Nutrient dynamics associated with leaf litter decomposition of three agroforestry tree species (*Azadirachta indica*, *Dalbergia sissoo*, and *Melia azedarach*) of bangladesh. *J. For. Res.* **2011**, *22*, 577–582. [CrossRef]

51. Ghestem, M.; Cao, K.; Ma, W.; Rowe, N.; Leclerc, R.; Gadenne, C.; Stokes, A. A framework for identifying plant species to be used as 'ecological engineers' for fixing soil on unstable slopes. *PLoS ONE* **2014**, *9*, e95876. [CrossRef] [PubMed]

52. Gaston, K.J. The importance of being rare. *Nature* **2012**, *487*, 46–47. [CrossRef] [PubMed]

53. Leitão, R.P.; Zuanon, J.; Villeger, S.; Williams, S.E.; Baraloto, C.; Fortunel, C.; Mendonca, F.P.; Mouillot, D. Rare species contribute disproportionately to the functional structure of species assemblages. *Proc. R. Soc. B Biol. Sci.* **2016**, *283*, 20160084. [CrossRef] [PubMed]

54. Mouillot, D.; Bellwood, D.R.; Baraloto, C.; Chave, J.; Galzin, R.; Harmelin-Vivien, M.; Kulbicki, M.; Lavergne, S.; Lavorel, S.; Mouquet, N.; et al. Rare species support vulnerable functions in high-diversity ecosystems. *PLoS Biol.* **2013**, *11*, e1001569. [CrossRef] [PubMed]

55. Gascon, C.; Brooks, T.M.; Contreras-MacBeath, T.; Heard, N.; Konstant, W.; Lamoreux, J.; Launay, F.; Maunder, M.; Mittermeier, R.A.; Molur, S.; et al. The importance and benefits of species. *Curr. Biol.* **2015**, *25*, R431–R438. [CrossRef] [PubMed]

56. O'Sullivan, O.S.; Holt, A.R.; Warren, P.H.; Evans, K.L. Optimising uk urban road verge contributions to biodiversity and ecosystem services with cost-effective management. *J. Environ. Manag.* **2017**, *191*, 162–171. [CrossRef] [PubMed]

57. Xie, Z. Characteristics and conservation priority of threatened plants in the yangtze valley. *Biodivers. Conserv.* **2003**, *12*, 65–72. [CrossRef]

58. Karim, M.N.; Mallik, A.U. Roadside revegetation by native plants: I. Roadside microhabitats, floristic zonation and species traits. *Ecol. Eng.* **2008**, *32*, 222–237. [CrossRef]

59. Hua, K.; Zhu, B.; Wang, X. Dissolved organic carbon loss fluxes through runoff and sediment on sloping upland of purple soil in the Sichuan Basin. *Nutr. Cycl. Agroecosyst.* **2014**, *98*, 125–135. [CrossRef]

60.  Wang, C.; Li, S.; Xia, J.; Wu, J. Clay composition and fertilizer conservation property of red soil and yellow soils in the southwest. *J. Sichuan Agric. Univ.* **1996**, *2*, 211–218.

61.  Li, Q.; Ma, M.; Liu, Y.; Ding, H.; Chen, M.; Chen, Y. Study on soil carbon and nutrients pools of several evergreen broad-leaved forest types in northwest sichuan. *J. Soil Water Conserv.* **2007**, *6*, 114–125.

62.  Peng, W.; Wang, G.; Zhou, J.; Zhao, J.; Yang, C. Studies on the temporal and spatial variations of urban expansion in chengdu, western china, from 1978 to 2010. *Sustain. Cities Soc.* **2015**, *17*, 141–150. [CrossRef]

63.  Jennings, S.B.; Brown, N.D.; Sheil, D. Assessing forest canopies and understorey illumination: Canopy closure, canopy cover and other measures. *Forestry* **1999**, *72*, 59–74. [CrossRef]

64.  Yang, Y.S.; Guo, J.F.; Chen, G.S.; Xie, J.S.; Cai, L.P.; Lin, P. Litterfall, nutrient return, and leaf-litter decomposition in four plantations compared with a nautral forest in subtropical china. *Ann. For. Sci.* **2004**, *61*, 465–476. [CrossRef]

65.  McCune, B.; Grace, J.B. *Principal Components Analysis*; MjM Software Design: Gleneden Beach, OR, USA, 2002; pp. 114–121.

# An Improved Syringe Agroinfiltration Protocol to Enhance Transformation Efficiency by Combinative Use of 5-Azacytidine, Ascorbate Acid and Tween-20

**Huimin Zhao [1], Zilong Tan [2], Xuejing Wen [2] and Yucheng Wang [1,2,\*]**

[1]    State Key Laboratory of Tree Genetics and Breeding, Northeast Forestry University, Harbin 150040, China; huimin0230@163.com

[2]    Key Laboratory of Biogeography and Bioresource in Arid Land, Xinjiang Institute of Ecology and Geography, Chinese Academy of Sciences, Urumqi 830011, China; smilebaobao@outlook.com (Z.T.); wxj-329@163.com (X.W.)

\*    Correspondence: wangyucheng@ms.xjb.ac.cn

Academic Editor: Milan S. Stankovic

**Abstract:** Syringe infiltration is an important transient transformation method that is widely used in many molecular studies. Owing to the wide use of syringe agroinfiltration, it is important and necessary to improve its transformation efficiency. Here, we studied the factors influencing the transformation efficiency of syringe agroinfiltration. The pCAMBIA1301 was transformed into *Nicotiana benthamiana* leaves for investigation. The effects of 5-azacytidine (AzaC), Ascorbate acid (ASC) and Tween-20 on transformation were studied. The β-glucuronidase (*GUS*) expression and GUS activity were respectively measured to determine the transformation efficiency. AzaC, ASC and Tween-20 all significantly affected the transformation efficiency of agroinfiltration, and the optimal concentrations of AzaC, ASC and Tween-20 for the transgene expression were identified. Our results showed that 20 μM AzaC, 0.56 mM ASC and 0.03% (*v/v*) Tween-20 is the optimal concentration that could significantly improve the transformation efficiency of agroinfiltration. Furthermore, a combined supplement of 20 μM AzaC, 0.56 mM ASC and 0.03% Tween-20 improves the expression of transgene better than any one factor alone, increasing the transgene expression by more than 6-fold. Thus, an optimized syringe agroinfiltration was developed here, which might be a powerful method in transient transformation analysis.

**Keywords:** ascorbate acid; 5-azacytidine; syringe agroinfiltration; transformation efficiency; Tween-20

## 1. Introduction

Genetic transformation is a powerful method used in a variety of molecular studies, such as gene function analysis, protein production, protein–protein interaction and promoter activity. There are two kinds of genetic transformation: stable transformation and transient transformation. Stable transformation is a labor-intensive low-throughput process. Comparatively, transient transformation is an easy and efficient method for gene transformation, which avoids the drawbacks of the stable transformation process, such as transformation efficiency, transformants selection, and regeneration [1]. Additionally, transient expression represents a rapid method to analyze the function of certain genes, and the analysis can be completed within a few days of transformation [2]. Among the transient transformation methods, agroinfiltration is a simple, rapid, versatile and widely used technique. "Syringe infiltration" is the most popular method for agroinfiltration, which is a simple procedure where a needleless syringe is used to introduce *Agrobacterium* into plant leaves, with no need for specialized equipment [3].

Syringe infiltration has many advantages, and can either transfer one target gene into the whole leaf, or introduce multiple genes into different areas of one leaf, allowing multiple assays to be conducted on a single leaf [4]. These advantages of syringe infiltration have led to its application in a wide range of studies, such as those concerned with transgenic complementation, promoter analysis, plant pathogen interaction study, abiotic stress tolerance assay, protein production, gene functional analysis with transient overexpressing or silencing assay, protein–protein interactions and protein localization assays [5–8].

As an important technique, some studies have been performed to improve the transformation efficiency of syringe agroinfiltration. For instance, to defend against viral attack, plant species have developed a post-transcriptional gene silencing (PTGS) system that produces small interfering RNAs to silence non-native genes. Meanwhile, many viruses have developed a system, such as tomato bushy stunt virus (TBSV) p19, which could interfere with the plant [9]. Therefore, *Agrobacterium* cells carrying the p19 protein are frequently co-infiltrated with the *Agrobacterium* carrying the construct of the gene of interest to improve the transgenic transformation efficiency. Dugdale et al. [6] provide a protocol for the design and construction of a split-gene in-plant activation (INPACT). INPACT enables the expression of recombinant proteins at up to 10% of total soluble protein in the leaf within 6 to 9 months. Wroblewski et al. [2] developed a protocol for efficient transient transformation of lettuce, Arabidopsis and tobacco. They found that *Agrobacterium tumefaciens* strain C58C1 did not elicit a necrotic response in plants and was the best strain for these plant species. Similarly, *A. tumefaciens* strain 1D1246 was found to provide high transient expression levels in solanaceous plants without a necrotic response, enabling routine transient expression in solanaceous species. Fujiuchi et al. [10] found that residual water from bacterial suspension in the intercellular space of detached leaves could significantly reduce the yield of recombinant protein expression in the syringe agroinfiltration process, and removal of bacterial suspension water in detached leaves after agroinfiltration significantly improved recombinant protein expression.

Methylation of DNA is found to be involved in the gene expression regulation, and there is an inverse correlation between the methylation level and the transcriptional activity of a gene [11,12]. In addition, DNA methylation level is closely associated with genetic transformation efficiency and transgene expression [12–14]. AzaC plays a role in reducing DNA methylation, and could increase the expression of transgenes by decreasing DNA methylation of the transgene [11,15,16]. Therefore, does AzaC also play a positive role in agroinfiltration transformation efficiency? This deserves to be further studied. The genetic transformation of plants mediated by *A. tumefaciens* is a type of pathogenic infection, which normally induces an oxidative burst, with rapid and transient production of reactive oxygen species (ROS) [17]. Excess ROS could be sufficiently toxic to both plant cells and also attack Agrobacterium cells, reducing the efficiency of transformation [17]. Therefore, scavenging ROS is important for genetic transformation, and the antioxidants, such as Ascorbate acid (ASC), may also be involved in enhancing agroinfiltration transformation efficiency. Neutral surfactants, such as Tween-20, Triton X-100 and Silwet L-77, play important roles in reducing surface tension and enhancing the entry of bacteria into plant tissues, and are usually used in genetic transformation, including floral dip transformation or other vacuum infiltration transformation methods [18,19]. However, it is still not known whether it plays a role in enhancing transformation efficiency in syringe agroinfiltration.

Previously, we had built a transient transformation method based on the Agrobacterium mediated method [20,21]. When improving the transformation efficiency of that method, we found that AzaC, ASC and Tween-20 all play roles in enhancing the expression of transgene. Therefore, we supposed that these reagents might also play positive roles in syringe agroinfiltration, and they were further studied in the present study. Our studies showed that 5-azacytidine (AzaC), Ascorbate acid (ASC) and Tween-20 all affect the transformation efficiency of syringe agroinfiltration, and can improve the expression of transgene at a certain level. Furthermore, we found that these factors could coordinately work on genetic transformation. The improvement of syringe infiltration was performed, which will be useful in transient transformation study.

## 2. Results and Discussion

### 2.1. The Effects of AzaC on Transformation Efficiency

Different concentrations of AzaC were used for study of the transformation efficiency. Both *GUS* expression and GUS activity measurements showed that AzaC from 10 to 30 μM increased the transformation efficiency, but 20 μM of AzaC increased both the *GUS* expression and GUS activity highest (Figure 1). Therefore, although AzaC at 10–30 μM can affect the expression of transgene, there is an optimal concentration of AzaC for improving the transformation efficiency, and the medium concentration (20 μM) of AzaC improves transformation efficiency more highly than the low (10 μM) or high (30 μM) concentration of AzaC did.

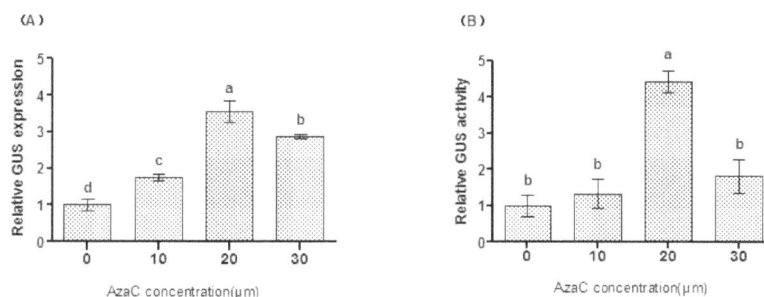

**Figure 1.** Effects of 5-azacytidine (AzaC) on the transformation efficiency. The expression (**A**) and activity (**B**) of the β-glucuronidase (*GUS*) transgene were analyzed to determine syringe agroinfiltration efficiency. (**A**) *GUS* expression analysis to study the transformation efficiency. The expression of *GUS* in the control (no AzaC supplement) was used as a calibrator to normalize the expression of *GUS* at different concentrations of AzaC (the ratio of the expression of *GUS* at different concentrations of AzaC was divided by the *GUS* expression without AzaC supplement); (**B**) GUS activity analysis to determine the effects of AzaC on syringe agroinfiltration efficiency. GUS activity without AzaC was used as a calibrator to normalize the results of different AzaC levels (the ratio of GUS activity at different concentrations of AzaC was divided by the *GUS* expression without AzaC supplement). a, b, c, d means not sharing a common superscript differ significantly according to ANOVA Tukey's test ($p < 0.05$).

Previous studies showed that AzaC inhibits and decreases DNA methylation, and treatment with AzaC was found to increase the expression of transgenes by reducing methylation in transferred T-DNA [11,14–16]. Additionally, AzaC also inhibits the inactivation of transgene expression. For example, kanamycin-resistant transgenic plants usually lose this resistance over time. However, treatment with AzaC could restore kanamycin-resistance activity and improve the growth of neomycin phosphotransferase-negative plants in the presence of kanamycin [15]. These results indicated that DNA methylation status is quite important in genetic transformation.

Supplement of 20 μM of AzaC significantly increased the transient transformation efficiency (Figure 1), suggesting that AzaC plays an important role in increasing the efficiency of genetic transformation. Previously, Palmgren et al. [12] showed that *Agrobacterium* cells treated with AzaC prior to transformation showed increased transient expression efficiency, which relies on the hypothesis that methylated Agrobacterium DNA will reduce its infection capability, and DNA demethylation could increase T-DNA transformation. Additionally, AzaC could inhibit the methylation-dependent inactivation of the reporter gene in the cells [12]. Therefore, this increased transformation efficiency at 20 μM of AzaC might be the reason that AzaC treatment demethylated the T-DNA, leading to increased transgene expression; and/or AzaC also demethylated Agrobacterium DNA that enhanced its infection capability. Our results also showed that a high level of AzaC decreased the expression of the transgene (Figure 1), perhaps because high AzaC is toxic to plant cells and/or cells of *A. tumefaciens*, which will reduce the transformation efficiency.

## 2.2. ASC Significantly Affects the Expression of Transgene

GUS activity and expression analysis both indicated that ASC at the concentrations from 0.28 to 1.68 mM all could significantly increase the expression of transgene (Figure 2A), but 2.24 mM of ASC did not affect transformation efficiency (Figure 2A,B). In addition, both *GUS* expression and activity assay showed that there was an optimal concentration for ASC to increase the expression of transgene (Figure 2). The optimum concentration of ASC for increasing the expression of transgene is 0.56 mM, and low or high ASC concentrations were not best for increasing the expression of transgene. These results indicated that the supply of ASC could significantly improve syringe agroinfiltration efficiency at certain concentrations.

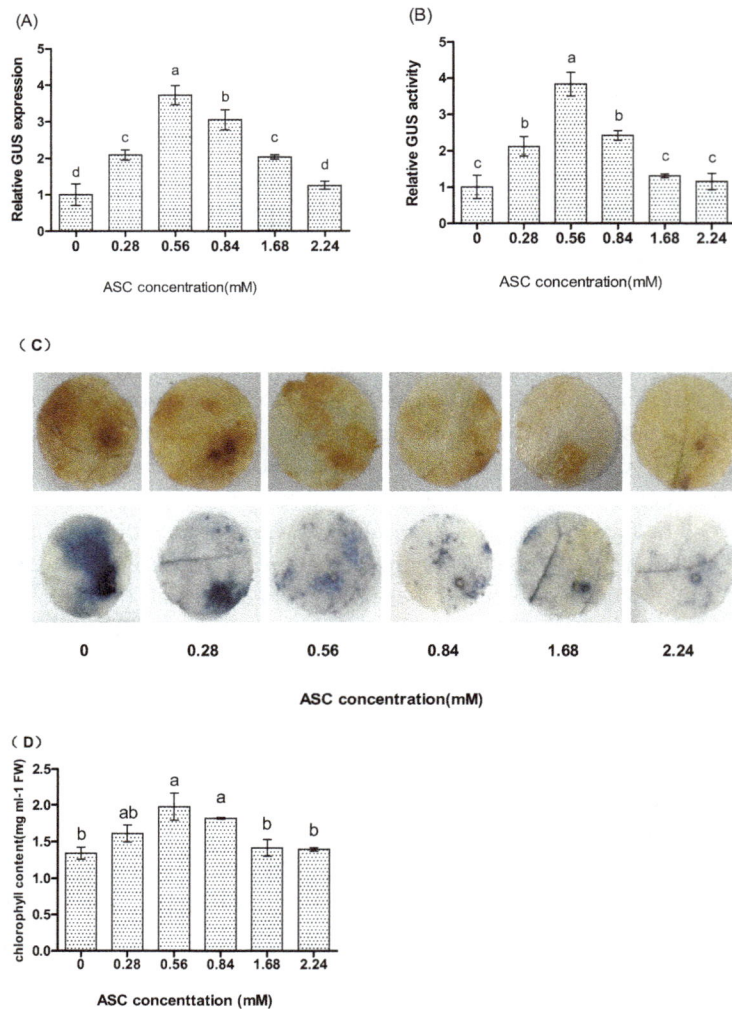

**Figure 2.** The effects of Ascorbate acid (ASC) on transformation efficiency by monitoring the expression of the *GUS* reporter gene (**A**) and GUS activity (**B**). (**A**) *GUS* expression analysis to study the transformation efficiency. *GUS* expression at different levels of ASC treatment was normalized by that in the control experiment (no ASC was supplied); (**B**) Determination of the effects of ASC on syringe agroinfiltration efficiency by GUS activity analysis. The activity of GUS in the control (without ASC supplement) experiment was used to normalize the activity of GUS at different concentrations of ASC. a, b, c, d means not sharing a common superscript differ significantly according to ANOVA Tukey's test ($p < 0.05$); (**C**) Analysis of reactive oxygen species (ROS) scavenging by supply of ASC. Detection of ROS by 3-diaminobenzidine (DAB) and nitroblue tetrazolium (NBT) staining that respectively indicate the level of $H_2O_2$ and $O^{2-}$; (**D**) Analyses of chlorophyll contents in the infiltrated leaves supplied with different concentrations of ASC.

Syringe agroinfiltration in this study is a kind of pathogenic infection in tobacco plants, which will lead to an oxidative burst, with rapid and transient production of reactive oxygen species (ROS) [17]. Excess ROS will cause tissue/cell necrosis, leading to inhibiting regeneration of the transformed cells/tissues, and will inhibit the potential of *Agrobacterium* to colonize plant cells and transfer T-DNA [22,23]. Moreover, the generated ROS could be sufficiently toxic to kill the attacking *Agrobacterium* directly, preventing *Agrobacterium* from transferring T-DNA into plants during attempted transfection [17]. All these will have a negative effect on the transformation efficiency. Therefore, it is important to scavenge the excess ROS during the syringe agroinfiltration process.

In the present study, we found that the addition of ASC to a certain level could increase the expression and activity of the transformed *GUS* gene significantly during agroinfiltration (Figure 2). For investigating whether ASC increases the transformation efficiency through scavenging ROS, nitroblue tetrazolium (NBT) and 3-diaminobenzidine (DAB) staining were performed. The result showed that $O^{2-}$ and $H_2O_2$ were both reduced, accompanied with the increasing of ASC level (Figure 2C), which reflects the protective effect of ASC against oxidative damage caused by excess ROS accumulation. Abiotic stress will induce the breakdown of chlorophyll, and therefore chlorophyll content can be used as an indicator of plant damage [24]. We further determined the chlorophyll contents in the infiltrated tobacco leaves. The results showed that the infiltrated leaves supplied with 0.28–0.84 mM ASC can retain higher chlorophyll contents than the infiltrated leaves without ASC supplied (Figure 2D). However, the leaves supplied with higher concentrations of ASC (1.68 and 2.24 mM) had similar chlorophyll contents to the leaves without ASC supplied (Figure 2D). This might be the reason why ASC at low and medium concentrations can protect plant cells by reducing excess ROS and is not toxic to plants, resulting in the reduced chlorophyll breakdown. Although a high level of ASC can effectively scavenge ROS, high ASC level is toxic to plant cells, which finally leads to decreased chlorophyll contents. Therefore, these results together might suggest that only a moderate ASC level (0.56 mM) can most highly improve transformation efficiency (Figure 2A,B), which effectively scavenges ROS and is not toxic to plant cells (Figure 2C,D); on the contrary, high ASC level (1.68 mM or more) can effectively reduce ROS accumulation (Figure 2C), but it will be toxic to plant cells, resulting in the breakdown of chlorophyll (Figure 2D), which will cancel out the reduced ROS accumulation, and fail to enhance the transformation efficiency (Figure 2A,B).

## 2.3. Tween-20 Could Increase the Transformation Efficiency of Syringe Agroinfiltration

The influence of Tween-20 on syringe agroinfiltration efficiency was studied. Tween-20 at 0.015% to 0.03% (*v/v*) significantly increased *GUS* expression and activity (Figure 3), indicating that Tween-20 affects the efficiency of syringe agroinfiltration. In particular, 0.03% Tween-20 could highly increase the transformation efficiency. However, concentrations of Tween-20 at 0.06% failed to increase the efficiency of syringe agroinfiltration (Figure 3).

Neutral surfactants such as Tween-20, Triton X-100 and Silwet L-77 are usually used in genetic transformation, because they play important roles in reducing surface tension and enhancing the entry of bacteria into plant tissues [18]. In the present study, the supply of Tween-20 in syringe infiltration improved the transformation efficiency (Figure 3), indicating that Tween-20 can also be used in improving the transformation efficiency. Therefore, Tween-20 in agroinfiltration might increase transformation efficiency through reducing the surface tension, which could enhance the entry of bacteria into plant tissues as shown by the findings of Clough and Bent [18]. However, 0.06% Tween-20 failed to enhance the expression of transgene, and this may be due to the fact that a high level of Tween-20 will damage the plant cells, leading to decreased transformation.

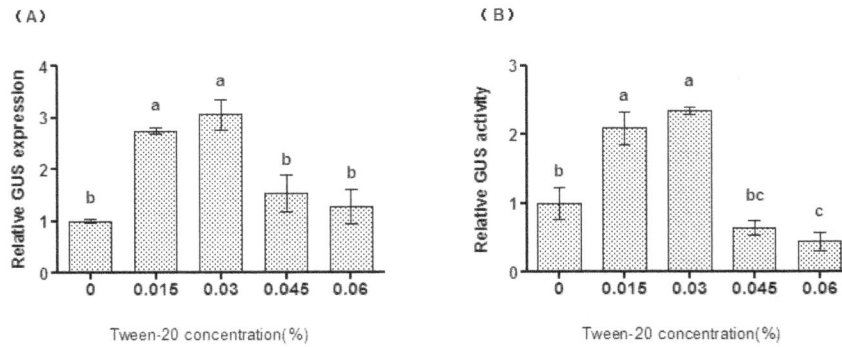

**Figure 3.** The effects of Tween-20 on syringe agroinfiltration efficiency by determination of *GUS* expression (**A**) and GUS activity (**B**). (**A**) Agroinfiltration efficiency analyzed by *GUS* expression. The expression of *GUS* in the control experiment (no Tween-20 was supplied) was used to normalize the expression of *GUS* at different concentrations of Tween-20; (**B**) The effects of Tween-20 on agroinfiltration efficiency were determined by GUS activity analysis. GUS activity at different levels of Tween-20 treatments was normalized by the level in the control. a, b, c means not sharing a common superscript differ significantly according to ANOVA Tukey's test ($p < 0.05$).

## 2.4. A Combination of AzaC, ASC and Tween-20 Highly Improves Transformation Efficiency

The above studies showed that AzaC, ASC and Tween-20 could improve syringe agroinfiltration efficiency at certain concentrations (20 μM of AzaC, 0.56 mM of ASC and 0.03% Tween-20). To study whether a combination of these factors could improve the transformation efficiency further, 20 μM AzaC, 0.56 mM ASC and 0.03% Tween-20 were supplied together in the infiltration buffer. Both GUS activity and *GUS* expression analyses showed that these factors together could improve the syringe agroinfiltration efficiency to a greater extent than any of the factors supplied alone (Figure 4), suggesting that AzaC, ASC and Tween-20 could increase the efficiency of syringe agroinfiltration synergistically.

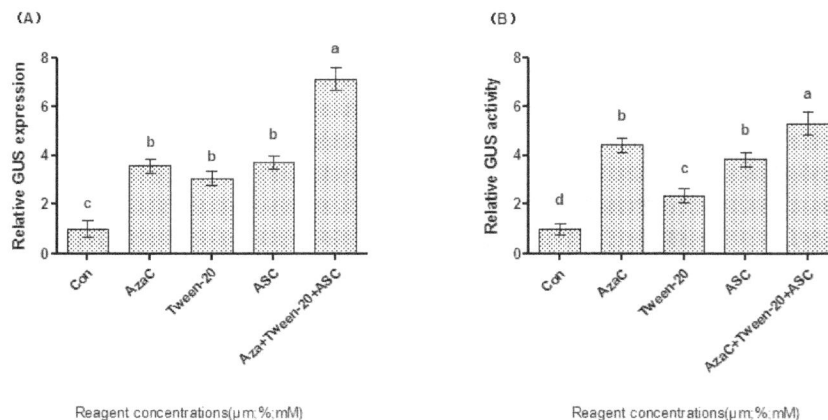

**Figure 4.** Analysis of the effects of Tween-20, ASC and AzaC together on syringe agroinfiltration efficiency by determination of *GUS* expression (**A**) and GUS activity (**B**). (**A**) Analysis of the effects of 0.03% Tween-20, 20 μM ASC and 0.56 mM AzaC separately or together on syringe agroinfiltration efficiency by *GUS* expression. The expression of *GUS* supplied with Tween-20, ASC and AzaC together was normalized by the expression of *GUS* from the control experiment (no AzaC, ASC and Tween-20 supplied); (**B**) Determination of the effects of AzaC, ASC and Tween-20 together on syringe agroinfiltration efficiency using GUS activity analysis. GUS activity, when supplied with AzaC, ASC and Tween-20 together, was normalized by the activity in the control experiment. a, b, c, d means not sharing a common superscript differ significantly according to ANOVA Tukey's test ($p < 0.05$).

From the above studies, we can see that although the change profiles of *GUS* expression and GUS activities are similar, there were also some differences between them. This should be due to the following reasons. One is that GUS is a very stable enzyme and this might mask some of the differences in gene expression; the other is that GUS activity reflects the post-transcriptional translation of the *GUS* gene that is not the reflection of transcripts of GUS directly.

## 3. Materials and Methods

### 3.1. Plant Materials and Growth Conditions

The seeds of *N. benthamiana* were planted into the pots containing the mixture of sands and soil (2:1) in a greenhouse under the conditions of 70%–75% relative humidity, 16 h light/8 h darkness photocycle at 25 °C. The 4-week-old plants were used for agroinfiltration. The plasmid pCAMBIA1301 that harbors a *GUS* gene under the control of 35S CaMV promoter was transformed into *Agrobacterium tumefaciens* (EHA105) and was used for the syringe agroinfiltration study.

### 3.2. Infiltration Procedures

Syringe agroinfiltration was performed according to the method of Broghammer et al. [25] as follows: The *Agrobacterium* strain harboring pCAMBIA1301 was cultured in lysogeny broth (LB) medium containing kanamycin (50 mg/L) and rifampicin (100 mg/L) with rotation at 200 rpm at 28 °C. After the culture reached an optical density at 600 nm ($OD_{600}$) of 0.8, the culture was diluted 50-fold with fresh LB medium, and was cultured at 200 rpm and 28 °C until the $OD_{600}$ reached 0.6. The *Agrobacterium* cells were centrifuged at $3500 \times g$ for collection, resuspended in infiltration buffer (10 mM $MgCl_2$, 10 mM MES, 150 µM acetosyringone, pH = 5.6), adjusted to an $OD_{600}$ of 0.3 and used for syringe agroinfiltration. The mixture (200 µL) was infiltrated into *N. benthamiana* leaves using a needless syringe. The infiltrated regions in leaves were harvested after infiltration for 72 h and used for study. The above procedure was the standard syringe agroinfiltration method, which was used for the following studies.

### 3.3. Factors Influencing Syringe Infiltration

To improve the transformation efficiency of syringe infiltration, some factors that might affect the transformation efficiency were investigated. We use the infiltration buffer (10 mM $MgCl_2$, 10 mM MES, 150 µM acetosyringone, pH = 5.6) as control, and different concentrations of AzaC, ASC and Tween-20 were added in the infiltration buffer for study their effects. To determine the influence of AzaC on syringe agroinfiltration efficiency, 10, 20 and 30 µM of AzaC were added to the infiltration buffer. For determination of the effects of ASC, the stock liquid of ASC (100 mM) was added into the infiltration buffer to reach the concentration at 0.28, 0.56, 0.84, 1.68 and 2.24 mM. To determine the effect of Tween-20 on syringe agroinfiltration efficiency, 0.015, 0.03, 0.045 and 0.06% (*v/v*) was supplied in the infiltration buffer. The varying concentrations of these three factors in the infiltration buffer were studied for their effects on agroinfiltration efficiency. To ensure the reliability in experiments, the following measures were performed. To eliminate the physiological difference among the different leaves, the experiment respectively had its control experiment, and the experiment and its control were performed respectively at two sides of the main vein within one leaf.

To reduce the variations of pCAMBIA1301 amount injected by syringe, the same amount of Agrobacterium cells (i.e., the same OD concentration and same volume, which is shown as Infiltration Procedures) in each experiment had been used. Three independent experiments were performed, and each experiment contains at least five leaves.

### 3.4. Determination of β-Glucuronidase (GUS) Activity

GUS activity expressed from pCAMBIA1301 was determined according to the method of Jefferson et al. [26]. In brief, leaves were ground into a fine powder in liquid nitrogen and homogenized in extraction buffer (50 mM $NaH_2PO_4$-$Na_2HPO_4$, pH 7.0, 10 mM ethylenediaminetetraacetic acid (EDTA), 10 mM β-mercaptoethanol, 0.1% Triton X-100, 0.1% sodium lauryl sarcosine). Enzyme reactions were performed in extraction buffer supplied with 1 mM 4-methylumbelliferyl-β-D-glucuronide (MUG) at 37 °C and the reaction was stopped by adding 450 µL of 0.2 M $Na_2CO_3$. The fluorescence of 4-methylumbelliferone was monitored using a DyNA Quant fluorometer (Hoefer Pharmacia, San Francisco, CA, USA). A protein standard curve was generated by the Bradford assay.

### 3.5. Quantitative Reverse Transcription PCR (qRT-PCR)

Total RNA was isolated from each sample using the Trizol reagent (Promega), and then treated with DNaseI to remove any DNA contamination. About 1 µg of total RNA was reverse-transcribed into cDNA with oligo deoxythymidine primers using PrimeScript™ RT reagent Kit (Takara, Dalian, China) in a reaction volume of 10 µL, which was subsequently diluted to 100 µL and used as the template for real-time PCR. The *Actin* (GenBank number: AB158612.1) and *α-tubulin* (GenBank number: AB052822.1) genes were used as internal references (see Additional file 1: Table S1 for primers used). The mean Ct of the two internal references was used to normalize the cDNA amount in each sample. Real-time PCR was carried out with an Opticon 2 System (Bio-Rad, Richmond, CA, USA). The reaction mixture contained 10 µL of Synergy Brands (SYBR) Green Real-time PCR Master Mix (Toyobo Co., Ltd., Osaka, Japan), 0.5 µM each of forward and reverse primers, and 2 µL of cDNA template (equivalent to the transcript from 20 ng of total RNA) in a total volume of 20 µL. PCR was performed with the following cycling parameters: 94 °C for 30 s; followed by 45 cycles at 94 °C for 12 s, 60 °C for 30 s, 72 °C for 40 s; and 1 s at 79 °C for plate reading. Melting curves were generated from the samples at the end of each run to assess the purity of the amplified products. Three independent biological replicates were performed, and expression levels were calculated from the cycle threshold according to the $2^{-\Delta\Delta Ct}$ method [27].

### 3.6. DAB and NBT Staining and Chlorophyll Content Assay

The infiltrated leaves were used for histochemical staining analysis. Infiltration of leaves with 3,3′-diaminobenzidine (DAB) or nitroblue tetrazolium (NBT), which allowed the detection of hydrogen peroxide and superoxide respectively, was performed following the method of Fryer et al. [28]. To measure chlorophyll contents, the infiltrated leaves were ground into fine powder under liquid nitrogen, added with 2 mL of 95% ethanol, incubated in the dark until the material became white, and centrifuged with 12,000 $g$ for 2 min. The supernatant was measured at 663 nm and 646 nm, using 95% ethanol as a blank. The concentration (mg/L) of chlorophyll a and b respectively was calculated according to the following formula. Ct: total chlorophyll concentration. $Ca = 12.72\ D_{663} - 2.59\ D_{646}$, $Cb = 22.88\ D_{646} - 4.67\ D_{663}$, $Ct = Ca + Cb = 20.29\ D_{646} + 8.05\ D_{663}$ (Ca and Cb indicate the concentrations of chlorophyll a and b respectively; OD663 and OD645 are the absorbance at wavelength of 663 nm and 646 nm). The Chlorophyll content (mg/g) was calculated as: NCtV/W (where N is the dilution factor; Ct is the total concentration of chlorophyll a and b (mg/L); V is the volume of the extract (mL); W is the fresh weight or dry weight of the sample (g)).

### 3.7. Statistical Analyses

Statistical analyses were conducted using SPSS 22.0 (SPSS Inc., Chicago, IL, USA) software. Data were compared using ANOVA Tukey's test (factor analysis of variance). Significant differences from each treatment compared with the control were considered when $p$-value $< 0.05$.

## 4. Conclusions

As an important and widely use method, it is important to improve the transformation efficiency of syringe agroinfiltration. We studied the factors that significantly influenced the transformation of syringe agroinfiltration. pCAMBIA1301 is used in the transformation study, because the *GUS* gene of pCAMBIA1301 was under the control of the CaMV 35S constitutive promoter that can constantly express the *GUS* gene in different plant tissues at different growth stages; and GUS protein can only be expressed in the eukaryotic cell as introns were present in its sequence. Therefore, the use of pCAMBIA1301 could reflect the transformation efficiency accurately. AzaC, ASC and Tween-20 can all improve the genetic transformation significantly at certain levels. In addition, AzaC, ASC and Tween-20 used together could improve transformation efficiency to a greater extent than any of the three compounds used alone (Figure 4), indicating that these factors could work synergistically to improve the transformation efficiency. Therefore, the optimized conditions for syringe agroinfiltration include 20 µM AzaC, 0.56 mM ASC and 0.03% Tween-20 together, which increases the expression of the transgene by more than 6-fold.

Due to a RNA-dependent RNA polymerase (RDR1) mutation, *N. benthamiana* is known to be compromised in its ability to silence foreign nucleic acids, which may affect the extrapolation of these results in other species. However, as for the mechanism of these factors in enhancing transformation efficiency and their applications in other genetic transformation methods, AzaC, ASC and Tween-20 could also be used in other plant species. The optimal concentrations may be varied in different plant species, and these need to be optimized.

**Acknowledgments:** This work was supported by the project of the culture of the young scientists in scientific innovation in Xinjiang Province (2013711001), and the National Natural Science Foundation of China (No. 31270703).

**Author Contributions:** Huimin Zhao and Yucheng Wang conceived and designed the experiments; Huimin Zhao performed the experiments and analyzed the data; Zilong Tan and Xuejing Wen contributed reagents/materials/analysis tools; and Huimin Zhao and Yucheng Wang wrote the paper.

**Conflicts of Interest:** The authors declare no conflict of interest.

## References

1. Guidarelli, M.; Baraldi, E. Transient transformation meets gene function discovery: The strawberry fruit case. *Front. Plant Sci.* **2015**, *6*, 444. [CrossRef] [PubMed]
2. Wroblewski, T.; Tomczak, A.; Michelmore, R. Optimization of Agrobacterium-mediated transient assays of gene expression in lettuce, tomato and Arabidopsis. *Plant Biotech. J.* **2005**, *3*, 259–273. [CrossRef] [PubMed]
3. Santi, L.; Batchelor, L.; Huang, Z.; Hjelm, B.; Kilbourne, J.; Arntzen, C.J.; Chen, Q.; Mason, H.S. An efficient plant viral expression system generating orally immunogenic Norwalk virus-like particles. *Vaccine* **2008**, *26*, 1846–1854. [CrossRef] [PubMed]
4. Vaghchhipawala, Z.; Rojas, C.M.; Senthil-Kumar, M.; Mysore, K.S. Agroinoculation and agroinfiltration: Simple tools for complex gene function analyses. *Methods Mol. Biol.* **2011**, *678*, 65–76. [PubMed]
5. Chen, Q.; Lai, H.; Hurtado, J.; Stahnke, J.; Leuzinger, K.; Dent, M. Agroinfiltration as an Effective and Scalable Strategy of Gene Delivery for Production of Pharmaceutical Proteins. *Adv. Tech. Biol. Med.* **2013**. [CrossRef]
6. Dugdale, B.; Mortimer, C.L.; Kato, M.; James, T.; Harding, R.M.; Dale, J.L. Design and construction of an in-plant activation cassette for transgene expression and recombinant protein production in plants. *Nat. Protoc.* **2014**, *9*, 1010–1027. [CrossRef] [PubMed]
7. Yang, Y.; Li, R.; Qi, M. In vivo analysis of plant promoters and transcription factors by agroinfiltration of tobacco leaves. *Plant J.* **2000**, *22*, 543–551. [CrossRef] [PubMed]
8. Johansen, L.K.; Carrington, J.C. Silencing on the spot Induction and suppression of RNA silencing in the Agrobacteriummediated transient expression system. *Plant Physiol.* **2001**, *126*, 930–938. [CrossRef] [PubMed]
9. Takeda, A.; Sugiyama, K.; Nagano, H.; Mori, M.; Kaido, M.; Mise, K.; Tsuda, S.; Okuno, T. Identification of a novel RNA silencing suppressor, NSs protein of Tomato spotted wilt virus. *FEBS Lett.* **2002**, *532*, 75–79. [CrossRef]

10. Fujiuchi, N.; Matsuda, R.; Matoba, N.; Fujiwara, K. Removal of bacterial suspension water occupying the intercellular space of detached leaves after agroinfiltration improves the yield of recombinant hemagglutinin in a N benthamiana transient gene expression system. *Biotechnol. Bioeng.* **2016**, *113*, 901–906. [CrossRef] [PubMed]

11. Bochardt, A.; Hodal, L.; Palmgren, G.; Mattsson, O.; Okkels, F.T. DNA methylation is involved in maintenance of an unusual expression pattern of an introduced gene. *Plant Physiol.* **1992**, *99*, 409–414. [CrossRef] [PubMed]

12. Palmgren, G.; Mattson, O.; Okkels, F.T. Treatment of Agrobacterium or leaf disks with 5-azacytidine increases transgene expression in tobacco. *Plant Mol. Biol.* **1993**, *21*, 429–435. [CrossRef] [PubMed]

13. Jiang, J.; Wing, V.; Xie, T.; Shi, X.; Wang, Y.P.; Sokolov, V. DNA methylation analysis during the optimization of agrobacterium-mediated transformation of soybean. *Genetika* **2016**, *52*, 66–73. [CrossRef] [PubMed]

14. Zhen, Z.; Karen, W.H.; Leaf, H. Effects of 5-azacytidine on transformation and gene expression in *Nicotiana tabacum. Cell. Dev. Biol. Plant* **1991**, *27*, 77–83.

15. Christman, J.K. 5-Azacytidine and 5-aza-2′-deoxycytidine as inhibitors of DNA methylation: Mechanistic studies and their implications for cancer therapy. *Oncogene* **2002**, *21*, 5483–5495. [CrossRef] [PubMed]

16. Weber, H.; Ziechmann, C.; Graessmann, A. In vitro DNA methylation inhibits gene expression in transgenic tobacco. *EMBO J.* **1990**, *9*, 4409–4415. [PubMed]

17. Wojtaszek, P. Oxidative burst: An early plant response to pathogen infection. *Biochem. J.* **1997**, *322*, 681–692. [CrossRef] [PubMed]

18. Dan, Y. Biological functions of antioxidants in plant transformation. *Cell. Dev. Biol. Plant* **2008**, *44*, 149–161. [CrossRef]

19. Kuta, D.D.; Tripathi, L. Agrobacterium-mduced hypersensitive necrotic reaction in plant cells: A resistance response against Agrobacterium-mediated DNA transfer. *Afr. J. Biotechnol.* **2005**, *4*, 752–757.

20. Ji, X.; Zheng, L.; Liu, Y.; Nie, X.; Liu, S.; Wang, Y. A Transient Transformation System for the Functional Characterization of Genes Involved in Stress Response. *Plant Mol. Biol. Rep.* **2014**, *3*, 732–739. [CrossRef]

21. Zang, D.; Wang, C.; Ji, X.; Wang, Y. *Tamarix hispida* zinc finger protein ThZFP1 participates in salt and osmotic stress tolerance by increasing proline content and SOD and POD activities. *Plant Sci.* **2015**, *235*, 111–121. [CrossRef] [PubMed]

22. Clough, S.J.; Bent, A.F. Floral dip: A simplified method for Agrobacterium-mediated transformation of *Arabidopsis thaliana. Plant J.* **1998**, *16*, 735–743. [CrossRef] [PubMed]

23. Tague, B.W.; Mantis, J. In planta Agrobacterium-mediated transformation by vacuum infiltration. *Methods Mol. Biol.* **2006**, *323*, 215–223. [PubMed]

24. Gous, P.W.; Gilbert, R.G.; Fox, G.P. Drought-proofing barley (*Hordeum vulgare*) and its impact on grain quality: A review. *J. Inst. Brew.* **2015**, *121*, 19–27. [CrossRef]

25. Broghammer, A.; Krusell, L.; Blaise, M.; Sauer, J.; Sullivan, J.T.; Maolanon, N.; Vinther, M.; Lorentzen, A.; Madsen, E.B.; Jensen, K.J.; et al. Legume receptors perceive the rhizobial lipochitin oligosaccharide signal molecules by direct binding. *Proc. Natl. Acad. Sci. USA* **2012**, *109*, 13859–13864. [CrossRef] [PubMed]

26. Jefferson, R.A.; Kavanagh, T.A.; Bevan, M.W. GUS fusions: β-glucuronidase as a sensitive and versatile gene fusion marker in higher plants. *EMBO J.* **1987**, *6*, 3901–3907. [PubMed]

27. Livak, K.J.; Schmittgen, T.D. Analysis of Relative Gene Expression Data Using Real-Time Quantitative PCR and the $2^{-\Delta\Delta CT}$ Method. *Methods* **2001**, *25*, 402–408. [CrossRef] [PubMed]

28. Fryer, M.J.; Oxborough, K.; Mullineaux, P.M.; Baker, N.R. Imaging of photo-oxidative stress responses in leaves. *J. Exp. Bot.* **2002**, *53*, 1249–1254. [CrossRef] [PubMed]

# Patterns of Growth Costs and Nitrogen Acquisition in *Cytisus striatus* (Hill) Rothm. and *Cytisus balansae* (Boiss.) Ball are Mediated by Sources of Inorganic N

**María Pérez-Fernández** [1,*], **Elena Calvo-Magro** [1], **Irene Ramírez-Rojas** [1], **Laura Moreno-Gallardo** [1] and **Valentine Alexander** [2]

[1] Department of Physical, Chemical and Natural Systems, University Pablo de Olavide, Carretera de Utrera Km, Seville 141013, Spain; ecmagro79@gmail.com (E.C.-M.); iramroj@alu.upo.es (I.R.-R.); lmorgal@alu.upo.es (L.M.-G.)

[2] Botany and Zoology Department, University of Stellenbosch, Private Bag X1, Matieland 7602, South Africa; alexvalentine@mac.com

\* Correspondence: maperfer@upo.es

Academic Editor: Maurizio Chiurazzi

**Abstract:** Nitrogen-fixing shrubby legumes in the Mediterranean area partly overcome nutrient limitations by making use of soil N and atmospheric $N_2$ sources. Their ability to switch between different sources lets them adjust to the carbon costs pertaining to N acquisition throughout the year. We investigated the utilization of different inorganic N sources by *Cytisus balansae* and *Cytisus striatus*, shrubby legumes under low and a sufficient (5 and 500 μM P, respectively) levels of P. Plants grew in sterile sand, supplied with N-free nutrient solution and inoculated with effective *Bradyrhizobium* strains; other treatments consisted of plants treated with (i) 500 μM $NH_4NO_3$; and (ii) 500 μM $NH_4NO_3$ and inoculation with effective rhizobial strains. The application of $NH_4NO_3$ always resulted in greater dry biomass production. Carbon construction costs were higher in plants that were supplied with mineral and symbiotic N sources and always greater in the endemic *C. striatus*. Photosynthetic rates were similar in plants treated with different sources of N although differences were observed between the two species. Non-fertilized inoculated plants showed a neat dependence on $N_2$ fixation and had more effective root nodules. Results accounted for the distribution of the two species with regards to their ability to use different N sources.

**Keywords:** legume; $N_2$ fixation; mineral N; C construction costs

---

## 1. Introduction

In natural stands where trees are absent, shrubby leguminous species play an important role in sustaining stand productivity and environmental values by regulating water uptake, the root environment and nutrient cycling [1,2]. This is of particular relevance in arid areas of the world, where nutrients are impoverished and plant growth is strongly hindered by limiting factors such as water, extreme temperatures and excessive solar radiation [3,4]. In such ecosystems, legumes contribute to global fertility by introducing nitrogen to the soils via their nitrogen-fixing symbiosis with legume-nodulating bacteria (for a review see [5,6]).

Legumes are the entryway through which nitrogen (N) enters ecosystems [6]. This provides legumes with a complementary N source as compared to non-legumes. However, little is known about the plant's preferences with respect to N sources. It has been proven that the Cape Fynbos legume, *Virgilia divaricata* (Adamson), is able to switch N sources for its growth, depending on the environmental cues [7]. Accordingly, Neff *et al*. [8] suggested that leguminous plants can absorb and assimilate nitrogenous compounds such as nitrate, ammonium, or amino acids directly from soil in response to both the plant's needs and the environmental restrictions imposed on organic matter decomposition. Changes in the sources of nitrogen are expected to induce differing responses in different plants species that would translate into contrasting photosynthetic rates and biomass production [9,10].

Biological nitrogen fixation (BNF) is not free and the plant must contribute a significant amount of energy in the form of photosynthates (photosynthesis-derived sugars) and other nutritional factors for the bacteria. However, some legumes are more efficient than others in fixing nitrogen. The process requires 160 Kcal·$mol^{-1}$ for a molecule of $N_2$ to be reduced [11,12]; hence, plants would only become involved in such a reaction when there are no sources of nitrogen other than the atmosphere [13].

The quantity of nitrogen fixed depends, amongst other factors, on the level of soil nitrogen, the rhizobia strain infecting the legume, the amount of legume plant growth, and the length of the growing season. Increased soil nitrogen availability results in decreased nodulation rates and N-fixing efficiency [14–17]. If given a choice, a legume plant will remove nitrogen from the soil before obtaining nitrogen from the air through $N_2$-fixation, thus reducing the benefits of the nodulation. A legume growing on a sandy soil, very low in nitrogen, will get most of its nitrogen from the air while a legume growing on a fertile river-bottom soil will get most of its nitrogen from the soil [18,19].

In the central western area of the Iberian Peninsula, water is the primary factor that limits plant growth. In addition to water scarcity, soils in the best part of this area are infertile due to deep soil erosion that drives losses of N, phosphorus (P) and other nutrients that hinder plant establishment [20]. At the same time, soils are characterized by low pH values, which are known to reduce the ability of legumes to establish effective symbiosis with their rhizobial symbionts, hence reducing BNF [21,22].

These soils typically harbor low concentrations of N and P, in amounts that are generally available for plant use in micro-molar concentrations, compromising metabolic process [23]. The proportion of N:P and co-limitation are important in explaining N-P relationships in plants, and can be used as a tool to diagnose both plant growth and dynamics with respect to nutrient availability in soils [24]. Legumes are highly dependent on the P concentration in the growing media in terms of nodulation and BNF [25–27]. P micro-molar concentrations are extremely low to drive the P-requiring metabolic processes [19], compromising the wellbeing of plants; however, different species may response differently to N:P ratio changes under altered growing conditions, which may then account for the species distribution with regards to nutrient availability.

*Cytisus striatus* (Hill) Rothm. is a shrubby legume endemic to the Iberian Peninsula that has colonized other parts of the world [28]. It grows in siliceous soils from 450–750 m a.s.l., avoiding cold distributions. In ecotonal areas, it can form loose mixed population with *Cytisus balansae* (Boiss.) Ball. The latter is well represented in the Iberian Peninsula and northern Morocco, forming dense populations on siliceous soils from 750–1300 m a.s.l. Both *C. balansae* and *C. striatus* have been reported to nodulate with *Bradyrhizobium* spp. [2,29,30]. The main objective of this work is to identify possible ways in which inorganic N is used by *C. balansae* and *C. striatus* in relation to P availability, and how it affects carbon construction costs, photosynthetic rates and efficiency of N-fixing in these two shrubs. Our working hypothesis is that the distribution of these two species in soils with low concentrations of nutrients, under harsh climatic conditions, is explained by the plants' ability to change their sources of N, either from the atmosphere or from the soil, during growth. Should this hypothesis be proven, it would be possible (i) to explain why these two species do not form mixed populations and (ii) to relate *C. striatus*' ability to colonize new areas to its greater plasticity in terms of N use under low construction costs.

## 2. Materials and Methods

### 2.1. Plant Material and Experimental Design

Seeds of *C. balansae* (cba) and *C. striatus* (cst) were hand harvested in the summer of 2014 from natural populations in central-west Spain. The strains cba and cst had been previously obtained from nodules of *C. balansae* and *C. striatus* plants in monospecific natural populations [29]. Strains were identified as *Bradyrhizobium* isolates with the accession numbers AF461191 and AF461194 for cba and cst, respectively. The strains were maintained on yeast extract mannitol (YEM) agar [31] at 4 °C. For inoculation of seedlings, cultures were grown for 6 days in YEM broth at 26 °C in an orbital shaker at 100 rpm before dilution to the required concentration of cells.

Seeds were hand scarified using an emery board. This treatment was followed by surface-sterilization in 70% ethanol for 5 min and 1% sodium hypochlorite for 3 min and then washed six times in sterile distilled water. Twenty-five seedlings per species were transplanted to 10-cm diameter pots containing sterile sand-river and were given the appropriate treatment (all seedlings were supplied with 25% Hoagland's solution—pH 5.8) [32], modified with either high P (500 μM) or low P (5 μM) as $NaH_2PO_4\ 2H_2O$). Plants were maintained in a glasshouse at the University Pablo de Olavide (Seville, Spain) under natural light and temperature, with a 12-h photoperiod (24 °C day and 18 °C night) and a photon flux density at the top of the plants of approximately 700 $\mu mol \cdot m^{-2} \cdot s^{-1}$ for 22 weeks (February until July 2015). Pots with different treatments were randomly distributed on benches in the glass house, 1 m apart from any other treatment, to prevent cross contamination; a total of 25 replicates per combination of species and treatments were maintained.

The control treatment consisted of un-inoculated Hoagland's solution from which nitrogenous compounds had been removed (−N−R). One of the treatments consisted of nitrogen-free Hoagland's solution and rhizobial inoculation (−N+R). A second treatment consisted of the application of 500 μM $NH_4NO_3$ as an N source with no rhizobial inoculation (+N−R). In the last experiment, plants received the same amount of $NH_4NO_3$ as before and were simultaneously inoculated (+N+R). All treatments were subjected to both high and low P levels.

Inoculation treatments consisted of growth phase broth-cultured inoculant at $1 \times 10^8$ cells $mL^{-1}$. Each plant species was inoculated with 100 mL of its own rhizobia, *i.e.*, cba (AF461191) and cst (AF461194). The surface of the pots was covered with sterile polyurethane beds and watering was conducted weekly through a watering pipe.

## 2.2. Harvesting and Nutrient Analysis

At harvest, plants were assessed for root nodule number, shoot and root dry matter, total nitrogen accumulation in shoots and biologically-fixed nitrogen ($\delta^{15}N$). The dry mass of shoot, root and nodules was obtained as the dry weight of plant material after drying in an oven at 50 °C for 48 h–72 h. The dried material was ground and analyzed for C, N and P concentrations. The nitrogen accumulated in shoots was calculated by multiplying the weight of dry shoots by the nitrogen content as measured by the semi micro-Kjeldahl method [33]. Milled dry shoots were sent for isotopic analysis to the UIB (University of the Balearic Islands, Balearic Islands, Spain) and for total N analyses to the Laboratório Químico Agrícola Rebelo da Silva (Lisbon, Portugal).

## 2.3. Calculations of %Ndfa

The isotopic ratio of $\delta^{15}N$ was calculated as $\delta$ = 1000‰ (Rsample/Rstandard), where $R$ is the molar ratio of the heavier to the lighter isotope of the samples and standards are defined by [34].

The fraction of N derived entirely from $N_2$ fixation (Ndfa) in the nodulated plants [35] was calculated as:

$$\%Ndfa = (\delta^{15}N\text{reference plant} - \delta^{15}N\text{legume})/(\delta^{15}N\text{reference plant} - B) \times 100$$

where: $\delta^{15}N$ref— is the $\delta^{15}N$ from a non-fixing $N_2$ reference plant (*Lolium perenne* in this study); B is the $\delta^{15}N$ natural abundance of the N derived from biological N-fixation of the above-ground tissue of *C. balansae* and *C. striatus*, grown in an N-free culture (plants only N source was $N_2$). The B value of *C. balansae* was determined in this study as −3.94‰ and that of *C. striatus* was −2.96‰. The total amount of N in the plant derived from $N_2$ fixation (*Nfix*) was determined as N*fix* = Ndfa × N content.

## 2.4. Carbon and Nutrition Cost Calculations

Construction costs, $C_W$ (mmolCg$^{-1}$DW), were calculated according to the methods proposed by [36], modified from the equation used by [37]:

$$C_W = (C + kN/14 \times 180/24) \times (1/0.89) \times (6000/180)$$

where $C_W$ is the construction cost of the tissue (mmolCg$^{-1}$DW), C is the carbon concentration (mmolCg$^{-1}$), k is the reduction state of the N substrate (k = + 5 for $NO_3$) and N is the organic nitrogen content of the tissue (g$^{-1}$DW) [38]. The constant (1/0.89) represents the fraction of the construction costs that provide reductant that is not incorporated into the biomass [37,38] and (6000/180) converts units of g glucose DW$^{-1}$ to mmolCg$^{-1}$DW.

Belowground allocation is the fraction of new biomass formed in terms of roots and nodules over the growth period. This was calculated according to [39]:

$$df/dt = RGR \times (\partial - B_r/B_t)$$

RGR is the relative growth rate (mg$\cdot$ g$^{-1}\cdot$ day$^{-1}$) and $\partial$ is the fraction of new biomass gained during the growth period. $B_r/B_t$ is the root weight ratio, based on total plant biomass ($B_t$) and root biomass ($B_r$).

## 2.5. Photosynthetic Rates

Photosynthesis was measured using a Licor 6200 Photosynthetic System (LICOR, Lincoln, NE, USA), equipped with a quarter-liter chamber. Measurements were made between 08:00 and 16:30 h when light quality was optimum in the growing area. As leaves of the study species are particularly small, full branches (also photosynthetically active) were enclosed in the chamber. Ten measurements were performed per treatment. Light during the measurements remained steady at saturation ($\pm$1400 mol$\cdot$ m$^{-2}\cdot$ s$^{-1}$) at photosynthetic biomass temperature of 24 $°$C and humidity level of about 40%–60%. Surface area of photosynthetically active parts was measured using a Licor 3000 leaf area meter (LICOR). Branches were dried to constant mass and weighed for calculating leaf mass per area (LMA, g nr$^2$).

## 2.6. Statistical Analysis

All data were tested for normality and homogeneity of variances using the Levene and Cochran tests. The effects of the factors and their interactions were tested using analysis of variance (ANOVA). When the ANOVA results revealed significant differences between treatments, the means (6–8) were separated using a *post-hoc* t-Student test ($p \leqslant 0.05$). Statistical analysis was computed using the SPSS software version 15.0 for Windows.

## 3. Results

### 3.1. Biomass Production

High seedling mortality was observed in the −N−R treatment both at low and high P levels; the remaining seedlings had yellow leaves showing the lack of nitrogen nutrition and poor biomass production. Plants of the two species grown at the high level of phosphate always had greater biomass production in all treatments, except for those in the −N−R. The addition of $NH_4NO_3$ (+N treatments) always triggered biomass accumulation (Table 1). Under the two levels of phosphate, the application of selected inoculants (+R treatments) resulted in increased biomass production compared with mass accumulation in the control plants. There was a differential biomass production in the +R treatments under the low and high levels of P. When P was scarce, the simultaneous addition of chemical N and inoculation significantly increased biomass production. Under high phosphate, rhizobial inoculation with or without mineral-supplied N induced a biomass decrease, especially for cst compared with plants only supplied with N (Table 1).

**Table 1.** Biomass production of *C. balansae* and *C. striatus* seedlings under four treatments of N acquisition under two levels of P nutrition.

| Biomass (g) | Plant sp. | −N−R | −N+R | +N−R | +N+R |
|---|---|---|---|---|---|
| | | **Low Phosphate** | | | |
| Shoot | cba | 0.165 ± 0.07 a | 0.332 ± 0.07 b | 0.485 ± 0.08 c | 0.654 ± 0.03 d |
| | cst | 0.312 ± 0.04 a | 0.367 ± 0.01 b | 0.645 ± 0.10 c | 0.640 ± 0.05 c |
| Root | cba | 0.050 ± 0.04 a | 0.114 ± 0.008 b | 0.189 ± 0.04 b | 0.202 ± 0.008 c |
| | cst | 0.115±0.02 a | 0.141 ± 0.008 b | 0.160 ± 0.02 c | 0.186 ± 0.008 c |
| Nodules | cba | Θ | 0.0279 ± 0.001 a | Θ | 0.036 ± 0.005 a |
| | cst | Θ | 0.0216 ± 0.003 a | Θ | 0.031 ± 0.001 a |
| Whole plant | cba | 0.215 ± 0.05 a | 0.446 ± 0.03 b | 0.674 ± 0.06 b | 0.856 ± 0.01 c |
| | cst | 0.327 ± 0.03 a | 0.508 ± 0.04 b | 0.805 ± 0.05 c | 0.826 ± 0.06 c |
| | | **High Phosphate** | | | |
| Shoot | cba | 0.213 ± 0.04 a | 0.403 ± 0.03 b | 0.882 ± 0.07 d | 0.828 ± 0.03 c |
| | cst | 0.124 ± 0.01 a | 0.566 ± 0.02 b | 0.985 ± 0.06 d | 0.623 ± 0.06 c |
| Root | cba | 0.069 ± 0.006 a | 0.124 ± 0.01 b | 0.129 ± 0.03 b | 0.142 ± 0.05 b |
| | cst | 0.217 ± 0.007 a | 0.173 ± 0.01 b | 0.191 ± 0.04 c | 0.136 ± 0.02 b |
| Nodules | cba | Θ | 0.011 ± 0.003 a | Θ | 0.010 ± 0.001 a |
| | cst | Θ | 0.014 ± 0.001 a | Θ | 0.012 ± 0.01 a |
| Whole plant | cba | 0.329 ± 0.03 a | 0.546 ± 0.01 b | 1.49 ± 0.07 d | 0.902 ± 0.01 c |
| | cst | 0.241 ± 0.01 a | 0.733 ± 0.06 b | 0.931 ± 0.08 d | 0.822 ± 0.07 c |

Values are means ± standard deviation. Different letters indicate significant differences amongst treatments ($p < 0.05$). All means are values obtained from 10 plants, except for treatment −N−R where only 5 seedlings survived. Θ indicates no nodulation.

### 3.2. Carbon Construction Cost and Photosynthetic Rate

There were significant differences in values of carbon construction costs between plant species and treatments. *C. balansae* always showed greater carbon costs than *C. striatus*. Carbon construction costs for the two plants species were significantly greater at the low phosphate level, with the greatest values in the presence of inoculants (+R treatments) (Figure 1a). Inoculated plants supplied with N in the low P study showed the lowest C construction cost, in contrast to those with no nitrogen added. Despite the high carbon construction costs in the +N+R treatment, both cba and cst resulted in the greatest amounts of biomass (Table 1). In addition, the greatest biomass production in *C. striatus* at the high P level was achieved under the treatment +N−R that is the one for which plants showed the greatest C construction cost. At the high level of P, no differences in carbon construction costs were detected in the species except for those under the +N−R treatment, which were significantly lower (Table 1 and Figure 1a). Under the low P level, the two species' allocation of resources to the roots was less pronounced in plants grown under the −N−R treatment. When the level of P was high, cst showed significantly greater root allocation in treatments −N−R and +N+R (Figure 1b). Nodule allocation was higher for the inoculated plants that relied solely on $N_2$ fixation, compared with plants with combined N sources (Figure 1c). Under low P, nodule allocation of cst was significantly greater than that of cba; however, under the high P conditions, no statistical differences in nodule allocation were observed between cba and cst (Figure 1c). The photosynthetic rate was not influenced by any of the treatments nor by the P levels (Table 2) except for the plants in treatment −N−R. Under the high level of P, the photosynthetic rate was always greater for plants in any of the four treatments, showing a clear positive effect of this nutrient on plant performance (Table 2).

**Table 2.** Leaf area, leaf area:plant dry weight and photosynthetic rate of seedlings of *C. balansae* and *C. striatus* under four treatments of N acquisition under two levels of P nutrition.

| | Plant sp. | Low Phosphate | | | |
|---|---|---|---|---|---|
| | | −N−R | +N−R | −N+R | +N+R |
| Leaf area (cm$^2$) | cba | 0.703 ± 0.07 a | 2.063 ± 0.06 a | 2.067 ± 0.14 a | 2.125 ± 0.10 a |
| | cst | 0.986 ± 0.08 a | 1.999 ± 0.08 a | 2.097 ± 0.05 a | 2.130 ± 0.12 a |
| Leaf area/DW | cba | 1.954 ± 0.11 a | 1.605 ± 0.20 b | 2.432 ± 0.08 b | 2.035 ± 0.05 b |
| | cst | 3.520 ± 0.21 a | 1.495 ± 0.25 b | 3.616 ± 0.25 b | 1.836 ± 0.06 b |
| Photosynthetic rate (μmol CO$^2$·m$^{-2}$·s$^{-1}$) | cba | 1.386 ± 0.13 a | 2.717 ± 0.13 b | 3.062 ± 0.09 b | 3.448 ± 0.07 b |
| | cst | 1.469 ± 0.16 a | 2.924 ± 0.12 b | 3.435 ± 0.17 b | 3.848 ± 0.14 b |
| | | High Phosphate | | | |
| Leaf area | cba | 2.104 ± 0.13 a | 2.045 ± 0.1 a | 2.111 ± 0.10 a | 2.057 ± 0.05 a |
| | cst | 2.083 ± 0.03 a | 1.089 ± 0.06 a | 2.068 ± 0.02 a | 2.143 ± 0.03 a |
| Leaf area/DW | cba | 4.178 ± 0.08 a | 1,062 ± 0.01 b | 2.426 ± 0.13 b | 2.007 ± 0.40 b |
| | cst | 5.786 ± 0.13 a | 1.061 ± 0.06 b | 3.132 ± 0.15 b | 1.514 ± 0.33 b |
| Photosynthetic rate (μmol CO$^2$·m$^{-2}$·s$^{-1}$) | cba | 2.786 ± 0.16 a | 4.303 ± 0.15 b | 3.503 ± 0.06 b | 4.538 ± 0.14 b |
| | cst | 2.717 ± 0.17 a | 3.683 ± 0.13 b | 3.269 ± 0.18 b | 3.752 ± 0.07 b |

Values are means ± standard deviation. Different letters indicate significant differences amongst treatments ($p < 0.05$). All means are values obtained from 10 plants, except for treatment −N−R where only 5 seedlings survived.

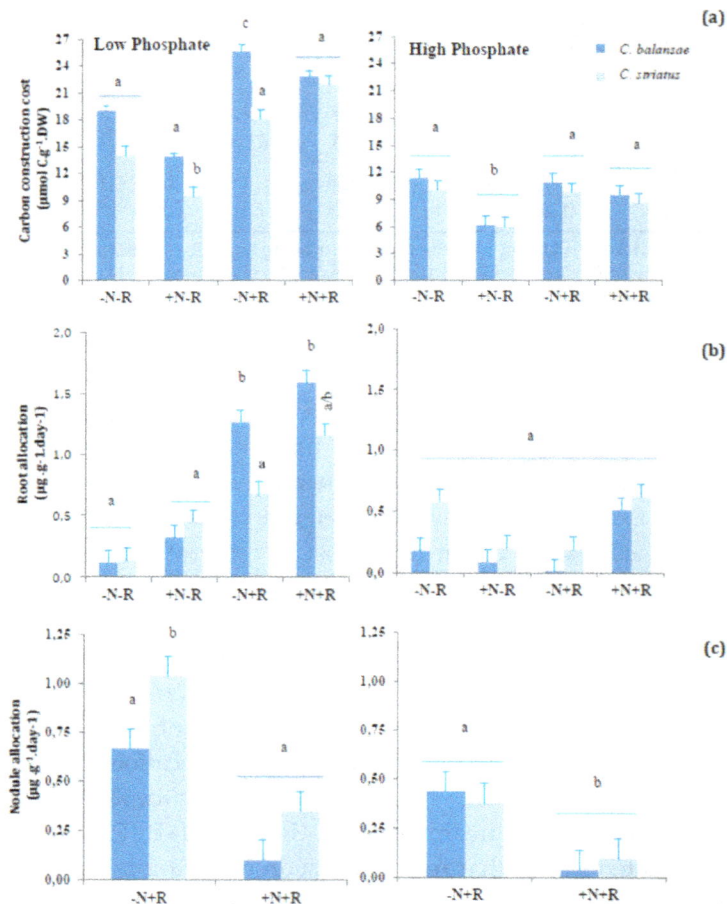

**Figure 1.** (**a**) Plant construction costs; (**b**) Root allocation and (**c**) Nodule allocation of 22-week-old *Cytisus balansae* (cba) and *Cytisus striatus* (cst) seedlings, grown in sand culture treated with −N−R, +N−R, +N−R and +N+R, under two levels of phosphate (Low and High). Values are means ($n = 10$, except for −N−R where $n = 6$) ± standard deviation. Different letters indicate significant differences among treatments (* $p \leqslant 0.05$).

### 3.3. Nitrogen Fixation

Total $N_2$ fixation varied between species and amongst treatments. $N_2$ was significantly lower in cba than in cst; overall for the two species, $N_2$ fixation efficiency was greater in the high P treatment than in the low P treatment (Figure 2a). The amount of $N_2$ fixed biologically was significantly lower in plants supplied with $NH_4NO_3$, as indicated by the decline in %Ndfa in plants in the +N+R treatment compared with plants grown in the inoculated treatment ($-N+R$) (Figure 2a). With the exception of the $-N-R$ treatment, for which N concentration was significantly low ($1.07 \pm 0.06$; $p = 0.038$), there were no differences for this variable in the +N$-$R ($1.91 \pm 0.23$; $p < 0.05$), $-N+R$ ($2.08 \pm 0.09$; $p < 0.043$) and +N+R ($2.46 \pm 0.11$; $p < 0.021$) treatments. $N_2$ fixation efficiency was greater in cst than in cba. Plants solely reliant on $N_2$ fixation were more efficient at fixing N at the two levels of P according to the amounts of N fixed per nodule (Figure 2b).

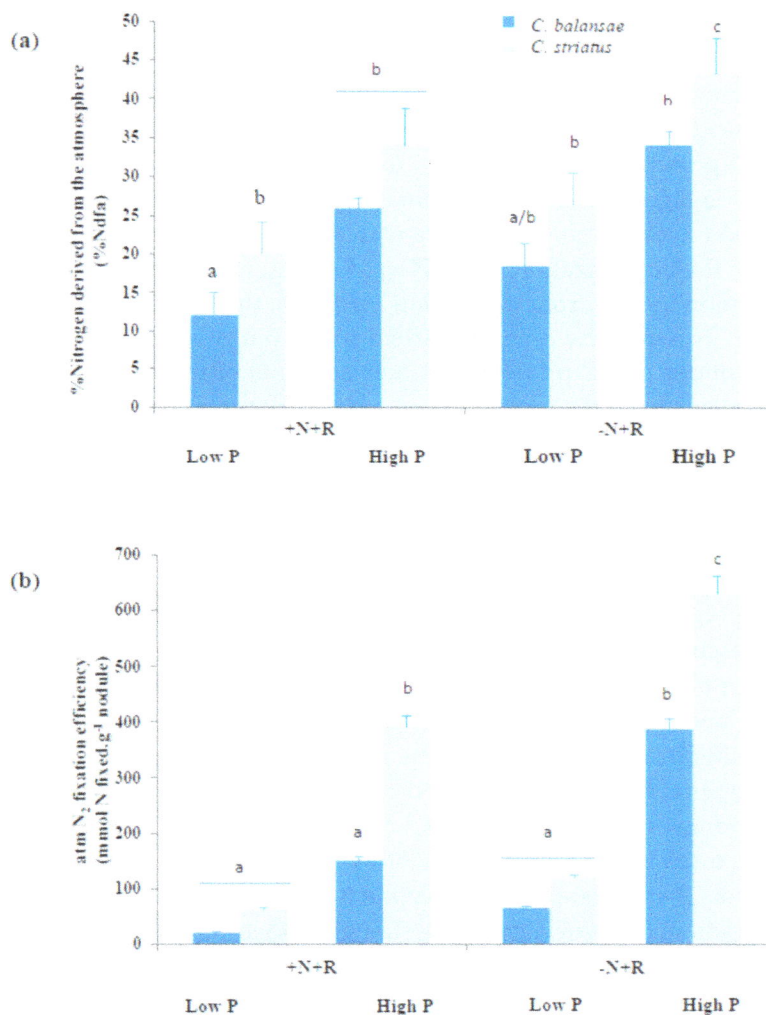

**Figure 2.** (a) Percentage N derived from the atmosphere (%Ndfa) and (b) $N_2$ fixation efficiency of 22-week-old *Cytisus balansae* (cba) and *Cytisus striatus* (cst) seedlings, grown in sand culture treated with $-N-R$, +N$-$R, +N$-$R and +N+R, under two levels of phosphate (Low and High). Values are means ($n = 10$, except for $-N-R$ where $n = 6$) $\pm$ standard deviation. Different letters indicate significant differences among treatments (* $p \leqslant 0.05$).

## 4. Discussion

Legumes are able to change the sources of N they use to meet their metabolic needs [40,41]. The two studied species in the present research confirm this fact, and the reported behavior in terms of N use matches their current distribution in nature. Under control conditions, we tested the responses of *C. balansae* and *C. striatus* to changes in P and N supplies as well as the role of rhizobial inoculation in plant growth and biomass allocation. Both species show shifts from organic to inorganic forms of N when P in the growing media is present, which allows them to adjust to changing environmental conditions. Strong differences in plant performance under $-N+R$, $+N-R$, $+N+R$ with significantly greater biomass production compared with plants under the $-N-R$ treatment prove the strong dependence of the two species on N and P availability. Similarly, the dependence of both species on N was clearly moderated by the micro-molar concentrations of P in the growing media.

Under sufficient levels of phosphorus in the growing media, both *C. balansae* and *C. striatus* were more efficient at incorporating $NH_4NO_3$ than at fixing atmospheric N. This can be explained by the fact that it is less expensive to acquire mineral sources of N than to fix them from the atmosphere [42]. The immediate result is a noticeable increase in biomass production when N and P are sufficient in the media. That would translate into a profuse colonization of soils by either of the two species. However, not all species are likely to colonize all soils because the amounts of nutrients needed for satisfactory plant growth would vary from one species to another. Differences in nutritional needs are linked to the legume-rhizobia combination as well as the inorganic source of N [42,43]. Similarly, legumes differ in their P requirements and in their ability to assimilate P from the soil [44,45], which correlates with their colonization status [46]. Most legumes from Western Australia would be killed by P concentration in soils from the Iberian Peninsula (toxic effect), whereas the latter would show P deficiencies if grown in the P-impoverished soils from Western Australia [47]. In our study, plants of cst grown under low P produced greater amounts of biomass and fixed more $N_2$ than those of cba. As all plants from the two species were experimentally maintained under exactly the same glasshouse conditions and nutrient availability, we explain the greater biomass production by cst in terms of greater efficiency of the legume-rhizobia interaction [42], which can simultaneously explain its ability to effectively colonize soils beyond its natural area of distribution [48,49]. It has been demonstrated that legumes under low or zero concentrations of P and N are forced to acquire N through symbiotic $N_2$ fixation; on the other hand, when N is present in the soil, legumes avoid the expensive process of $N_2$ reduction [50,51]; this very same scheme is depicted by cba and cst in this experiment, which resembles the behavior of the Fynbos legume *Vigiglia divaricata* [7]. Nevertheless shifts in the use of N are actually mediated by levels of P. Under limiting levels of P for plant growth, both atmospheric $N_2$ and $NH_4NO_3$ supplies contribute to increase the C sink strength of cba and cst plants in order to maintain enough carbon in the plant tissues to maintain both N fixation and soil N acquisition. Concomitantly, carbon costs and root allocations of plants in the $+N+R$ treatments were the greatest amongst all treatments; that fact proves that the plants of the two species have to maintain the structures for N acquisition [7,50,51]. We observed differences in plant biomass accumulation under low P and N supply between the two species, with a marked biomass production by cba, which we explain in terms of carbon sink strength and lower photosynthetic rate of *C. balansae*. This behavior, which might have to be related to evolutionary processes through which cba selectively occupies areas with limited resources, also accounts for the restricted distribution of this species and the extended distribution of *C. striatus* [48]. It is interesting that plants under the $+N+R$ and $-N+R$ treatments showed the greatest values of C construction costs regardless of the level of P they were supplied with. Plants in these treatments also showed the greatest leaf area per plant mass. This can only be interpreted as a way to increase the leaf area ratio that the two species need to meet for the photosynthetic requirements to build up the nodules, as has been demonstrated in *Glycine max* (L. Merr.) [52] and *Virgilia divaricata* [7].

An external supply of mineral N exerts inhibiting effects on nodulation and nitrogen fixation [13,42], which are dependent on the combination of plant-rhizobia and seem to be driven by the bacterial strains [42,53]. This was clearly shown by *C. balansae* and *C. striatus* in our experiment, where a reduction in %Ndfa was observed in plants supplied with the combined sources of N; the plant species that achieved the greatest efficiency in the use of different sources of N was cst, which has allowed this species to expand its area of distribution. The two Iberian shrubs have shown behavior similar to that previously described for *V. divaricata* [7], *i.e.*, a decrease in %Ndfa when plants have enough P and inorganic sources of N, regardless of the presence of inoculants in the media This proves that plants tend to use less energy demanding sources of N (mineral sources).

Overall, these results support the initial hypothesis that the ability of the shrubby legume *C. striatus* to change sources of N plays a role in its distribution and that limitations of *C. balansae* to efficiently fix $N_2$ has restricted its distribution. Similarly, the broader tolerance of cst to P and N concentrations in the soil account for its extended distribution. We have shown that *C. striatus* is the species that can make better use of any available source of N and at the time, is the one with the lowest carbon costs (at a constant photosynthetic rate). *C. balansae* plants are more reliant on inorganic sources of N, and the maintenance of nodules corresponds to the greatest carbon constructions costs, which represents a strong limit to its growth. Differences in the behavior of the two species and restricted ability to quickly and efficiency change the use of N might be the reasons why *C. striatus* continues to expand and *C. balansae* is restricted in its area of distribution.

**Author Contributions:** All authors have equally contributed to all stages of the research and manuscript production.

**Conflicts of Interest:** The authors declare no conflict of interest.

## References

1. Rejili, M.; Lorite, M.J.; Mahdhi, M.; Sanjuan Pinilla, J.; Ferchichi, A.; Mars, M. Genetic diversity of rhizobial populations recovered from three *Lotus* species cultivated in the infra-arid Tunisian soils. *Prog. Nat. Sci.* **2009**, *19*, 1079–1087. [CrossRef]
2. Ruiz-Díez, B.; Fajardo, S.; Puertas-Mejía, M.A.; de Felipe, M.R.; Fernández-Pascual, M. Stress tolerance, genetic analysis and symbiotic properties of root-nodulating bacteria isolated from Mediterranean leguminous shrubs in Central Spain. *Arch. Microbiol.* **2009**, *191*, 35–46. [CrossRef] [PubMed]
3. Forti, M.; Lavie, Y.; Ben-Dov, Y.; Pauker, R. Long-term plant survival and development under dryland conditions in an experimental site in the semi-arid Negev of Israel. *J. Arid Environ.* **2006**, *65*, 1–28. [CrossRef]
4. Laranjo, M.; Oliveira, S. Tolerance of *Mesorhizobium* type strains to different environmental stresses. *Antonie van Leeuwenhoek* **2011**, *99*, 651–662. [CrossRef] [PubMed]
5. Azan, F. Legume-bacterium (*Rhizobium*) association-symbiosis, a marriage of convenience, necessary evil or bacterium taken hostage by the legume. *Pak. J. Biol. Sci.* **2001**, *4*, 757–761.
6. Sprent, J.I.; Gehlot, H.S. Nodulated legumes in arid and semi-arid environemnts: Are they important? *Plant Ecol. Divers.* **2010**, *3*, 211–219. [CrossRef]
7. Magadlela, A.; Pérez-Fernández, M.A.; Kleinert, A.; Dreyer, L.L.; Valentine, A.J. Source of inorganic N affects the cost of growth in a legume tree species (*Virgilia divaricata*) from the Mediterranean-type Fynbos ecosystem. *J. Plant Ecol.* **2016**. [CrossRef]
8. Neff, J.C.; Chapin, F.S., III; Vitousek, P.M. Breaks in the cycle: Dissolved organic nitrogen in terrestrial ecosystems. *Front. Ecol. Environ.* **2003**, *1*, 205–211. [CrossRef]
9. Crawfor, N.M.; Glass, A.D.M. Molecular and physiological aspects of nitrate uptake in plants. *Trends Plant Sci.* **1998**, *3*, 389–395. [CrossRef]
10. Miller, A.J.; Smith, S.J. Nitrate transport and compartmentation in cereal root cells. *J. Exp. Bot.* **1996**, *47*, 843–854. [CrossRef]
11. Postgate, J. *Nitrogen Fixation*; Cambridge University Press: Cambridge, UK, 1998; p. 377.
12. Minchin, F.; Witty, J.F. Respiratory/carbon costs of symbiotic nitrogen fixation in legumes. In *Plant Respiration*; Lambers, H., Ribas-Carbó, M., Eds.; Springer: Dordrecht, The Netherlands, 2005; pp. 195–205.
13. Simms, E.L.; Taylor, D.L. Partner choice in nitrogen-fixation mutualisms of legumes and Rhizobia. *Integr. Comp. Biol.* **2002**, *42*, 369–380. [CrossRef] [PubMed]

14. Caetano-Anolles, G.; Gresshoff, P.M. Plant genetic control of nodulation. *Ann. Rev. Microbiol.* **1991**, *45*, 345–382. [CrossRef] [PubMed]

15. Lang, P.; Martin, R.; Golvano, M.P. Effect of nitrate on carbon metabolism and nitrogen fixation in root nodules of *Lupinus albus*. *Plant Physiol. Biochem.* **1993**, *31*, 639–648.

16. Rubio Arias, H.O.; de la Vega, L.; Ruiz, O.; Wood, K. Differential nodulation response and biomass yield of Alexandria clover as affected by levels of inorganic nitrogen fertilizer. *J. Plant Nutr.* **1999**, *22*, 1233–1239. [CrossRef]

17. Thomas, R.B.; Bashkin, M.A.; Ritcher, D.D. Nitrogen inhibition of nodulation and $N_2$ fixation of a tropical $N_2$-fixing tree (*Gloricidia sepium*) grown in elevated atmospheric $CO_2$. *New Phytol.* **2000**, *145*, 233–243. [CrossRef]

18. Thrall, P.H.; Laine, A.L.; Broadhurst, L.M.; Bagnall, D.J.; Brockwell, J. Symbiotic effectiveness of rhizobial mutualists varies in interactions with native Australian legume genera. *PLoS ONE* **2011**, *6*, e23545. [CrossRef] [PubMed]

19. Sulieman, S.; Ha, C.V.; Schulze, J.; Tran, L.S. Growth and nodulation of symbiotic *Medicago truncatula* at different levels of phosphorus availability. *J. Exp. Bot.* **2013**, *64*, 2701–2712. [CrossRef] [PubMed]

20. Pérez-Fernández, M.A.; Calvo-Magro, E.; Valentine, A. Benefits of the Symbiotic Association of Shrubby Legumes to Re-vegetate Heavily Damaged Soils. *Land Degrad. Dev.* **2015**. [CrossRef]

21. Dilworth, M.J.; Howieson, J.G.; Reeve, W.G.; Tiwari, R.P.; Glenn, A.R. Acid tolerance in legume root nodule bacteria and selecting for it. *Aust. J. Exp. Agri.* **2001**, *41*, 435–446. [CrossRef]

22. Pérez-Fernández, M.A.; Hill, Y.J.; Calvo-Magro, E.; Valentine, A. Competing *Bradyrhizobia* strains determine niche occupancy by two native legumes in the Iberian Peninsula. *Plant Ecol.* **2015**, *216*, 1537–1549.

23. López-Mosquera, M.E.; Moirón, C.; Carral, E. Use of dairy-industry sludge as fertilizer for grasslands in northwest Spain: Heavy metal levels in the soil and plants. *Resour. Conserv. Recycl.* **2000**, *30*, 95–109. [CrossRef]

24. Xu, B.; Gao, Z.; Wang, J.; Xu, W.; Palta, J.A.; Chen, Y. N:P ratio of the grass *Bothriochloa ischaemum* mixed with the legume *Lespedeza davurica* under varying water and fertilizer supplies. *Plant Soil* **2016**, *400*, 67–79. [CrossRef]

25. De Oliveira, W.S.; Meinhardt, L.W.; Sessitsch, A.; Tsai, S.M. Analysis of *Phaseolus*-Rhizobium interactions in a subsistence farming system. *Plant Soil* **1998**, *204*, 107–115. [CrossRef]

26. Taiwo, L.B.; Nworgu, F.C.; Adatayo, O.B. Effect of bradyrhizobium inoculation and phosphorus fertilization on growth, nitrogen fixation and yield of promiscuity nodulating soybean (*Glycine max* (L.) Merr.) in a tropical soil. *Crop Res.* **1999**, *18*, 169–177.

27. Vance, C.P. Symbiotic nitrogen fixation and phosphorus acquisition. Plant nutrition in a world of declining renewable resources. *Plant Physiol.* **2011**, *127*, 390–397. [CrossRef]

28. Herrera-Reddy, A.M.; Carruthers, R.I.; Mills, N.J. Integrated management of Scotch broom (*Cytisus scoparius*) using biological control. *Invasive Plant Sci. Manag.* **2012**, *5*, 69–82. [CrossRef]

29. Rodríguez-Echeverría, S.; Pérez-Fernéndez, M.A.; Vlaar, S.; Finnan, T. Analysis of the legume-rhizobia symbiosis in shrubs from central western Spain. *J. Appl. Microbiol.* **2003**, *95*, 1367–1374. [CrossRef] [PubMed]

30. Rodríguez-Echeverría, S.; Pérez-Fernández, M.A. Potential use of Iberian shrubby legumes and rhizobia inoculation in re-vegetation projects under acidic soil conditions. *Appl. Soil Ecol.* **2005**, *29*, 203–208. [CrossRef]

31. Vincent, J.M. *A Manual for the Practical Study of Root Nodule Bacteria*. IBP Handbook 15; Blackwell: Oxford and Edinburgh, UK, 1970.

32. Hoagland, D.R.; Arnon, D.I. *The Water-Culture Method of Growing Plants without Soil*; California Agricultural Experimental Station: Berkeley, CA, USA, 1950; p. 347.

33. Sarruge, J.R.; Haag, H.P. *Análises Químicas em Plantas*; ESALQ/USP: São Paulo, Brazil, 1979. (In Portuguese)

34. Farquhar, G.D.; Ehleringer, J.R.; Hubick, K.T. Carbon isotope discrimination and photosynthesis. *Ann. Rev. Physiol. Plant Mol. Biol.* **1989**, *40*, 503–537. [CrossRef]

35. Högberg, P. Tansley Review No. 95. [15]N natural abundance in soil-plant systems. *New Phytol.* **1997**, *137*, 179–203. [CrossRef]

36. Mortimer, P.E.; Archer, E.; Valentine, A.J. Mycorrhizal C costs and nutritional benefits in developing gravevines. *Mycorrhiza* **2005**, *15*, 159–165. [CrossRef] [PubMed]

37. Peng, S.; Eissenstat, D.M.; Graham, J.H.; Williams, K.; Hodge, N.C. Growth depression in mycorrhizal citrus at high-phosphorus supply: Analysis of carbon costs. *Plant Physiol.* **1993**, *101*, 1063–1070. [PubMed]

38. Williams, K.; Percival, F.; Merino, J.; Mooney, H.A. Estimation of tissue construction cost from heat of combustion and organic nitrogen content. *Plant Cell Environ.* **1987**, *10*, 725–734.

39. Bazzaz, F.A. Allocation of resources in plants: state of science and critical questions. In *Plant Resource Allocation*; Bazzaz, F.A., Grace, J., Eds.; Academic Press: San Diego, CA, USA, 1997; pp. 1–37.

40. Paoli, G.D.; Curran, L.M.; Zak, D.R. Phosphorus efficiency of Bornean rain forest productivity: Evidence against the unimodal efficiency hypothesis. *Ecology* **2005**, *86*, 1548–1561. [CrossRef]

41. Lü, X.T.; Reed, S.; Yu, Q.; He, N.P.; Wang, Z.W.; Han, X.G. Convergent responses of nitrogen and phosphorus resorption to nitrogen inputs in a semiarid grassland. *Glob. Chang. Biol.* **2013**, *19*, 2775–2784. [CrossRef] [PubMed]

42. Becana, M.; Minchin, F.R.; Sprent, J.I. Short-term inhibition of legume $N_2$ fixation by nitrate: I. Nitrate effects on nitrate-reductase activities of bacteroids and nodule cytosol. *Planta* **1989**, *180*, 40–45. [CrossRef] [PubMed]

43. Nebiyu, A.; Huygens, D.; Upadhayay, H.R.; Diels, J.; Boeckx, P. Importance of correct *B* value determination to quantify biological $N_2$ fixation and N balances of faba beans (*Vicia faba* L.) via $^{15}N$ natural abundance. *Biol. Fertil. Soils.* **2013**. [CrossRef]

44. Chisholm, R.H.; Blair, G.J. Phosphorus efficiency in pasture species. I. Measures based on total dry weight and P content. *Aust. J. Agric. Res.* **1988**, *39*, 807–816. [CrossRef]

45. Sanginga, N.; Bowen, G.D.; Danso, S.K.A. Intra-specific variation in growth and $N_2$ fixation of *Leucaena leucocephala* and *Gliricidia sepium* at low levels of soil P. *Plant Soil* **1991**, *127*, 169–178. [CrossRef]

46. Sanginga, N. Role of biological nitrogen fixation in legume based cropping systems; a case study of West Africa farming systems. *Plant Soil* **2003**, *252*, 25–39. [CrossRef]

47. Groom, P.G.; Lamont, B.B. *Plant Life of Southwestern Australia—Adaptations for Survival*; De Gruyter Open: Warsaw, Poland, 2015; pp. 63–67.

48. Shaben, J.; Myers, J.H. Relationship between Scotch broom (*Cytisus scoparius*), soil nutrients, and plant diversity in the Garry oak savannah ecosystem. *Plant Ecol.* **2010**, *207*, 81–91. [CrossRef]

49. Pérez-Fernández, M.; Lamont, B.B. Competition and facilitation between Australian and Spanish legumes in seven Australian soils. *Plant Species Biol.* **2015**. [CrossRef]

50. Kaschuk, G.; Kuyper, W.T.; Leffelaar, P.A.; Hungria, M.; Giller, K.E. Are the rates of photosynthesis stimulated by the carbon sink strength of rhizobial and arbuscular mycorrhizal symbioses. *Soil Biol. Biochem.* **2009**, *41*, 1233–1244. [CrossRef]

51. Kaschuk, G.; Xinyou, Y.; Hungria, M.; Leffelaar, P.A.; Giller, K.E.; Kuyper, W.T. Photosynthetic adaptation of soy bean due to varying effectiveness of $N_2$ fixation by two distinct *Bradyrhizobium japonicum* strains. *Environ. Exp. Bot.* **2012**, *76*, 1–6. [CrossRef]

52. Harris, D.; Pacovsky, R.S.; Paul, E.A. Carbon economy of soybean-*Rhizobium-Glomus* associations. *New Phytol.* **1985**, *101*, 427–440. [CrossRef]

53. He, T.; Lamont, B.B. Species versus genotypic diversity of a nitrogen-fixing plant functional group in a metacommunity. *PopEcol* **2010**, *52*, 337–345. [CrossRef]

# Comparative Phenotypical and Molecular Analyses of Arabidopsis Grown under Fluorescent and LED Light

**Franka Seiler, Jürgen Soll and Bettina Bölter \***

Department Biologie I-Botanik, Ludwig-Maximilians-Universität, Großhadernerstr. 2-4, Planegg-Martinsried 82152, Germany; f.andersch@lmu.de (F.S.); soll@lmu.de (J.S.)
\* Correspondence: boelter@bio.lmu.de

Academic Editor: Milan S. Stankovic

**Abstract:** Comparative analyses of phenotypic and molecular traits of *Arabidopsis thaliana* grown under standardised conditions is still a challenge using climatic devices supplied with common light sources. These are in most cases fluorescent lights, which have several disadvantages such as heat production at higher light intensities, an invariable spectral output, and relatively rapid "ageing". This results in non-desired variations of growth conditions and lowers the comparability of data acquired over extended time periods. In this study, we investigated the growth behaviour of Arabidopsis Col0 under different light conditions, applying fluorescent compared to LED lamps, and we conducted physiological as well as gene expression analyses. By changing the spectral composition and/or light intensity of LEDs we can clearly influence the growth behaviour of Arabidopsis and thereby study phenotypic attributes under very specific light conditions that are stable and reproducible, which is not necessarily given for fluorescent lamps. By using LED lights, we can also roughly mimic the sun light emission spectrum, enabling us to study plant growth in a more natural-like light set-up. We observed distinct growth behaviour under the different light regimes which was reflected by physiological properties of the plants. In conclusion, LEDs provide variable emission spectra for studying plant growth under defined, stable light conditions.

**Keywords:** Arabidopsis; light quality; LED; phenotype; gene expression

---

## 1. Introduction

Light is the most important parameter for plant growth and development. Not only the intensity of light, but also the spectral quality has a great effect on many aspects of plant life such as photosynthetic performance, differentiation, and flowering [1]. Therefore, plants feature several sensors for light quantity, day length (photoperiod), and spectral quality [2]. These sensors (photoreceptors) are at the root of complex signalling networks, which control developmental, physiological, and morphological processes. The range of wavelengths that can be utilized by plants for different purposes ranges from 280 nm to 750 nm [3]. Within this spectrum, 380–730 nm covers the visible light, which is absorbed mainly by chlorophyll a, b, and carotenoids. Violet, blue, and red light play a great role in photosynthesis, whereas red and far-red light (730 nm) also influence germination, vegetative growth, budding, and flowering [4]. Several classes of photoreceptors have been described: phytochromes (in Arabidopsis there are five family members called phyA-phyE) absorb red and far-red light, whereas blue light is perceived by cryptochromes (cry.1 and cry.2), phototropins (phot.1 and phot.2), and Zeitlupes (ZKL, FKF1 and LKP2) [2]. UV-B radiation can be sensed via specific receptors called UVR8. Downstream signalling first and foremost leads to stress-related adaptations which results in the protective measures against the harmful UV-B radiation [5]. Virtually all processes within a life cycle of a plant are initiated and/or regulated by light via perception by photoreceptors and their downstream signalling cascades. This often involves the (de-) activation of transcription factors, which influence the expression of genes e.g., associated with hormone synthesis/transport.

One prominent example is the red and far red light-absorbing phytochrome family that is critically involved in photomorphogenesis, the developmental response of an organism to the information contained in light. This process includes slowed stem elongation, cotyledon/leaf expansion, and greening and straightening of the apical hook. Characteristically, phytochromes occur in an inactive (Pr) and active (Pfr) form, which are interconvertible via conformational changes upon absorption of red or far red light, respectively. Red light (R) induces the active conformation (Pfr), which is translocated to the nucleus, where phytochromes inhibit two different classes of repressive transcription factors. PIFs (phytochrome interacting factors) directly bind to Pfr forms of phytochromes upon which these are phosphorylated followed by ubiquitination and degradation by the proteasomal system. The removal of these and other repressive transcription factors allows the initiation of photomorphogenesis. The molecular details of the complex signalling cascades include massive differential gene regulation and are still not fully understood [6]. Further processes influenced by red/far red light are the shade avoidance response and flowering initiation [2]. Induction of shade avoidance is mediated by a lowered R:FR light ratio, which in fact represents reflected far red light by neighbouring vegetation. Since the red light amount remains constant, the ratio drops and that signals probable imminent shading to the receiving plant which then initiates the avoidance response [7].

Blue light acts via cryptochromes, which also control many developmental processes in plants, specifically de-etiolation, elongation, flowering, and maintenance of the circadian clock. Cryptochromes also play an important role in photomorphogenesis. They derive from ancient DNA repair enzymes known as photolyases. Though cryptochromes no longer fulfil DNA repair-related functions, they are structurally still very similar to their progenitors and have kept FAD as a co-factor. Illumination with blue light seems to induce continuous cycling between different redox states of FAD, which correlates with biological activity. In Arabidopsis, both cryptochrome isoforms (Cry.1 and Cry.2) were localised to the nucleus, but Cry.1 was also found in the cytosol where it seemed to be involved in different regulatory processes than the nuclear localised form [8].

Natural sun light covers the complete range of usable wavelengths, representing an ideal light source, whereas the usually applied illuminants in plant breeding and research are fluorescent, metal-halide, high-pressure sodium, or incandescent lamps. These not only contain wavelengths dispensable for the plant, mainly in the green light area, but also consume huge amounts of energy and generate a great measure of unwanted heat. In comparison, light-emitting diodes (LEDs) have several advantageous features, such as the possibility to control spectral composition and therefore mimic the natural light as closely as possible, a long lifespan, small mass and volume, and negligible heat emission combined with low energy consumption [9]. Though the first reports about the influence of LED light on plant growth appeared already two decades ago [10] and several studies on the topic were published over the years especially concerning legumes or crop plants [9,11–14], no comprehensive investigation on the behaviour of the model plant *Arabidopsis thaliana* grown under LED lights in comparison to fluorescent lights has been conducted to our knowledge.

In the present study, we aimed to compare the morphological and physiological traits of *Arabidopsis thaliana* Columbia 0 plants grown under LED lights of different intensity and spectral composition with plants cultivated under fluorescent lights in a climatic chamber. In addition, we performed gene expression analysis of all plants at day 18 after sowing (18 DAS). We observed extensive differential expression of many RNAs under different light intensities, which are partially correlated to the analysed phenotypic parameters. This in turn makes it quite clear that analysis of gene expression profiles is a good first step, but to clearly and unequivocally correlate this to phenotypic traits, in-depth biochemical investigations are mandatory.

## 2. Results

### 2.1. Plants Grown under Different Light Regimes Show Distinct Phenotypes

We compared the growth behaviour of *Arabidopsis thaliana* Col 0 plants cultivated on soil under either fluorescent or LED lights at different intensities, as well as plants grown under LED light of diverse spectral composition (Figure 1b–f).

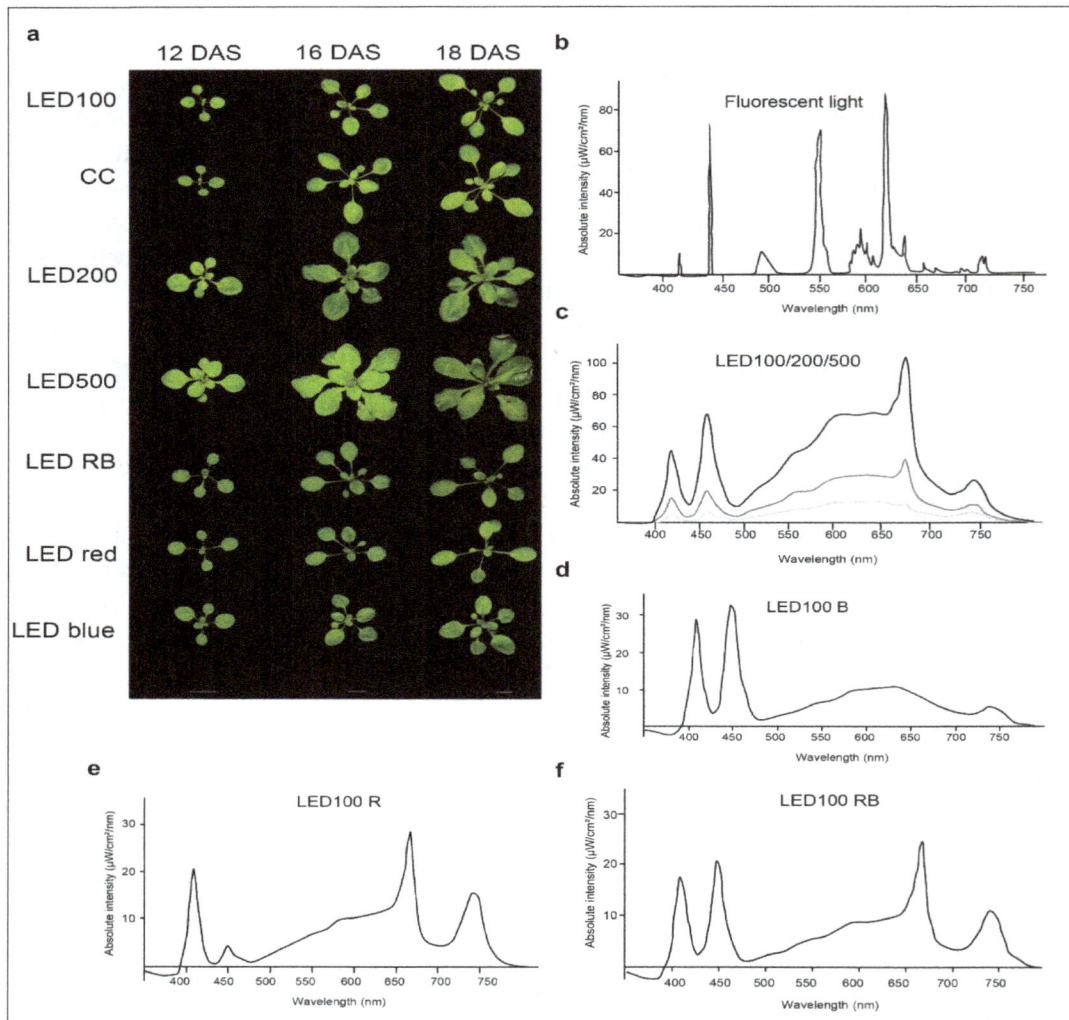

**Figure 1.** *Arabidopsis thaliana* shows distinct growth behaviour under different light and climate regimes. (**a**) Phenotypes of plants grown at either 100 μmol m$^{-2}$ s$^{-1}$ under white light-emitting diode (LED) light (LED100) or fluorescent light in a climatic chamber, or LED light at 200 μmol m$^{-2}$ s$^{-1}$ or 500 μmol m$^{-2}$ s$^{-1}$ LED red, LED blue or LED red/blue. Plants were grown on soil and photographed at 12, 16, and 18 days after sowing (DAS). The scale bar represents 1 cm; (**b–f**) Spectra of the different light regimes. Please note that the total light intensity represents the area below the spectral curves. Thus, the peak heights at the different wavelengths do not reach the value of the total output; (**b**) Spectrum of the fluorescent lamp in the climate chamber; (**c**) LED spectra at 100 (lower light grey curve), 200 (middle dark grey curve), and 500 μmol m$^{-2}$ s$^{-1}$ (upper black curve); (**d**) LED spectrum for LED100 with a raised portion of blue light (B); (**e**) LED spectrum for LED100 with a raised portion of red light (A); (**f**) LED spectrum for LED100 with a raised portion of red/blue light R/B.

Please note that the total light intensity represents the area below the spectral curve. Thus, the peak heights at the different wavelengths do not reach the value of the total output. Pictures were taken at days 12, 16, and 18 after sowing (DAS) (Figure 1a). Our reference point in all experiments was LED100, where the wavelength distribution was adapted according to the sunlight emission spectrum as much as technically possible with our LED set-up (please refer to Figure 6b). Plants grown at the same light intensity of 100 $\mu$mol m$^{-2}$ s$^{-1}$ under fluorescent (Figure 1b, light grey line) or LED light in a climatic chamber (Figure 1b, dark grey line) do not show dramatically different phenotypes (Figure 1a, top and second row). Surprisingly to us, the development of plants cultivated under fluorescent lamps was faster, judged by the number of true leaves, which is significantly higher by 20–30% after 16 and 18 DAS, respectively (Figure 1a, row one and two, Figure 2c indicated by asterisks). The rosette area of plants grown in the climatic chamber differs even more at all days measured from plants cultivated in the LED chamber: 120% after 12 DAS, 160% after 16 DAS, and even 270% after 18 DAS (Figure 2a). However, the fresh weight was not significantly different from plants grown under fluorescent or LED light (Figure 2b), indicating that even though there are less leaves with a smaller area, their thickness is higher in plants grown under LED lamps, which could be due to the comparatively higher amount of blue light.

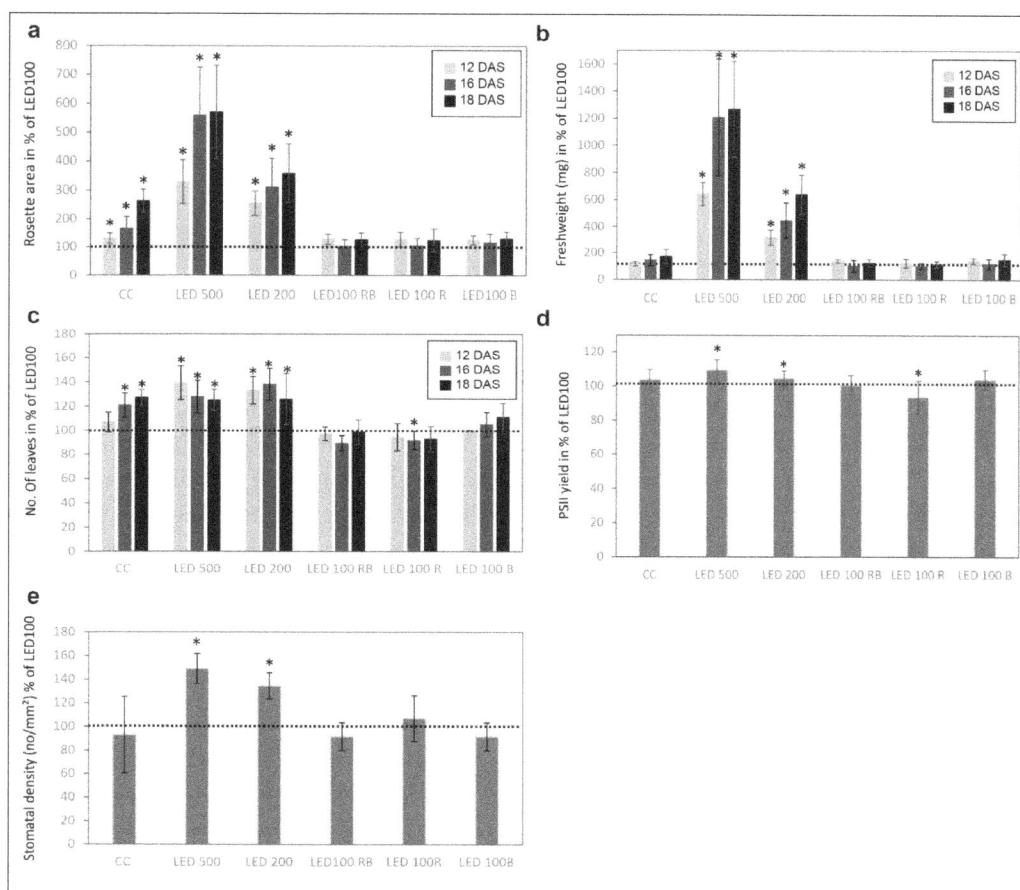

**Figure 2.** Physiological parameters of plants grown under different light regimes. (**a**) Mean rosette area in % compared to LED100 from $n = 6$ plants measured at 12, 16, and 18 DAS. The dotted line represents the value of plants grown at LED100 in all graphs; (**b**) Mean fresh weight (mg) from $n = 6$ plants at 12, 16, and 18 DAS; (**c**) Number of leaves in % compared to LED100 from $n = 6$ plants measured at 12, 16, and 18 DAS; (**d**) PSII yield from plants at 18 DAS in % compared to LED100; (**e**) Stomatal density given in number/mm$^2$ from plants at 18 DAS. Significant changes compared to LED100 according to the student's T test are indicated by an asterisk (*).

Raising the light intensity in the LED chamber to 200 μmol or 500 μmol, respectively, under constant spectral distribution led as expected to faster development (more leaves, bolting visible) and growth (bigger leaf areas and higher fresh weight) (Figure 1a, third and fourth row, Figure 2a–c). Plants subjected to high light intensities accumulated much more fresh weight: in the case of LED200 about 600%, at LED500 even up to 1300% compared to LED100. To exclude any adverse effects of heat emitted by the LED lamps under high light intensities, we measured the temperature at the leaf level and found it to be at 23 °C, thus we can rule out that our observations on phenotypic or gene expression levels are caused by elevated temperatures.

In parallel, we analysed the effects of spectral quality while keeping the total light intensity at a constant 100 μmol. Therefore, white light (3 K) was lowered in all three conditions. To achieve a light quality with a relatively high portion of red light (for simplicity we call it "red, R"), the electrical output of the 660 nm LEDs was enhanced and the 730 nm LEDs were kept unchanged, while the UV (395 nm) and blue light (440 nm) LEDs were lowered in their output. To bring about a high portion of blue light (named "blue, B"), UV and blue LED output was raised, while red and far red light LEDs were lowered in their output. In a combination of both settings (we call it "red blue, RB"), we achieved spectral peaks at 440 nm as well as 660/730 nm. An overview about the spectral composition, measured as fluence rates, is depicted in Figure 1d–f. By this we ensured that the total light intensity (meaning the fluence rate) was kept at 100 μmol in all conditions. Plants grown under a combination of red and blue (RB) LED light compared to LED100 were growing very similarly (Figure 1a, top and fifth panel, Figure 2). Red light seemingly led to marginally slower growth, however this was not represented by significant differences in the measured phenotypical characteristics (Figures 1a and 2). Blue light resulted in a more compact-appearing phenotype and shorter petioles, whereas neither rosette area nor fresh weight nor number of leaves were different from LED100. Thus, we could clearly demonstrate that by using LED lamps with distinct spectral characteristics we can influence the growth behaviour concerning the appearance of Arabidopsis plants.

## 2.2. Physiological Traits Differ in Plants Grown under Different Light Regimes

In addition to visual analyses, we investigated several physiological traits of the differently cultivated plants at 18 DAS (Figure 2d,e). The photosynthetic performance showed small but clear differences between the plant populations (Figure 2d). The PSII yield was higher in plants from LED200 and 500 and red light. Thus, the higher number of leaves and greater leaf area in the plants from the climate chamber with fluorescent lights does not result in a higher photosynthesis rate compared to plants from the same light intensity under LED lamps. Consequently, these leaves seem to feature either a lower number of photosystems per area or less active ones. In contrast, high light intensity leads to overall bigger plants and better photosynthetic performance.

Since stomata development is known to be light dependent [15], we analysed the stomatal density in plants from all conditions at 18 DAS (Figure 2e). LED200 and LED500 plants had an evidently higher stoma density (140 and 130% of LED100, respectively). All other plants were similar to the ones from LED100 concerning the number of stomata, indicating that for this physiological trait light fluence rate is more important than wavelength composition. This observation is in line with previously reported data that high irradiation increases the stomatal density [16]. It was found in the same study that monochromatic red light also influences this trait, but probably since we did not apply monochromatic light, but rather increased the portion of red light with still white and blue light present, we did not detect a similar effect.

## 2.3. Affymetrix Analysis Reveals Differentially Regulated RNA Expression

In an attempt to correlate the monitored morphological and physiological features with gene expression, we performed Affymetrix analysis with RNA isolated from leaves harvested at 18 DAS. Genes were defined as differentially regulated the if the p-value was lower than 0.05. As for the data in Figure 2, LED100 was applied as the reference point in all evaluations. We then compiled separate groups for comparative analyses: 1. climate chamber with fluorescent lamps and LED lights; 2. LED200 and LED500 as well as 3. R, B, and RB (Figures 3–5). For each group except the first, we generated Venn diagrams of total, up- and down-regulated genes (Figures 4a–c and 5a–c), respectively. We executed MapMan analyses for all groups in which the bins are represented as bar charts (Figures 3, 4d,e and 5d,e).

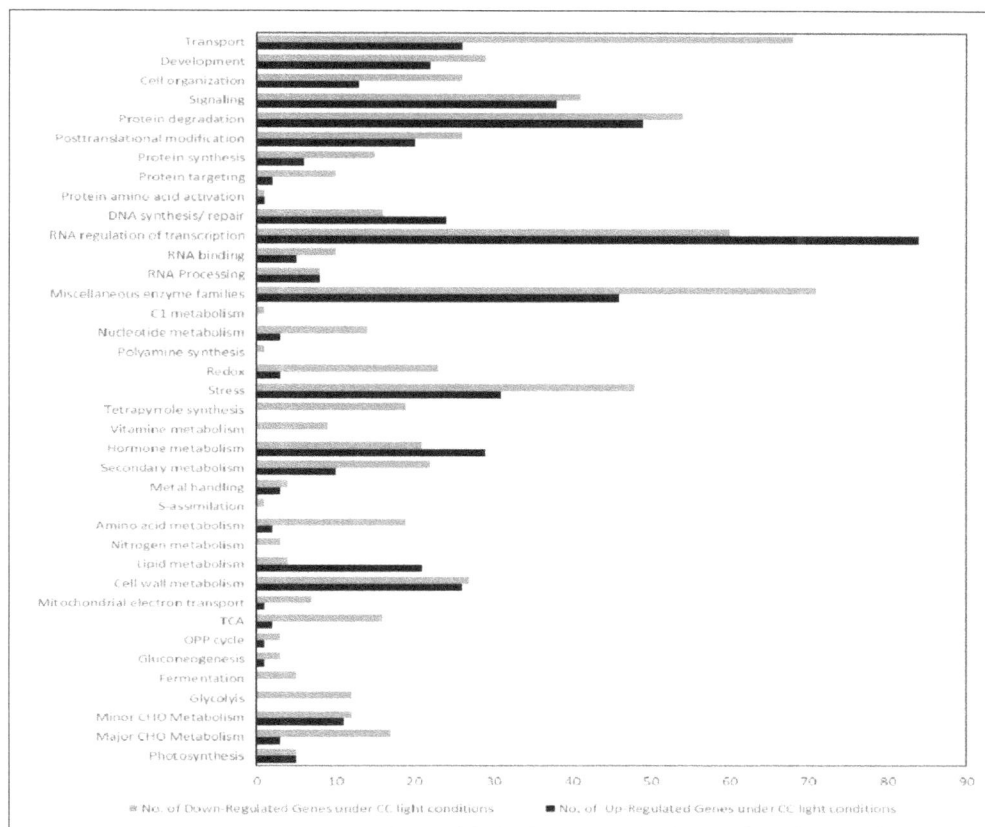

**Figure 3.** Comparative gene expression analysis of RNA isolated from plants grown in LED100, climate chamber. Bar chart of up (black bars) and down (grey bars) regulated genes from plants grown in the climate chamber.

For the first group (climate chamber with fluorescent lights compared to LED100), we found about 200 genes differentially expressed with a fold change (FC) >2 (Supplemental Table S1). We can, however, not completely exclude that some effects on gene expression between plants in the climate chamber with fluorescent lamps and plants in the chamber with LED lights are due to the different climate chambers. These are of extremely similar build, volume, and set-up, but not of identical make.

Noticeably, the highest number of genes regulated is to be found in the MapMan bin RNA regulation and transcription, indicating that gene expression in general is clearly influenced by the different light quality. Apart from that, many genes in the groups of transport, protein homeostasis, and enzymes are differentially expressed, though no single pathway stands out. Since the appearance of the plants is so similar, this indicates that diverse expression of genes is not necessarily reflected in a distinctive visible phenotype. Nevertheless, one needs to be aware that applying different light sources leads to fundamental changes in the transcriptome, even though the overall characteristics of plants grown under disparate types of lamps are not easily distinguishable.

In the second group, comprising plants from the different light intensities LED200 and LED500 (Figure 4), a total of 9500 RNAs is differentially expressed compared to LED100, demonstrating that many physiological activities are altered under higher light intensities, which is clearly reflected by the pronounced growth of the plants. From these, 1454 genes are down regulated in LED200 (Figure 4c) as well as in LED500 (Figure 4b) and 1481 are up regulated in both categories. In inference, this means that 2754 RNAs are less-expressed solely in plants from LED500, and 156 RNAs from LED200. Down-regulated exclusively in LED500 were 3485 genes, restricted to LED200 were 186 genes (Figure 4a–c). Thus, higher light intensity leads in general to a greater response in RNA level. The fact that about 30% of all genes are regulated under both conditions with the FC bigger at higher intensities demonstrates the reproducibility of our approach. Concerning the distribution of these genes across the MapMan bins, the picture is somewhat different from the first group (Figure 4d,e). Focussing on genes with a FC > 2 (Supplementary Table S1) reveals that most differentially expressed RNAs belong to the bin proteins. The largest portion of up regulated RNAs from this bin code for proteins involved in the processes of degradation, followed by modification. This implicates that under higher light intensities protein turnover might be faster. Surprisingly, among the lower expressed RNAs, most can be assigned to the class of synthesis of ribosomal proteins, which suggests that while plants grown under higher light intensities perform protein degradation to a larger extent, they do not counteract that by increasing the translation machinery to produce more proteins de novo. Since these plants clearly develop faster and have significantly more biomass, the phenotypic observations seem quite contradictory to the RNA expression patterns. However, it may well be that the ribosomal proteins are protected against degradation and are simply more stable than many other proteins. Thus, this RNA expression pattern does not necessarily indicate a lower protein synthesis activity per se. The highest fold change can be observed for BAM5 under LED500 which codes for a putative ß-amylase (FC = 157), while other genes from the starch degradation pathway are less drastically but still prominently up-regulated (FC = 4–13). Considering the accelerated growth under the applied conditions, this nicely reflects the elevated need for carbohydrates. In the same line, we can interpret the increased expression of genes related to transport functions, since many metabolites need to be transported within the plant to ensure growth. Since plant growth, among many other complex processes includes cellular expansion, the up-regulation of auxin-related genes fits into this picture [17]. High light intensities of especially 500 μmol also lead to increased generation of reactive oxygen species, consequently we find RNAs belonging to the bins stress and redox at a higher expression level in plants from LED200 and LED500. Since the plants have a generally healthy appearance, this up-regulation does not necessarily indicate stress, but acclimation mechanisms to prevent stress-related damage.

In the last group, we compared gene expression in plants grown under red light, blue light, or a combination of red and blue, all supplemented with white light at 25% of the default intensity LED100, thereby ensuring a total intensity of 100 μmol m$^{-2}$ s$^{-1}$. In contrast to our expectations, very few genes were differentially expressed compared to LED100 (Figure 5). It has been known for some time that red light induces the phytochrome system, whereas blue light results in the induction of phototropins and cryptochromes (for review, see [2]). However, all these previous studies were conducted after implementing pulses of the respective wavelengths, immediately followed by gene expression analysis, while we grew the plants continuously under the indicated spectral regimes for 18 days. This obviously leads to long-term acclimation and thereby results in a different pattern of gene expression, when the peak of light-induced expression might already have flattened. A total of only 121 RNAs was regulated in expression in relation to LED100 (Figure 5a). A single gene is commonly up-regulated in all conditions which belongs to the signalling bin. Nothing more is known about its function, though it is annotated as a receptor kinase (Supplementary table). This gene was found to be significantly regulated in many conditions, thus it's questionable if this represents a specific effect. 38 RNAs appear at a higher level solely in blue light, most of them distributed among the bins signalling, transcription, protein degradation, enzymes, and redox (Figure 5e). Expressed at a lower level, we found RNAs mainly concerning photosynthesis-related functions, which is surprising in view of our PSII yield measurements that show a slightly higher, though not significantly, activity of photosystem II (Figure 2d). Since the effected genes are coding for PSI, PSII, and NDH complex components, the RNA expression level obviously does not correspond to process performance in this case. Support for this conclusion comes from studies of translational activity compared to gene expression, which is not necessarily coupled [18]. Other down-regulated genes primarily participate in protein synthesis, transcription, stress, and cell wall metabolism (Figure 5e).

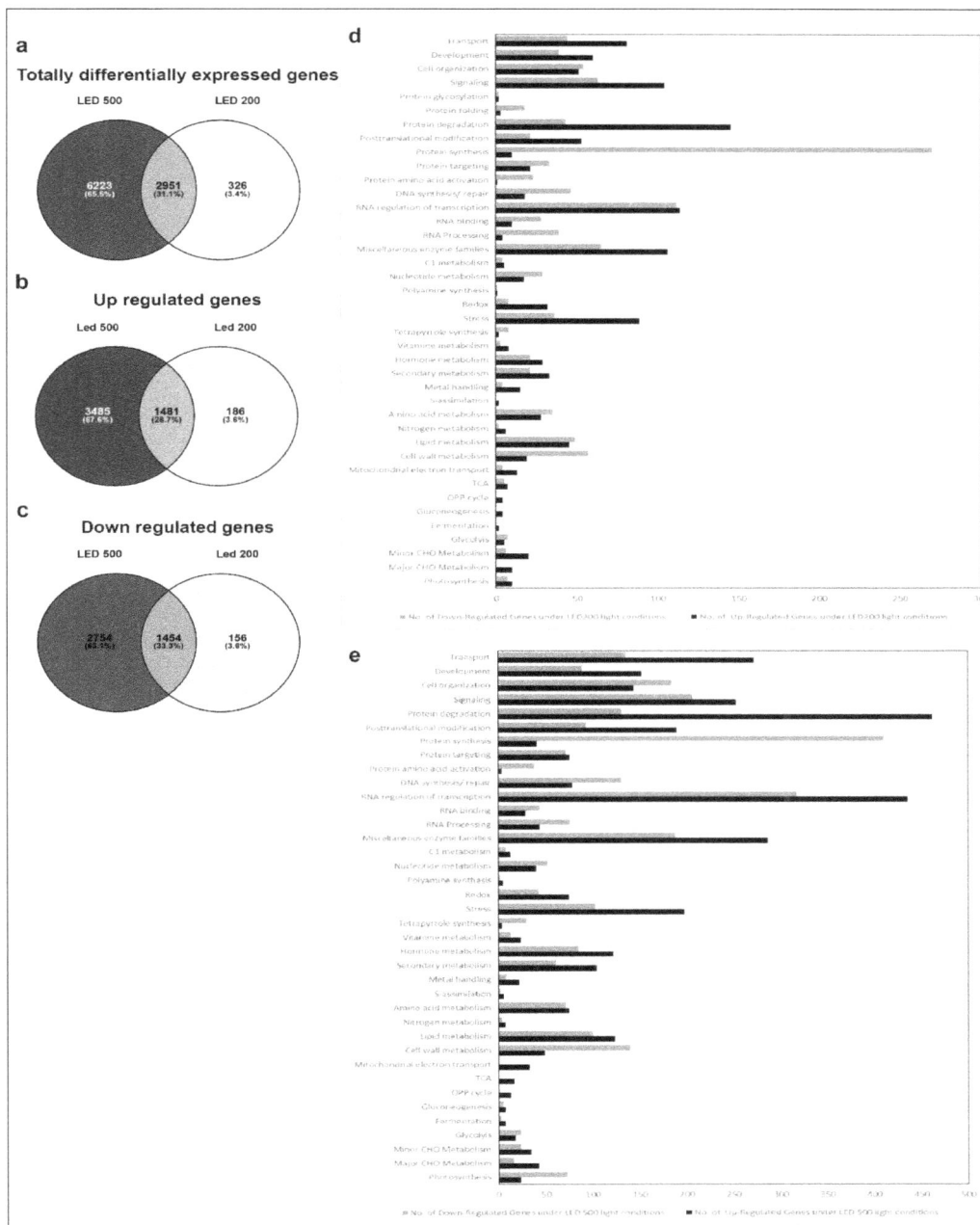

**Figure 4.** Comparative gene expression analysis of RNA isolated from plants grown in LED100, LED200 and LED500. (**a–c**) Venn diagrams of a total differentially expressed; (**b**) up regulated; (**c**) down regulated genes; (**d**) Bar chart of up (black bars) and down (grey bars) regulated genes from plants grown in LED200; (**e**) Bar chart of up (black bars) and down (grey bars) regulated genes from plants grown in LED500.

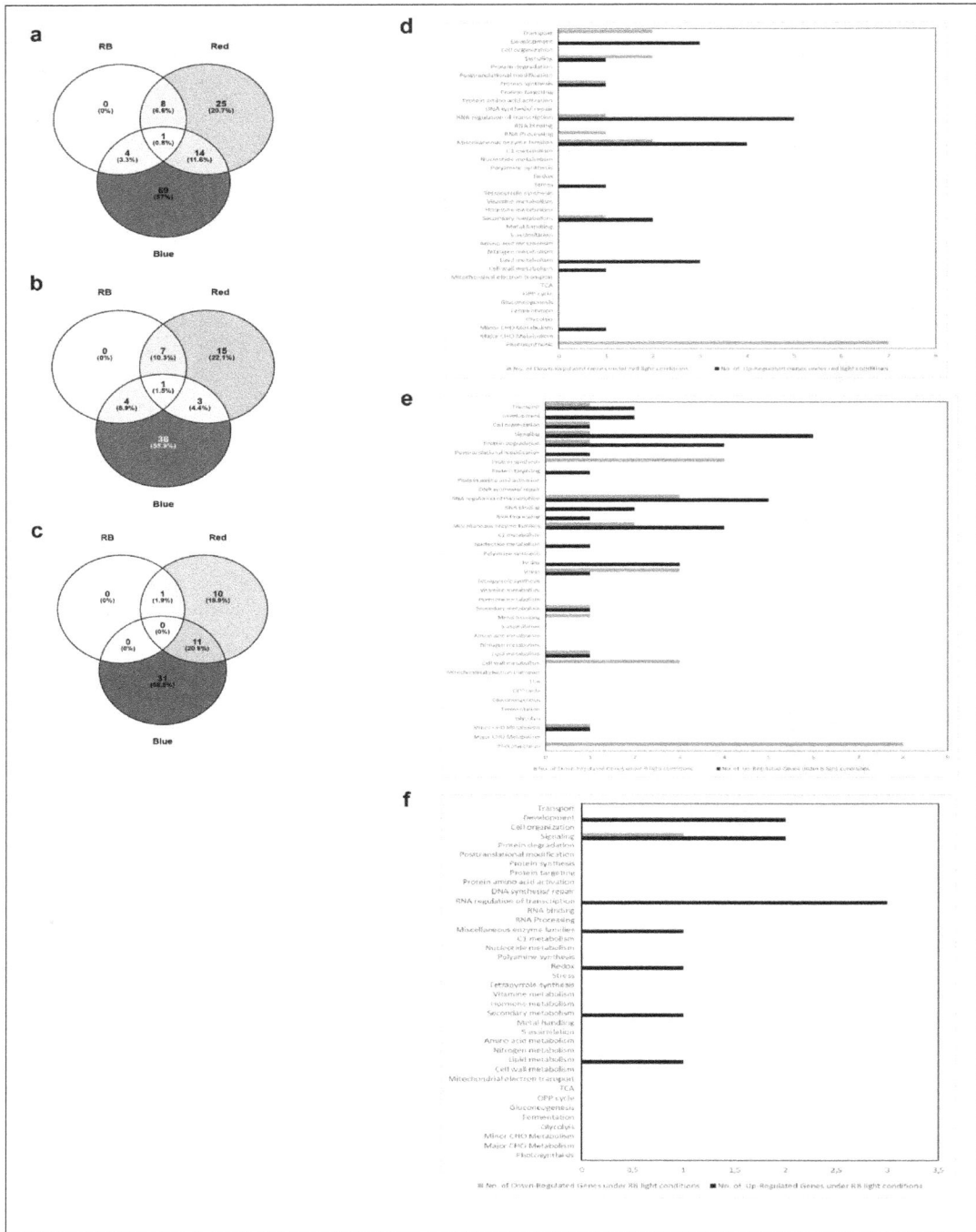

**Figure 5.** Comparative gene expression analysis of RNA isolated from plants grown in LED100, LED R, LED B, LED RB. (**a–c**) Venn diagrams of a total differentially expressed; (**b**) up regulated; (**c**) down regulated genes; (**d**) Bar chart of up (black bars) and down (grey bars) regulated genes from plants grown LED R; (**e**) Bar chart of up (black bars) and down (grey bars) regulated genes from plants grown in LED B; (**f**) Bar chart of up (black bars) and down (grey bars) regulated genes from plants grown in LED RB.

Red light leads to the elevated expression of genes from development, signalling, protein synthesis, transcription, miscellaneous enzymes, stress/acclimation, secondary metabolism, lipid metabolism, and cell wall and CHO metabolism (Figure 5d). Less transcribed under red light are predominantly genes belonging to the category of photosynthesis, concerning specifically PSII constituents (Supplementary table), which in this case is in line with the pulse-amplitude modulations (PAM) measurements (Figure 2d). The remaining down-regulated genes are involved in transport, signalling, protein synthesis, transcription and RNA processing, enzymes, and secondary and CHO metabolism.

Combining red and blue light results in a revocation of effects from red or blue light on gene expression level. While the plants phenotypically more resemble the ones cultivated under only red light, differential RNA synthesis overlaps with red and blue light treatment (Figure 5f). Interestingly, the number of genes that are regulated here is lower than under red or blue light, respectively (Figure 5a). The respective genes appear consistently as differentially expressed throughout all conditions analysed, indicating that these might represent key genes regulated upon specific treatments with red and blue light.

## 3. Discussion

Altogether, we can conclude that growth behaviour of *Arabidopsis thaliana* Col 0 is clearly influenced by light quantity and quality (Figure 1). This in itself is not a novel concept, but our conditions for plant growth were quite unusual, in that we did not treat the plants with short periods of different wave lengths and look for short-term effects, but kept them under high portions of red, blue, or red/blue light throughout the whole experiment.

Provided with sufficient nutrients to build cellular material, high light intensities led to about ten-times faster growth and development (LED500 compared to LED100). In spite of the high radiation, plants cultivated at 500 μmol appeared healthy and were obviously not stressed, which would have been reflected in anthocyanin accumulation and thus a reddish colour [19]. One should keep that in mind when analysing the effect of high light on plant physiology, so that general stress reactions are not confused with high light specific acclimatory processes.

Exposing the plants to higher doses of blue light also had visible effects on their phenotypes: it led to more stunted growth with short petioles but more leaves. In a combination of red and blue light, the effect of red light prevails and plants looked very similar to those raised under red light only. It was shown in several physiological studies that for the fine-tuning of light responses, the photoreceptors for red and blue light act synergistically [20,21]; under the conditions applied in our study, signalling from red light receptors seems to have a dominant effect.

Though LED light still has profound consequences for gene expression, phenotypically the plants are not dramatically different. This observation is in line with a previous publication where several laboratories strived to achieve comparability in their plant growth setups and found that, especially on the gene expression level, the variation was already quite high between individual plants from the same growth cabinet which seemed to originate from changes in the microenvironment [22]. Gene expression obviously reacts much faster than phenotypic appearance, which one needs to take into consideration upon interpreting gene expression data in correlation to phenotype or biochemical activity.

Plants incubated under 100 μmol m$^{-2}$ s$^{-1}$ LED light displayed slightly slower growth compared to plants from the climate chamber equipped with fluorescent lights (Figure 1a). Considering that the LED light features a nearly optimal wavelength composition with regard to comparability to the natural sun light and in contrast to the spectral distribution provided by fluorescent lamps (Figure 1b–d), we would have anticipated accelerated development under LED light. However, on closer scrutiny of the spectra and super-positioning of both emission curves (Figure 1b,c), it becomes clear that the very narrow but high peaks emitted by the fluorescent lamps in the blue and red light range cover a very similar area as the broader but lower peaks from the LED lights. This probably results in roughly the same or even higher amount of photosynthetically active radiation emitted by the fluorescent lamps at 100 μmol m$^{-2}$ s$^{-1}$ compared to our setup of the LEDs. Thus, in a next step we will further vary the spectral distribution of the LED lights at 100 μmol m$^{-2}$ s$^{-1}$ and compare those to fluorescent lamps.

In this study, we could clearly show that manipulation of the wavelengths and/or intensity of the applied light source can influence the growth behaviour and gene expression of Arabidopsis plants. By using LEDs, it is possible to roughly mimic the natural light conditions and study plant growth of different genotypes as closely as possible to what would happen in the field. This is a considerable advantage of LEDs over other commercial light sources, which feature fixed emission spectra. Furthermore, fluorescent lamps exhibit changes in emission quality after a relatively short life time, whereas LEDs provide stable light quality and quantity over extended time periods. In addition, using LEDs avoids the added stress factor of excessive heat produced by commercial lamps, which drastically influences gene expression and development. This is currently a common problem when studying plant response to high light treatment, because it leads unavoidably to a higher leaf temperature. In our setup, the temperature at leaf level was constantly 23 °C under 500 μmol LED light. Furthermore, the energy consumption of LEDs is much lower than that of any other available light source, thus substantially reducing the running costs of climate chambers/greenhouses.

## 4. Materials and Methods

### 4.1. Plant Materials and Growth Conditions

All experiments were performed on *Arabidopsis thaliana* ecotype Columbia 0. Seeds of Arabidopsis were sown on soil and vernalized for two days at 4 °C. Then, the plants were transferred to environmentally controlled growth chambers under different light treatments: climatic chamber (CC) refers to a growth chamber equipped with fluorescent lights and LED refers to a growth chamber supplied with light emitting diodes. In addition, the spectral output of the fluorescent lamps was determined for a new lamp compared to one of about six months of age (Supplemental Figure S1) to provide an idea of when to change lamps for comparable experiments.

The LED lights had the following specifications: OSRAM Osslon SSL Far red 730, OSRAM Osslon SSL Deep blue 450, OSRAM Osslon SSL hyper red 660, OSRAM Osslon SSL light colour 3.000 K (Osram, Munich, Germany), Edison Federal 3535-UVA 395–410 (LED-Tech, Moers, Germany). Environmental conditions in the chambers were set at a 22 °C/18 °C day/night temperature, air relative humidity of 50–65%, and a 16 h photoperiod. In total 50 seedlings per condition and light treatment were used. Plants were watered by subirrigation and fertilised upon every second watering, usually every 2 to 3 days, depending on the growth stage. The spectral distributions for the seven light treatments are shown in Figure 1b–f and were determined using a calibrated OceanOptics (Duiven, Netherlands ) spectrometer at the day of sowing. In short, we programmed the desired light intensities into the software of the LED chamber. The actual output is controlled by a built-in sensor within the chamber. The spectra were recorded at the setting for absolute intensities ($\mu W/cm^2/nm$), which is the most precise method. Therefore, we could monitor the total light output at the same time as the distribution over the wavelength spectrum.

### 4.2. Phenotypic Analysis

To closely monitor the growth of Arabidopsis plants under different light conditions, a detailed phenotypic analysis was conducted. For visual monitoring, eight plants per condition were photographed in two-day intervals. Depicted in Figure 1 are exemplary plants from 12 DAS, 16 DAS, and 18 DAS.

### 4.3. Spectroscopic Analysis

Chlorophyll a fluorescence of Arabidopsis leaves was measured using a pulse-modulated fluorimeter (Imaging PAM Mini; Walz) and the PSII yield was determined according to the manufacturer's instructions.

## 4.4. Stomata Density

Three fully-expanded rosette leaves from 3 individual plants were collected into 70% ethanol, cleared from chlorophyll overnight at room temperature, and stored at 4 °C as needed and then further cleared in chloral hydrate solution (chloral hydrate:water:glycerol (8:2:1, $w/v/w$)). Differential Interference Microscopy (DIC) images of the abaxial surface were captured with a Leica DM1000 microscope (Leica Microystems, Wetzlar, Germany)at $40\times$ magnification. Stomatal density ($mm^2$) was manually counted for all pictures and all leaves.

## 4.5. Fresh Weight Determination

Total fresh weight data were obtained by carefully removing all of the leaves including petioles from six Arabidopsis plants from each light condition. All of the individual leaves were immediately weighed to obtain fresh weight data. Data were subjected to statistical analyses (Student's $t$ Test).

## 4.6. Rosette Area Determination

Rosette areas from eight plants per condition were determined graphically with ImageJ and the data were statistically evaluated by Student's $t$ Test.

## 4.7. Transcriptomic Profiling Using Affymetrix ATH1 Microarray

For microarray analysis, leaves of 18-day-old Arabidopsis plants were used. To provide biological replicates, three samples were harvested from 10 individual plants. Total RNA was extracted using the Plant RNeasy Extraction kit (Qiagen, Hilden Germany). RNA concentration, purity, and integrity were determined. The purified RNA (200 ng) was used to produce biotinylated cRNA probes by using an Affymetrix 3'-IVT Express kit (Affymetrix, High Wycombe, UK) according to the manufacturer's instructions. A total of 12 μg biotinylated cRNA was fragmented and hybridized to GeneChip *Arabidopsis* ATH1 arrays containing 22,810 probe sets. Washing and staining were done on an Affymetrix GeneChip Fluidics Station 250. The array chips were scanned using an Affymetrix GeneArray Scanner 3000. Raw signal intensity values (CEL files) were computed from the scanned array images using the Affymetrix GeneChip Command Console 3.0 (Affymetrix, Santa Clara, USA). For quality checking and normalisation, the raw intensity values were processed with Robin software [23] using default settings. Specifically, for background correction, the robust multiarray average normalization method [24] was performed across all arrays. Statistical analysis of differential gene expression was carried out using the linear model-based approach developed by Smyth [25]. The obtained $p$ values were corrected for multiple testing using the strategy described by Benjamini and Hochberg [26] separately for each of the comparisons made. Genes that showed a $p$-value lower than 0.05 were considered significantly differentially expressed. The normalised log2 values where then used to compare the transcriptomic changes using MapMan. Based on MapMan BINs [27] the significantly expressed genes were functionally annotated. In total, we analysed the following comparisons: LED100 versus Climate chamber, LED100 versus LED500, LED100 versus LED200, LED100 versus LED RB, LED100 versus LED R, and LED100 versus LED B.

# 5. Conclusions

While plant growth under fluorescent lights leads to acceptable results and healthy plants, the spectral distribution provided by those lamps is far from comparable to natural sun light. If we strive to observe and analyse our plants as close as possible to the conditions outside in their natural habitat, we need to apply variable light sources such as LED lights where single wavelengths can be adjusted and freely combined. As a further subject for thought, in addition to our study conducted with plants cultivated in climatic chambers, we grew Arabidopsis outside under completely natural conditions (Figure 6a). These plants are obviously smaller and more compact than the plants from climatic chambers, while the leaf phenotypes concerning petiole length and leaf area are quite similar to the plants grown at LED500. The spectral distribution of LED500 was adjusted as close as technically possible to full sun light (Figure 6b). Of course, one needs to keep in mind that the plants grown in the field were subjected to constantly changing humidity and temperature as well as very variable light conditions due to weather changes, which we didn't monitor continuously. As depicted in Figure 6b,c, the intensity alone changed immensely from cloudy conditions to full sun light. Thus, the divergent growth behaviour is not solely due to different spectral quality. With the use of LED lights, we can move forward towards our goal of observing plant growth and behaviour under conditions mimicking those occurring in nature—especially when further technical advance will allow us to synchronise light conditions from outside with those in climatic chambers.

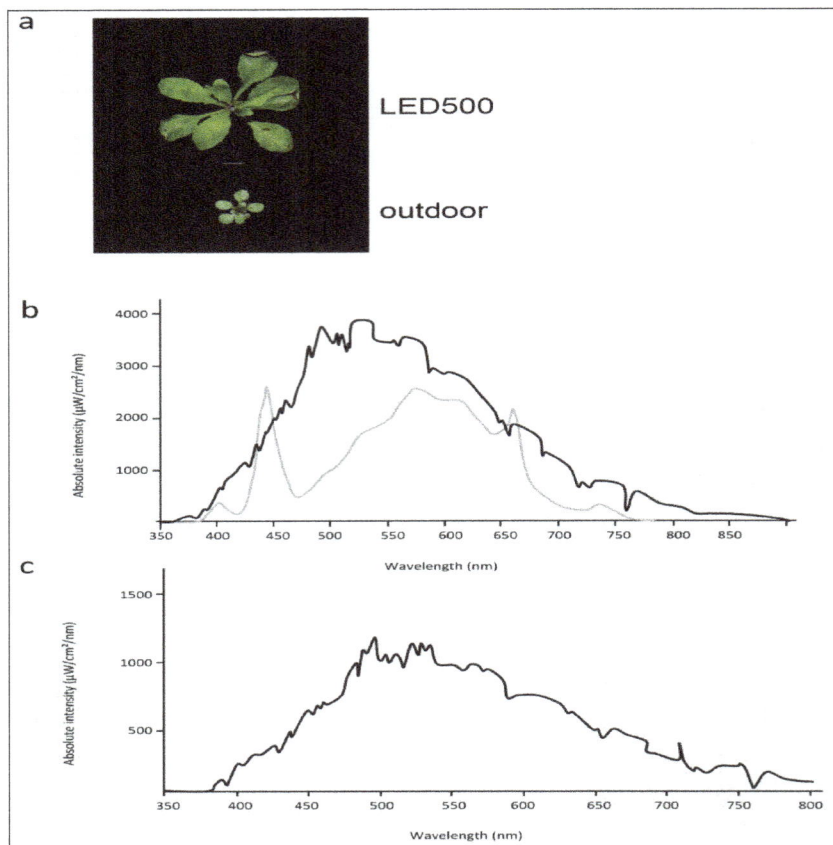

**Figure 6.**  Arabidopsis grows better under completely controlled environmental conditions. (**a**) Phenotypes of plants grown under LED500 or outdoor. Plants were grown on soil and photographed at 18 DAS or, in case of the plant grown outdoor, 38 DAS, respectively. The scale bar represents 1 cm. Please note that the picture of LED500 is identical to Figure 1; (**b**) Emission spectra of LED500 (lower grey line) and natural light at sunny conditions (upper black line); (**c**) Emission spectrum of natural light at cloudy conditions.

**Acknowledgments:** We thank K. Hoefig for critical reading of the manuscript. F.S. was financially supported by Rhenac Green Tec AG (Hennef, Germany) for some time of the conducted study.

**Author Contributions:** Franka Seiler did all the experimental work and analysed data, Jürgen Soll and Bettina Bölter analysed data and wrote the manuscript.

**Conflicts of Interest:** The authors declare no conflict of interest.

## Abbreviations

| | |
|---|---|
| LED | light emitting diode |
| DAS | days after sowing |
| R | high portion of red light |
| B | high portion of blue light |
| RB | high portions of red and blue light |

## References

1. Smith, H. Light quality, photoperception, and plant strategy. *Annu. Rev. Plant Physiol.* **1982**, *33*, 481–518. [CrossRef]
2. Kami, C.; Lorrain, S.; Hornitschek, P.; Fankhauser, C. Chapter two—Light-regulated plant growth and development. In *Current Topics in Developmental Biology*; Marja, C.P.T., Ed.; Academic Press: Cambridge, MA, USA, 2010; Volume 91, pp. 29–66.
3. Pocock, T. Light-emitting diodes and the modulation of specialty crops: Light sensing and signaling networks in plants. *HortScience* **2015**, *50*, 1281–1284.
4. Singh, D.; Basu, C.; Meinhardt-Wollweber, M.; Roth, B. Leds for energy efficient greenhouse lighting. *Renew. Sustain. Energy Rev.* **2015**, *49*, 139–147. [CrossRef]
5. Jenkins, G.I. The uv-b photoreceptor uvr8: From structure to physiology. *Plant Cell* **2014**, *26*, 21–37. [CrossRef] [PubMed]
6. Xu, X.; Paik, I.; Zhu, L.; Huq, E. Illuminating progress in phytochrome-mediated light signaling pathways. *Trends Plant Sci.* **2015**, *20*, 641–650. [CrossRef] [PubMed]
7. Pierik, R.; de Wit, M. Shade avoidance: Phytochrome signalling and other aboveground neighbour detection cues. *J. Exp. Bot.* **2014**, *65*, 2815–2824. [CrossRef] [PubMed]
8. Wu, G.; Spalding, E.P. Separate functions for nuclear and cytoplasmic cryptochrome 1 during photomorphogenesis of arabidopsis seedlings. *Proc. Natl. Acad. Sci. USA* **2007**, *104*, 18813–18818. [CrossRef] [PubMed]
9. Lin, K.-H.; Huang, M.-Y.; Huang, W.-D.; Hsu, M.-H.; Yang, Z.-W.; Yang, C.-M. The effects of red, blue, and white light-emitting diodes on the growth, development, and edible quality of hydroponically grown lettuce (lactuca sativa l. Var. Capitata). *Sci. Hortic.* **2013**, *150*, 86–91. [CrossRef]
10. Brown, C.S.; Schuerger, A.C.; Sager, J.C. Growth and photomorphogenesis of pepper plants under red light-emitting diodes with supplemental blue or far-red lighting. *J. Am. Soc. Hortic. Sci.* **1995**, *120*, 808–813. [PubMed]
11. Barreiro, R.; Guiamét, J.J.; Beltrano, J.; Montaldi, E.R. Regulation of the photosynthetic capacity of primary bean leaves by the red:Far-red ratio and photosynthetic photon flux density of incident light. *Physiol. Plant.* **1992**, *85*, 97–101. [CrossRef]
12. Johkan, M.; Shoji, K.; Goto, F.; Hahida, S.; Yoshihara, T. Effect of green light wavelength and intensity on photomorphogenesis and photosynthesis in lactuca sativa. *Environ. Exp. Bot.* **2012**, *75*, 128–133. [CrossRef]
13. Li, Q.; Kubota, C. Effects of supplemental light quality on growth and phytochemicals of baby leaf lettuce. *Environ. Exp. Bot.* **2009**, *67*, 59–64. [CrossRef]
14. Nanya, K.; Ishigami, Y.; Hikosaka, S.; Goto, E. *Effects of Blue and Red Light on Stem Elongation and Flowering of Tomato Seedlings*; International Society for Horticultural Science (ISHS): Leuven, Belgium, 2012; pp. 261–266.
15. Klermund, C.; Ranftl, Q.L.; Diener, J.; Bastakis, E.; Richter, R.; Schwechheimer, C. Llm-domain b-gata transcription factors promote stomatal development downstream of light signaling pathways in *Arabidopsis thaliana* hypocotyls. *Plant Cell* **2016**, *28*, 646–660. [CrossRef] [PubMed]
16. Casson, S.A.; Franklin, K.A.; Gray, J.E.; Grierson, C.S.; Whitelam, G.C.; Hetherington, A.M. Phytochrome b and pif4 regulate stomatal development in response to light quantity. *Curr. Biol.* **2009**, *19*, 229–234. [CrossRef] [PubMed]
17. Lavy, M.; Estelle, M. Mechanisms of auxin signaling. *Development* **2016**, *143*, 3226–3229. [CrossRef] [PubMed]
18. Oelze, M.-L.; Muthuramalingam, M.; Vogel, M.O.; Dietz, K.-J. The link between transcript regulation and de novo protein synthesis in the retrograde high light acclimation response of *Arabidopsis thaliana*. *BMC Genom.* **2014**, *15*, 320. [CrossRef] [PubMed]

19.  Mouradov, A.; Spangenberg, G. Flavonoids: A metabolic network mediating plants adaptation to their real estate. *Front. Plant Sci.* **2014**, *5*, 620. [CrossRef] [PubMed]

20.  Mohr, H. Coaction between pigment systems.  In *Photomorphogenesis in Plants*; Kendrick, R.E., Kronenberg, G.H.M., Eds.; Springer Netherlands: Dordrecht, The Netherlands, 1994; pp. 353–373.

21.  Usami, T.; Matsushita, T.; Oka, Y.; Mochizuki, N.; Nagatani, A. Roles for the n- and c-terminal domains of phytochrome b in interactions between phytochrome b and cryptochrome signaling cascades. *Plant Cell Physiol.* **2007**, *48*, 424–433. [CrossRef] [PubMed]

22.  Massonnet, C.; Vile, D.; Fabre, J.; Hannah, M.A.; Caldana, C.; Lisec, J.; Beemster, G.T.S.; Meyer, R.C.; Messerli, G.; Gronlund, J.T.; et al. Probing the reproducibility of leaf growth and molecular phenotypes: A comparison of three arabidopsis accessions cultivated in ten laboratories. *Plant Physiol.* **2010**, *152*, 2142–2157. [CrossRef] [PubMed]

23.  Lohse, M.; Nunes-Nesi, A.; Krüger, P.; Nagel, A.; Hannemann, J.; Giorgi, F.M.; Childs, L.; Osorio, S.; Walther, D.; Selbig, J.; et al. Robin: An intuitive wizard application for r-based expression microarray quality assessment and analysis. *Plant Physiol.* **2010**, *153*, 642–651. [CrossRef] [PubMed]

24.  Irizarry, R.A.; Hobbs, B.; Collin, F.; Beazer-Barclay, Y.D.; Antonellis, K.J.; Scherf, U.; Speed, T.P. Exploration, normalization, and summaries of high density oligonucleotide array probe level data. *Biostatistics* **2003**, *4*, 249–264. [CrossRef] [PubMed]

25.  Smyth, G.K. Linear models and empirical bayes methods for assessing differential expression in microarray experiments. *Stat. Appl. Genet. Mol. Biol.* **2004**, *3*, 3. [CrossRef] [PubMed]

26.  Benjamini, Y.; Hochberg, Y. Controlling the false discovery rate: A practical and powerful approach to multiple testing. *J. R. Stat. Soc.* **1995**, *57*, 289–300.

27.  Usadel, B.; Poree, F.; Nagel, A.; Lohse, M.; Czedik-Eysenberg, A.; Stitt, M. A guide to using mapman to visualize and compare omics data in plants: A case study in the crop species, maize. *Plant Cell Environ.* **2009**, *32*, 1211–1229. [CrossRef] [PubMed]

# Quantification of *Plasmodiophora brassicae* Using a DNA-Based Soil Test Facilitates Sustainable Oilseed Rape Production

Ann-Charlotte Wallenhammar [1,2,*], Albin Gunnarson [3], Fredrik Hansson [4] and Anders Jonsson [2,5]

[1] Rural Economy and Agricultural Society, HS Konsult AB, P.O. Box 271, SE-701 45 Örebro, Sweden
[2] Precision Agriculture and Pedometrics, Department of Soil and Environment, Swedish University of Agricultural Sciences, P.O. Box 234, SE-523 23 Skara, Sweden; Anders.Jonsson@jti.se
[3] Swedish Seed and Oilseed Growers, P.O. Box 53, SE-230 53 Alnarp, Sweden; Albin@svenskraps.se
[4] Rural Economy and Agricultural Society, Borgeby Slottsväg 11, SE-237 91 Bjärred, Sweden; Fredrik.Hansson@hushallningssallskapet.se
[5] Swedish Institute of Agricultural and Environmental Engineering, Green Tech Park, P.O. Box 63, SE-532 21 Skara, Sweden
[*] Correspondence: Ann-Charlotte.Wallenhammar@hushallningssallskapet.se

Academic Editors: Dilantha Fernando and Rishi R. Burlakoti

**Abstract:** Outbreaks of clubroot disease caused by the soil-borne obligate parasite *Plasmodiophora brassicae* are common in oilseed rape (OSR) in Sweden. A DNA-based soil testing service that identifies fields where *P. brassicae* poses a significant risk of clubroot infection is now commercially available. It was applied here in field surveys to monitor the prevalence of *P. brassicae* DNA in field soils intended for winter OSR production and winter OSR field experiments. In 2013 in Scania, prior to planting, *P. brassicae* DNA was detected in 60% of 45 fields on 10 of 18 farms. In 2014, *P. brassicae* DNA was detected in 44% of 59 fields in 14 of 36 farms, in the main winter OSR producing region in southern Sweden. *P. brassicae* was present indicative of a risk for >10% yield loss with susceptible cultivars (>1300 DNA copies g soil$^{-1}$) in 47% and 44% of fields in 2013 and 2014 respectively. Furthermore, *P. brassicae* DNA was indicative of sites at risk of complete crop failure if susceptible cultivars were grown (>50 000 copies g$^{-1}$ soil) in 14% and 8% of fields in 2013 and 2014, respectively. A survey of all fields at Lanna research station in western Sweden showed that *P. brassicae* was spread throughout the farm, as only three of the fields (20%) showed infection levels below the detection limit for *P.brassicae* DNA, while the level was >50,000 DNA copies g$^{-1}$ soil in 20% of the fields. Soil-borne spread is of critical importance and soil scraped off footwear showed levels of up to 682 million spores g$^{-1}$ soil. Soil testing is an important tool for determining the presence of *P. brassicae* and providing an indication of potential yield loss, e.g., in advisory work on planning for a sustainable OSR crop rotation. This soil test is gaining acceptance as a tool that increases the likelihood of success in precision agriculture and in applied research conducted in commercial oilseed fields and at research stations. The present application highlights the importance of prevention of disease spread by cleaning of farm equipment, footwear, *etc.*

**Keywords:** *Plasmodiophora brassicae*; qPCR assay; predictive soil tests

## 1. Introduction

Clubroot caused by *Plasmodiophora brassicae*, is recognized as a serious soil-borne disease in Brassica crops and is associated with appreciable yield losses [1,2]. Disease outbreaks have caused problems in winter oilseed rape (WOSR) growing regions in southern Sweden and more frequently also in fields of spring oilseed rape (SOSR) and WOSR in central Sweden in recent years. This economically important disease has also proliferated worldwide in oilseed rape (OSR) and vegetable brassicas [3]. Persistent resting spores produced in high numbers remain viable in the soil for up to 17 years [4]. Integrated management strategies including the use of fungicides and resistant cultivars have been implemented in cabbage crops [5]. Chemical treatment is not an economically viable option for commercial OSR production. However, partly resistant OSR cultivars have been available to WOSR growers for approximately 10 years [6] and the release of several new cultivars onto the Swedish market has allowed WOSR production to expand into fields where *P. brassicae* is present. Soil bioassay tests were first used for monitoring the widespread prevalence of *P. brassicae* in a region in central Sweden in the 1980s [4], and were later offered and adopted as a commercial soil advisory testing service for farmers [7]. More recently developed methods with higher throughput and accuracy enabled by molecular techniques based on polymerase chain reaction (PCR) have allowed rapid assessment of the infection potential of *P. brassicae*. A protocol using real-time PCR for direct detection and quantification of genomic DNA of *P. brassicae* from resting spores in the soil has been developed and used for naturally and artificially infested soil samples containing different concentrations of *P. brassicae* [8]. A DNA-based soil testing service that identifies fields where *P. brassicae* poses a significant risk is now also available to growers [9]. The Swedish Oilseed Growers' Association performs about 120 field experiments annually, mainly in oilseed growers' fields. As a consequence of the recent outbreaks of clubroot [9], several field experiments have been abandoned, and thus an analysis of experimental sites for the presence of *P. brassicae* DNA has become a necessary measure prior to establishing experiments. The objective of this paper is to report the first results of implementation of the Biological Soil Mapping (BioSoM) project, including a new service for farmers and researchers based on molecular analysis of field survey samples that quickly and reliably quantifies *P. brassicae* DNA and predicts the infection potential of clubroot in field soils intended for OSR production or OSR field experiments.

## 2. Results

### 2.1. Assessment of Farm Fields

Survey results from Scania in southern Sweden in 2013 are presented in Table 1, and show that *P. brassicae* is prevalent as DNA was detected in 60% of the 45 fields on 10 of 18 the farms sampled. On one farm, DNA was detected in 8 of 11 fields sampled.

**Table 1.** Results of *Plasmodiophora brassicae* analyses in soil samples from 45 fields in Scania, Sweden, 2013.

| DNA copies g$^{-1}$ soil | 0 * | <1300 ** | 1300–50,000 | 50,000–325,000 | >325,000 |
|---|---|---|---|---|---|
| Number of samples | 18 | 6 | 15 | 4 | 2 |
| % of all fields | 40 | 13 | 33 | 9 | 5 |

* *P. brassicae* DNA not detected (number of copies below the limit of detection); ** *P. brassicae* DNA detected, but at levels below the limit of quantification.

### 2.2. Assessment of Farm Fields Intended for Field Experiment on Winter Oilseed Rape

The results from sampling 59 fields in 2014, representing 36 farms in six counties, intended for field experiments in winter OSR are presented in Table 2. 44% of the fields, located on 16 farms were contaminated and in 8% of the fields *P. brassicae* DNA was found at levels >50,000 DNA copies g$^{-1}$ soil.

The fields infested with *P. brassicae* DNA were in the counties of Scania, Halland, Kalmar, East Gothia and West Gothia, while levels below the detection limit of *P. brassicae* DNA were found in the two fields tested on the island of Gotland. On seven farms, two to five fields had to be sampled in order to find a healthy field. On five of these farms *P. brassicae* DNA was found in 75% of the fields. No *P. brassicae* DNA was found in either of the sampled fields on two farms.

**Table 2.** Results of *Plasmodiophora brassicae* analyses in soil samples from 59 fields intended for field experiments in winter oilseed rape in southern and central Sweden, July 2014.

| DNA copies g$^{-1}$ soil | 0 * | <1300 ** | 1300–50,000 | 50,000–325,000 | >325,000 |
|---|---|---|---|---|---|
| Number of samples | 33 | 16 | 5 | 2 | 3 |
| % of all fields | 56 | 27 | 9 | 3 | 5 |

* *P. brassicae* DNA not detected (number of copies below the limit of detection); ** *P. brassicae* DNA detected, but at levels below the limit of quantification.

Soil samples analyzed from 15 fields at Lanna research station (Figure 1) are illustrated by a soil map. Fields with levels below the detection level and fields with a high risk of infection and multiplication of *P. brassicae* are shown. There are only three fields (20%) below the detection level of *P. brassicae* DNA. The highest amounts of *P. brassicae* DNA, 658,750 and 823,750 DNA copies g$^{-1}$ soil respectively, were found in fields with recent OSR field experiments.

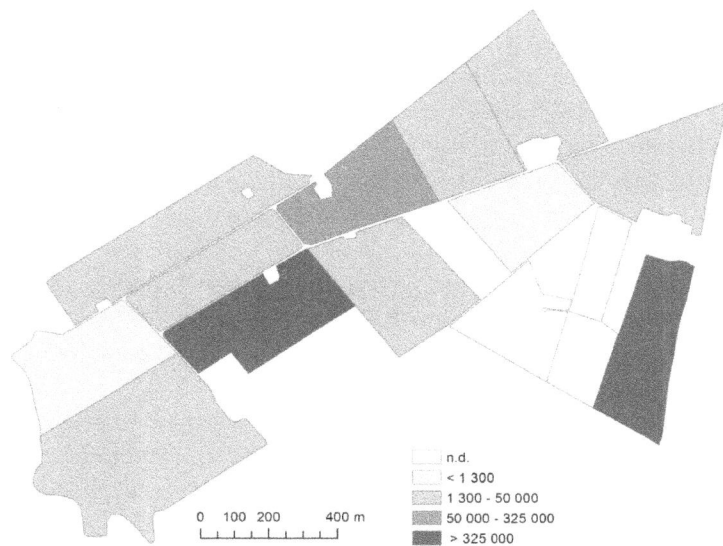

**Figure 1.** Field distribution of *Plasmodiophora brassicae* (DNA copies g$^{-1}$ soil) at Lanna Research Station in western Sweden. n.d. represents fields with no detection.

## 3. Discussion

We here report the first results of implementation of the Biological Soil Mapping (BioSoM) project, a new service for farmers and researchers based on molecular analysis of field survey samples that quickly and reliably quantifies *P. brassicae* DNA and predicts the infection potential of clubroot in field soils.

The results from growers' fields in the winter OSR districts of south-west Sweden show that *P. brassicae* is prevalent as DNA was detected in 60% of the 45 fields sampled. In 14% of the fields tested here (Table 1) levels >50,000 DNA copies g$^{-1}$ soil were found, indicating a high risk of complete crop failures if a susceptible OSR cultivar is sown.

Based on previous Swedish studies [8] the following temporary guidelines have been formulated for levels of DNA copies g$^{-1}$ soil [10]. For DNA-levels <1300 DNA copies, corresponding to approximately 3000 spores g$^{-1}$ soil, the risk of yield loss in susceptible crops is probably less than 10% [11]. At DNA levels ranging from 1300 to 325,000 DNA copies g$^{-1}$ soil (corresponding to 3000 and 750,000 spores g$^{-1}$ soil, respectively) resistant cultivars are recommended. At levels >325,000 DNA copies g$^{-1}$ soil the risk of multiplication of soil inoculum is considerable [8]. Yield loss caused by infections of *P. brassicae* depends not only on soil pre-plant DNA levels, but on other soil factors, and particularly on the time point of infection viz. on the part of growing season available after infection. Studies performed in spring OSR estimating yield loss based on soil inoculum levels assessed with bioassays showed a relationship ($R^2$ = 0.94) [1]. Recent studies on the performance of partly resistant cultivars of winter OSR confirm the variation in disease severity between years, and also point at an extremely severe yield loss of susceptible cultivars in a field experimental site with preplant inocula <10 million DNA copies g$^{-1}$ soil [12].

The results from sampling 59 fields in 2014, representing 36 farms intended for field experiments in winter oilseed rape crop, clearly show that *P. brassicae* has also proliferated throughout the growing districts of southern and central Sweden with 44% of the fields, located on 16 farms being contaminated (Table 2). In 8% of the fields *P. brassicae* DNA was found at levels (>50,000 DNA copies g$^{-1}$ soil) where a severe outbreak is likely to occur given optimal conditions for infection.

The presence of clubroot may cause problems at research stations where OSR field trials have been conducted over the years. The results from Lanna research station (Figure 1), clearly show an obvious spread, as there are only three fields (20%) below the detection level of *P. brassicae* DNA.

Thus, a predictive test indicating the infection potential of *P. brassicae* is essential when considering the choice of cultivar and identifying a healthy field for running field experiments. Routine testing of fields in contaminated areas is now recommended and offered by commercial laboratories [11] and the number of soil samples recently analyzed has reached a level that indicate a general adoption of this for routine clubroot detection prior to sowing. Increasing knowledge of fields where *P. brassicae* is prevalent will assist measures to prevent spread of the pathogen.

Clubroot was first discovered by assessing diseased plants in Swedish OSR crops in districts with high precipitation and a history of Brassica production in farm fields [8]. Previous findings in an extensive survey based on bioassays of 190 farm fields distributed in 18 farms where clubroot was assessed in central Sweden, showed that *P. brassicae* was present in 78% of the fields [4]. The investigations reported upon here (Tables 1 and 2) clearly show that the pathogen has also proliferated into the main regions of winter OSR production.

A patchy distribution of *P. brassicae* has been clearly demonstrated in earlier studies [8]. Therefore, when *P. brassicae* DNA is found in a field, further information can be obtained by conducting a point sampling, e.g., collecting 10 subsamples within a radius of 3 meters at different intensities e g one sampling point per hectare, [8] to identify the variation in occurrence of *P. brassicae* DNA. At Lanna Research Station, where OSR field experiments are routinely conducted, more detailed knowledge on the spread in selected fields is necessary in order to maintain high quality in field experiments on OSR crops.

In Swedish soils with a long history of cultivation of various Brassica crops [9] the pathogen was prevalent prior to the onset of OSR production in the 1940s. Breeding for clubroot resistance in turnips and swedes started in 1929 at a branch station of the Swedish Seed Association in western Sweden, where the disease was prevalent at that time [13], and resulted in several resistant cultivars. During the late1930s experiments were carried out on infested soil at Lönnstorp experimental farm in southwest Sweden (55°66′N, 13°10′E), which is now a SLU research station. This information is of great importance in indicating that there is most likely a history of *P. brassicae* distributed throughout Sweden, and that after repeated OSR production the pathogen multiplies, as shown in the long-term field experiments in southern Sweden [14] where the OSR crop was severely diseased due to clubroot after 10–12 rotations of spring OSR grown every four years.

The important role of machinery [15] and wind [16] in the dispersal of clubroot has been demonstrated in Canadian studies identifying the importance of equipment cleaning and sanitation [15]. While inspecting a field experiment site at Bollerup in southern Sweden with very high levels of *P. brassicae* DNA (58 million DNA copies corresponding to 133 million spores g$^{-1}$ soil), soil samples were taken from rubber boots. Samples scraped off the soles of the left and right rubber boots weighed 143.5 g and 135.5 g, respectively and contained *P. brassicae* DNA corresponding to 530 and 682 million spores g$^{-1}$ soil, respectively. This shows that spread even by footwear might be critically significant, bearing in mind that 1000 spores $g^{-1}$ soil is the limit of detection [9]. The potential risk of spread by footwear has to be prevented by routinely using shoe covers in experimental fields.

These results presented here suggest that *P. brassicae* is widely present in winter OSR regions and now poses a threat to OSR production in areas where clubroot was first reported only in recent years [8].

## 4. Experimental Section

### 4.1. Assessment of Farm Fields

Growers in the WOSR districts of south-west Sweden, in former Malmöhus county, were offered a soil analysis for *P. brassicae*. Analyses were performed in 2013 in samples from 45 fields on 18 farms. For some of the farms more than one field were sampled. Soil was sampled by field research staff from the Rural Economy and Agricultural Society. A soil auger with diameter 22 mm and volume 76 mL, was used to extract samples from the top 20 cm of the soil. The sampling procedure was performed according to standard procedures used by the Rural Economy and Agricultural Society in Scania, with four samples per hectare, taken within a plot of 50 × 50 m at points determined by GPS. The soil samples were sent to Eurofins Food and Agro Testing Sweden AB, Kristianstad, Sweden where qPCR analyses were performed.

### 4.2. Assessment of Farm Fields Intended for Field Experiment on Winter Oilseed Rape

WOSR trials are regularly being conducted in oilseed growers' fields. The fields intended for WOSR production in 2014 were sampled in July prior to sowing according to the instructions for Biological Soil Mapping (BioSoM) project. They were sent to Eurofins Food and Agro Testing Sweden AB, Kristianstad, Sweden where qPCR analyses were performed.

### 4.3. Biological Soil Mapping of Lanna Research Station

At Lanna research station in western Sweden (58°38′N, 13°16′E),which is owned and run by SLU, biological soil mapping was carried out to establish a general level of infection potential for each field. The fields were sampled according to BioSoM instructions, with 40 subsamples collected with an auger as described in 4.1 along a W-shaped sampling transect covering the field. The samples were handled and analyzed according to Eurofins protocol described in [8] at the laboratory of the Department of Soil and Environment, SLU, Skara.

## 5. Conclusions

The study demonstrates that the commercially available qPCR assay offers growers and researchers a fast and reliable predictive test for determining *P. brassicae* DNA levels in individual fields and can be used as a tool in integrated OSR production and in agricultural field research. The results clearly show that *P. brassicae* (as measured by DNA level in soil) is prevalent in the main winter OSR producing regions in Sweden. Soil analysis is therefore highly recommended by the Swedish Oilseed Grower' Association and advisory services as a management tool, and is an essential measure prior to OSR and Brassica production, research and development. Introduction of new research results in advisory work is often slow and time-consuming. However, we found that the soil test is already being adopted by commercial laboratories and by farmers.

**Acknowledgments:** We thank Katarzyna Marcez-Schmidt, Johanna Wetterlind and Mattias Gustafsson, SLU, for technical assistance. The study was performed within the framework of Biological Soil Mapping (BioSoM) project supported by the Faculty of Natural Resources and Agriculture (NL), Swedish Agricultural University (SLU), the Swedish Farmers' Foundation for Agricultural Research, the Farmers' Cooperation Foundations (VL and SL) and the Foundation for Swedish Oil Plant Research.

**Author Contributions:** Anders Jonsson conceived the project; Fredrik Hansson, Albin Gunnarson, Ann-Charlotte Wallenhammar and Anders Jonsson designed the study; Ann-Charlotte Wallenhammar and Anders Jonsson analysed the data and wrote the paper.

**Conflicts of Interest:** The authors declare no conflict of interest.

## References

1.   Wallenhammar, A.-C. Observations on yield loss from *Plasmodiophora brassicae* infections in spring oilseed rape. *Zeitschrift Pflanzenkrankheiten Pflanzenschutz* **1998**, *105*, 1–7.
2.   Strelkov, S.E.; Hwang, S.-F. Clubroot in the Canadian canola crop: 10 years into the outbreak. *Can. J. Plant Pathol.* **2014**, *36*, 27–36. [CrossRef]
3.   Dixon, G.R. Clubroot (*Plasmodiophora brassicae* Woronin)—An agricultural and biological challenge worldwide. *Can. J. Plant Pathol.* **2014**, *36*, 5–18. [CrossRef]
4.   Wallenhammar, A.-C. Prevalence of *Plasmodiophora brassicae* in a spring oilseed rape growing area in central Sweden and factors influencing soil infestation levels. *Plant Pathol.* **1996**, *45*, 710–719. [CrossRef]
5.   Donald, C.; Porter, I. Integrated control of clubroot. *J. Plant Growth Regul.* **2009**, *28*, 289–303. [CrossRef]
6.   Diedrichsen, E.; Frauen, M.; Linders, E.G.; Hatakeyama, K.; Hirari, M. Status and perspectives of clubroot resistance breeding in crucifer crops. *J. Plant Growth Regul.* **2009**, *28*, 265–281. [CrossRef]
7.   Engquist, L.G. Distribution of Clubroot in Sweden and the effect of infection on oil content of rape. *J. Swed. Seed Ass.* **1994**, *104*, 82–86.
8.   Wallenhammar, A.-C.; Almquist, C.; Söderström, M.; Jonsson, A. In field distribution of *Plasmodiophora brassicae* measured using quantitative real-time PCR. *Plant Pathol.* **2012**, *61*, 16–28.
9.   Wallenhammar, A.-C.; Almquist, C.; Schwelm, A.; Roos, J.; Marzec-Schmidt, K.; Jonsson, A.; Dixelius, C. Clubroot, a persistent threat to Swedish oilseed rape production. *Can. Plant Pathol.* **2014**, *36*, 135–141. [CrossRef]
10.  Almquist, C. Monitoring Important Soil-Borne Plant Pathogens in Swedish Crop Production Using Real-Time PCR. Ph.D. Thesis, Swedish University of Agricultural Sciences, Uppsala, Sweden, 2016.
11.  Eurofins. Klumprotsjuka. Analys av *Plasmodiophora brassicae* i Jord med Snabb och Specific Kvantifiering med DNA-Baserad Teknik. Available online: http://www.eurofins.se/dokument/lantbruk/Folder-rapssjukdomar20150707.pdf (accessed on 18 February 2016). (In Swedish)
12.  Gunnarsson, A. Höstrapssorter med klumprotresistens. In *Field Trial Report of Central Sweden*; Rural Economy and Agricultural Society Skaraborg: Falköping, Sweden, 2015; pp. 155–156. (In Swedish)
13.  Olsson, P.A. Club root disease (*Plasmodiophora brassicae* WOP.) in turnips and swedes and the means of control, especially by plant breeding. II. Further investigations, and experiments in breeding resistant strains. *J. Swed. Seed Ass.* **1940**, *50*, 287–360.
14.  Jonsson, A.J.; Marzec-Schmidt, K.; Börjesson, G.; Wallenhammar, A.-C. Quantitative PCR shows propagation of *Plasmodiophora brassicae* in Swedish long term trials. *Eur. J. Plant Pathol.* **2016**. [CrossRef]
15.  Cao, T.; Manolii, V.P.; Hwang, S.F.; Howard, R.J.; Strelkov, S.E. Virulence and spread of *Plasmodiophora brassicae* [clubroot] in Alberta, Canada. *Can. J. Plant Pathol.* **2009**, *31*, 321–329. [CrossRef]
16.  Rennie, D.C.; Holtz, M.D.; Turkington, T.K.; Lebouldus, J.M.; Hwang, S.-F.; Howard, R.J.; Strelkov, S.E. Movement of *Plasmodiophora brassicae* resting spores in windblown dust. *Can. J. Plant Pathol.* **2015**, *37*, 188–196. [CrossRef]

# Antibacterial Properties of Flavonoids from Kino of the Eucalypt Tree, *Corymbia torelliana*

Motahareh Nobakht, Stephen J. Trueman *, Helen M. Wallace, Peter R. Brooks, Klrissa J. Streeter and Mohammad Katouli

Centre for Genetics, Ecology and Physiology, University of the Sunshine Coast, Maroochydore DC, QLD 4558, Australia; mnobakht@research.usc.edu.au (M.N.); hwallace@usc.edu.au (H.M.W.); pbrooks@usc.edu.au (P.R.B.); kstreete@usc.edu.au (K.J.S.); mkatouli@usc.edu.au (M.K.)
* Correspondence: strueman@usc.edu.au

**Abstract:** Traditional medicine and ecological cues can both help to reveal bioactive natural compounds. Indigenous Australians have long used kino from trunks of the eucalypt tree, *Corymbia citriodora*, in traditional medicine. A closely related eucalypt, *C. torelliana*, produces a fruit resin with antimicrobial properties that is highly attractive to stingless bees. We tested the antimicrobial activity of extracts from kino of *C. citriodora*, *C. torelliana* × *C. citriodora*, and *C. torelliana* against three Gram-negative and two Gram-positive bacteria and the unicellular fungus, *Candida albicans*. All extracts were active against all microbes, with the highest activity observed against *P. aeruginosa*. We tested the activity of seven flavonoids from the kino of *C. torelliana* against *P. aeruginosa* and *S. aureus*. All flavonoids were active against *P. aeruginosa*, and one compound, (+)-(2S)-4′,5,7-trihydroxy-6-methylflavanone, was active against *S. aureus*. Another compound, 4′,5,7-trihydroxy-6,8-dimethylflavanone, greatly increased biofilm formation by both *P. aeruginosa* and *S. aureus*. The presence or absence of methyl groups at positions 6 and 8 in the flavonoid A ring determined their anti-*Staphylococcus* and biofilm-stimulating activity. One of the most abundant and active compounds, 3,4′,5,7-tetrahydroxyflavanone, was tested further against *P. aeruginosa* and was found to be bacteriostatic at its minimum inhibitory concentration of 200 μg/mL. This flavanonol reduced adhesion of *P. aeruginosa* cells while inducing no cytotoxic effects in Vero cells. This study demonstrated the antimicrobial properties of flavonoids in eucalypt kino and highlighted that traditional medicinal knowledge and ecological cues can reveal valuable natural compounds.

**Keywords:** antibiotic resistance; antimicrobial activity; cytotoxicity; ethnobotany; *Eucalyptus*; natural products; *Pseudomonas aeruginosa*; stingless bees; *Tetragonula*; traditional medicine

---

## 1. Introduction

The long-term use of antibiotics has led to widespread bacterial resistance, and so there has been a global drive to develop new antibiotics [1–4]. Traditional medicine based on natural products has led to the discovery of new drugs, and ecological cues can also reveal bioactive natural compounds [5–8]. Eucalypt kino is a trunk exudate produced by eucalypt trees (*Angophora*, *Corymbia* and *Eucalyptus* spp.) that contains high levels of potentially useful polyphenol compounds. Kino is characterised by its deep rich coloration, high tannin content, polyphenol composition and astringency [9,10]. Indigenous Australians have used kino from *Corymbia* and *Eucalyptus* trees to cure ailments such as diarrhoea, scabies, and haemorrhage [11]. Kino exudates from *Corymbia dichromophloia* trees have also been used as a treatment for toothache, cold and flu, and heart, lung, and bronchial diseases [12]. Kino from *C. citriodora* trees has been used traditionally as a treatment for chronic bowel inflammation [11], and kino from *C. intermedia* trees has been used to treat wounds [13]. Leaf or bark extracts from naturalized *C. torelliana* trees in Nigeria have been used to treat gastrointestinal disorders, wounds, and coughs [14,15].

The biological activity of eucalypt kinos used in traditional medicine has, until recent times, received little attention. Two compounds from *C. citriodora* and *C. maculata* kino, aromadendrin 7-methyl ether and ellagic acid, have long been known to possess antimicrobial activity against the Gram-positive bacterium, *Staphylococcus aureus* [16]. Aqueous kino extracts from 15 eucalypt species have recently been tested for their antimicrobial activity against *S. aureus*, *Bacillus subtilis*, *Kocuria rhizophila*, *Pseudomonas aeruginosa*, *Escherichia coli*, and *Saccharomyces cerevisiae*. Extracts from *C. maculata*, *C. ficifolia*, and *C. calophylla* kino exhibited strong activity against the Gram-positive bacteria, although no activity was observed from any of the eucalypt species against the Gram-negative bacteria [17]. Aqueous and ethanolic extracts from *C. intermedia* leaves also exhibit antimicrobial activity against *S. aureus* and the unicellular fungus, *Candida albicans* [13]. Volatile components from *C. citriodora* essential oil have strong activity against *Mycobacterium tuberculosis* [18]. Extracts of *C. torelliana* leaves or bark have antibacterial activity against a wide range of species including *M. tuberculosis* and non-tuberculous *Mycobacteria* spp., *S. aureus*, *E. coli*, *P. aeruginosa*, *Klebsiella* sp. and *Helicobacter pylori* [14,15,19,20]. *C. torelliana* also has an unusual mutualistic relationship with stingless bees that disperse its seeds and use its fruit resin to construct their nests [21–26]. The fruit resin from *C. torelliana* has recently been found to possess antimicrobial properties [27–29]. Bees prefer the fruit resin from *C. torelliana* to fruit resin from other species, and this fruit resin may protect their nest from pathogenic microbes [26–28]. However, extracts from the kino of *C. torelliana* and *C. citriodora* have not previously been tested for their antimicrobial activity.

In this study, we investigated the antimicrobial activity of extracts from kino of *C. citriodora*, *C. torelliana* and their widely planted hybrid, *C. torelliana* × *C. citriodora*. We examined their activity against three Gram-negative bacteria, *P. aeruginosa*, *E. coli*, and *Salmonella typhimurium*, two Gram-positive bacteria, *S. aureus* and *B. cereus*, and one fungus, *Candida albicans*. Furthermore, we investigated the antibacterial activity against *P. aeruginosa* and *S. aureus* of seven individual flavonoids (Figure 1) isolated from the kino of *C. torelliana* [30]. We also assessed whether the kino extracts and one of the most-abundant and active flavonoids in *C. torelliana* kino had cytotoxic effects.

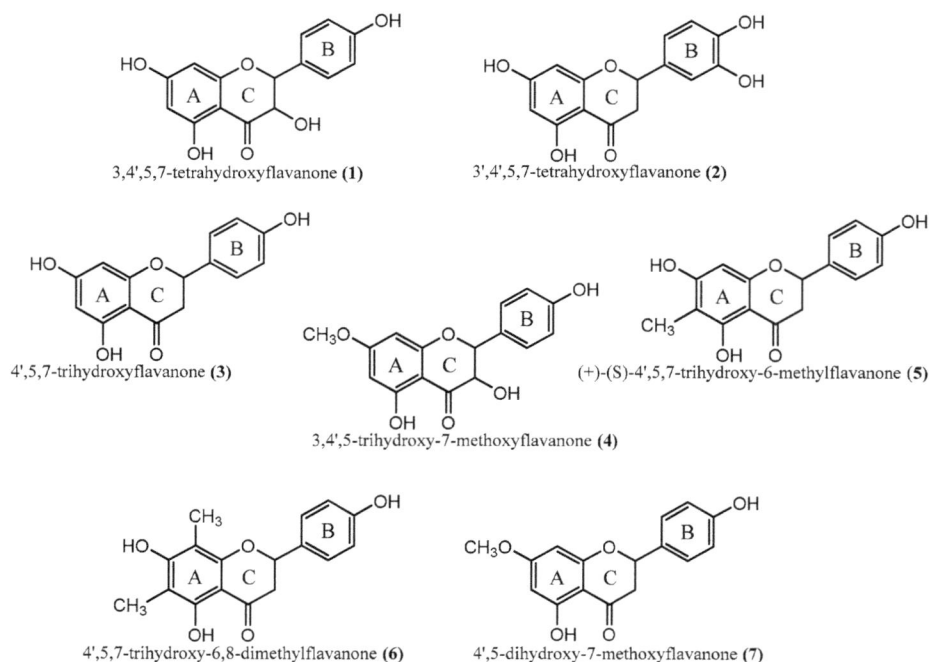

**Figure 1.** Structures of seven flavonoids isolated from the kino of *Corymbia torelliana*.

## 2. Results and Discussion

### 2.1. Antimicrobial Activity and Cytotoxicity of Crude Extracts from Corymbia Trees

Ethanolic extracts from kinos of *Corymbia citriodora*, *C. torelliana* × *C. citriodora*, and *C. torelliana* showed strong antimicrobial activity against all of the tested microorganisms (Figure 2). The extent of the inhibition zone varied among the extracts and microorganisms, but the highest inhibition with all three extracts was obtained against *P. aeruginosa*. This is the first report of antimicrobial activity of kino from *C. citriodora*, *C. torelliana* × *C. citriodora* and *C. torelliana* against Gram-negative bacteria, Gram-positive bacteria and *C. albicans*. These results are similar to findings that aqueous kino extracts from the closely related species, *C. maculata*, and the more-distantly related species, *C. ficifolia* and *C. calophylla*, are active against the Gram-positive bacteria, *S. aureus*, *B. subtilis*, and *K. rhizophila* [17]. Aqueous and ethanolic extracts from the leaves of another distantly-related species, *C. intermedia*, also have strong activity against *S. aureus* and *C. albicans* [13]. However, the extracts from *C. maculata*, *C. ficifolia*, *C. calophylla*, and *C. intermedia* had little or no activity against *P. aeruginosa* [13,17]. The use of aqueous rather than ethanolic extracts [17] and the sampling of leaves rather than kino [13] may explain the lack of activity of other *Corymbia* extracts against Gram-negative bacteria. Alternatively, the differences may be the result of specific chemical compounds that are present in *C. citriodora*, *C. torelliana* × *C. citriodora*, and *C. torelliana* kino.

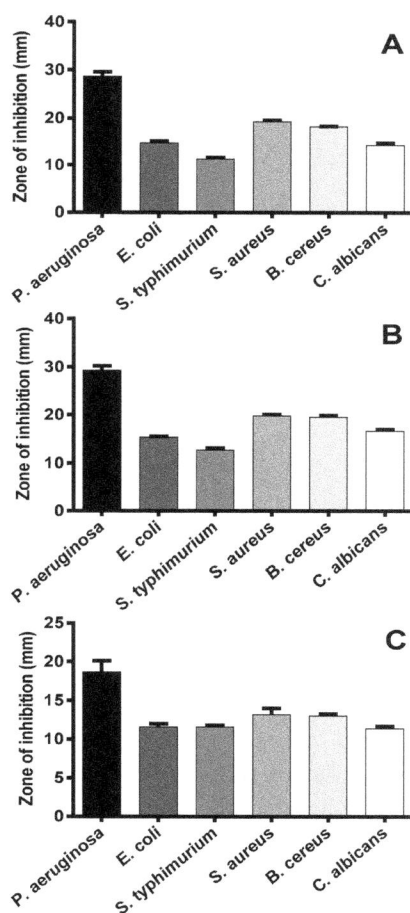

**Figure 2.** Antimicrobial activity of 400 μg of kino extract from (**A**) *Corymbia citriodora*, (**B**) *C. torelliana* × *C. citriodora*, and (**C**) *C. torelliana* against six microorganisms. Zones of inhibition are presented as mean + S.E. ($n$ = 21 trees for *C. citriodora* and the hybrid; $n$ = 3 trees for *C. torelliana*). Means among the three kino extracts do not differ significantly (ANOVA, $p > 0.05$).

Different strains of a bacterial species may exhibit different levels of susceptibility to an antimicrobial agent. The highest activity of the crude extracts was against *P. aeruginosa* and so we extended our screening by testing the extracts against four strains of *P. aeruginosa* (i.e., strains C1, C8, C11, and C19) that represented four different clonal groups, isolated recently from clinical cases in our laboratory [31]. The kino extracts showed strong activity against all four strains, except that C11 was resistant to the *C. torelliana* extract (Table 1). This strain was also highly resistant to ticarcillin and intermediately resistant to aztreonam and ticarcillin-clavulanic acid. Ticarcillin is a fourth generation of penicillin, a β-lactam antibiotic. This group of antimicrobial agents inhibits bacteria by penetrating the cytoplasmic membrane and attaching to penicillin binding proteins [32]. Resistance of bacteria to this antibiotic normally develops through a mechanism that inhibits the antibiotic from reaching this target. The resistance of the C11 strain to crude extract of *C. torelliana* suggests that the active component of *C. torelliana* kino might operate by a similar mechanism to these antibiotics. The MIC of kino extracts from *C. citriodora*, *C. torelliana* × *C. citriodora* and *C. torelliana* was 200 μg/mL against each of the bacteria in this study (data not presented). This suggests that, irrespective of the *Corymbia* species, the type and concentration of active compounds in the extracts were similar. At this MIC, 200 μg/mL, the extracts had bacteriostatic activity.

**Table 1.** Antimicrobial activity of a 1.0% (v/v) solution of kino extract from *Corymbia citriodora*, *C. torelliana* × *C. citriodora*, or *C. torelliana* against four clinical strains of *Pseudomonas aeruginosa* (C1, C8, C11, and C19) representing different clonal types. Antibiotic susceptibility profile of isolates is also given.

| *Corymbia* Species or Hybrid | Strain/Zone of Inhibition (mm) * | | | |
|---|---|---|---|---|
| | C1 | C8 | C11 | C19 |
| *C. citriodora* | 12 ± 0 | 18 ± 1 | 15 ± 1 | 11 ± 0 |
| *C. torelliana* × *C. citriodora* | 11 ± 0 | 16 ± 1 | 19 ± 1 | 12 ± 0 |
| *C. torelliana* | 11 ± 0 | 11 ± 1 | 1 ± 1 ** | 12 ± 0 |
| **Antibiotic Susceptibility Profile †** | | | | |
| Amikacin (30 μg) | S | S | S | S |
| Aztreonam (30 μg) | S | I | I | I |
| Ceftazidime (30 μg) | S | S | S | S |
| Cefepime (30 μg) | S | S | S | S |
| Piperacillin (100 μg) | S | I | S | R |
| Piperacillin-tazobactam (100/10 μg) | S | R | S | I |
| Ticarcillin (75 μg) | I | I | R | I |
| Gentamicin (10 μg) | S | I | S | S |
| Ciprofloxacin (5 μg) | S | S | S | S |
| Norfloxacin (10 μg) | S | S | S | S |
| Imipenem (10 μg) | R | S | S | S |
| Ticarcillin-clavulanic acid (75/10 μg) | I | R | I | R |

* Zones of inhibition are presented as mean ± S.E. ($n$ = 9 trees for *C. citriodora* and the hybrid; $n$ = 3 trees for *C. torelliana*). Means within a *P. aeruginosa* strain do not differ significantly (ANOVA, $p > 0.05$); ** Distance (mm) from the rim of the well; † Antibiotic susceptibility profile is classified as susceptible (S), intermediate (I) or resistant (R) according to CLSI guidelines [33].

The kino extracts were also tested for their ability to inhibit biofilm formation by the bacteria. All extracts increased biofilm formation of the bacterial strains (Figure 3). These results were unexpected, as we anticipated that extracts capable of inhibiting microbial growth would have also reduced their biofilm formation. However, similar results have been observed during testing of aqueous extracts from the neem tree, *Azadirachta indica*, against two yeast strains [34]. These researchers concluded that increased biofilm formation could be related to an increased level of hydrophobicity, which is a non-specific mechanism for adhesion of bacteria to surfaces. The use of ethanolic extracts could have partly effected the high level of biofilm formation in the current study, as ethanol has potential to increase hydrophobicity. The preparation of aqueous extracts was not feasible due to low solubility of *C. citriodora*, *C. torelliana* × *C. citriodora*, and *C. torelliana* kino in water. From a clinical perspective, biofilm formation is important for survival of bacteria that colonize the host and it assists in physical resistance to phagocytosis and tolerance to antibiotics [35,36]. Under these conditions, antibiotics are prevented from diffusing through the physical barrier formed by the exopolymeric substances in biofilms [37]. Nonetheless, the high antimicrobial activity of all extracts against *P. aeruginosa* indicates the presence of an active compound (or compounds) with anti-*Pseudomonas* activity in the kino from these *Corymbia* species.

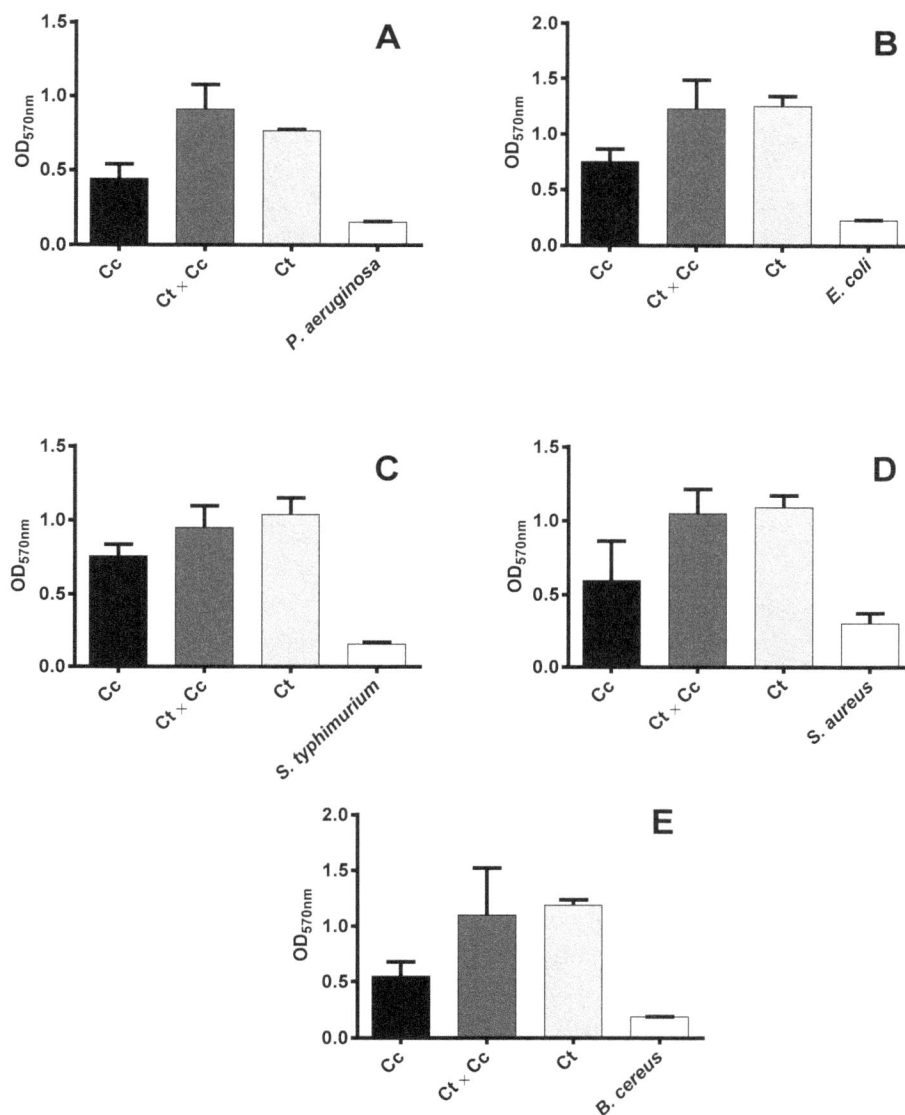

**Figure 3.** Biofilm formation in the presence of kino extracts from *Corymbia citriodora* (Cc), *C. torelliana* × *C. citriodora* (Ct × Cc), and *C. torelliana* (Ct) by (**A**) *Pseudomonas aeruginosa*, (**B**) *Escherichia coli*, (**C**) *Salmonella typhimurium*, (**D**) *Staphylococcus aureus*, and (**E**) *Bacillus cereus*. Optical densities at 570 nm wavelength ($OD_{570nm}$) are presented as mean + S.E. ($n$ = 3 trees for Cc, Ct × Cc, and Ct; $n$ = 3 for the kino-free control). Means among the three kino extracts do not differ significantly (ANOVA, $p > 0.05$).

The cytotoxicity of *C. citriodora*, *C. torelliana* × *C. citriodora*, and *C. torelliana* kino was tested using 1000 µg of the extracts against Vero cells. No cytopathic effect (CPE), such as detachment or rounding of the cells, was observed. However, cells showed morphological changes, characterized by shrinking, within 48 h. Studies investigating the CPE of bacterial toxins on eukaryotic cells, including Vero cells, have defined cytotoxicity as 50% or more of the cells showing CPE such as rounding or disruption of the cell monolayer within 4 h [38–40].

*2.2. Antibacterial Activity of Flavonoids from Corymbia torelliana*

Seven flavonoids isolated from the kino of *C. torelliana* (Figure 1) possessed antibacterial activity against *P. aeruginosa* (Table 2). The activity of these flavonoids against this Gram-negative bacterium is highly significant because the outer membrane of Gram-negative bacteria possesses narrow porin channels that slow the penetration of small hydrophilic solutes and increase their tolerance to antibiotics [41]. Our results suggest that flavonoids from the kino of *C. torelliana* have a mechanism that overcomes this barrier.

**Table 2.** Antimicrobial activity of seven flavonoids [3,4′,5,7-tetrahydroxyflavanone (**1**), 3′,4′,5,7-tetrahydroxyflavanone (**2**), 4′,5,7-trihydroxyflavanone (**3**), 3,4′,5-trihydroxy-7-methoxyflavanone (**4**), (+)-(2S)-4′,5,7-trihydroxy-6-methylflavanone (**5**), 4′,5,7-trihydroxy-6,8-dimethylflavanone (**6**), and 4′,5-dihydroxy-7-methoxyflavanone (**7**)] from *Corymbia torelliana* kino against *Pseudomonas aeruginosa* and *Staphylococcus aureus*.

| Bacterium | Zone of Inhibition (mm) | | | | | | |
|---|---|---|---|---|---|---|---|
| | Compound | | | | | | |
| | 1 | 2 | 3 | 4 | 5 | 6 | 7 |
| *P. aeruginosa* | 20.3 ± 1.8 | 6.7 ± 6.7 | 19.7 ± 1.9 | 12.3 ± 6.3 | 18.7 ± 1.2 | 24.7 ± 2.9 | 20.3 ± 2.8 |
| *S. aureus* | inactive | inactive | inactive | inactive | 12.7 ± 1.8 | inactive | inactive |

Means (± S.E.) among seven flavonoids within *P. aeruginosa* do not differ significantly (ANOVA, $p > 0.05$).

Only one of the seven compounds, (+)-(2S)-4′,5,7-trihydroxy-6-methylflavanone (**5**), was active against *S. aureus* (Table 2). The other six flavonoids might inhibit *S. aureus* growth at concentrations higher than 50 μg/well, although there can also be relationships between the structure of flavonoids and their antibacterial activity [42]. Hydroxyl groups in the chemical structure increase the activity of flavonoids against methicillin-resistant *S. aureus*, while methoxy groups reduce their activity [43]. However, we did not find the presence of hydroxyl groups to be a defining character in the anti-*Staphylococcus* activity of our flavonoids. For example, (+)-(2S)-4′,5,7-trihydroxy-6-methylflavanone (**5**) and 4′,5,7-trihydroxy-6,8-dimethylflavanone (**6**) both have three hydroxyl groups, located at positions 5 and 7 of the A ring and position 4′ of the B ring, while neither compound contains a methoxy group. The unique aspect of compound (**5**) is that it contains a single methyl group, located at position 6 of the A ring. We did not investigate possible relationships between the concentration of flavonoids and their antibacterial effects due to difficulties in obtaining sufficient quantities of flavonoids from *C. torelliana* extracts. However, there can be relationships between the concentration of flavonoids and their antibacterial activity [44].

One of the compounds, 4′,5,7-trihydroxy-6,8-dimethylflavanone (**6**), greatly increased biofilm formation by both *P. aeruginosa* and *S. aureus* (Figure 4). This compound is unique in possessing two methyl groups, located at positions 6 and 8 of the A ring (Figure 1). The stimulation of biofilm formation by this flavonoid was highly unexpected. However, extracts of *Azadirachta indica* increase biofilm formation by *C. albicans* [34] and some phenolics and aminoglycosides, at sub-inhibitory concentrations, increase biofilm formation by *P. aeruginosa* and *E. coli* [45,46]. A relationship between the existence of two methyl groups and enhanced biofilm formation has not been reported previously. However, it could be concluded that this phenomenon of increased biofilm formation is entirely, or at least partly, due to a hydrophobicity effect rather than the methyl groups.

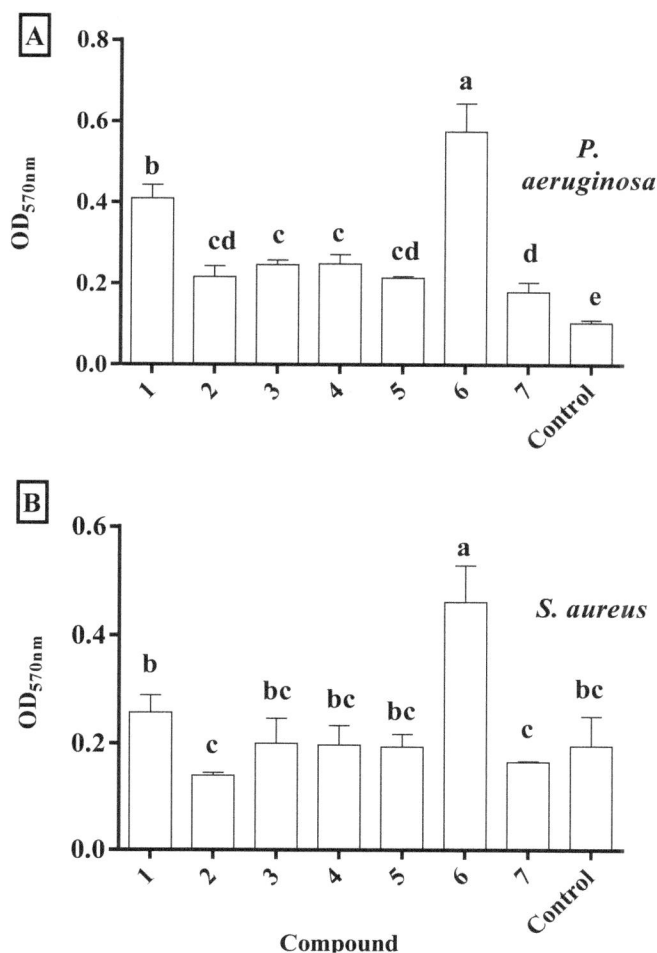

**Figure 4.** Biofilm formation ($OD_{570nm}$) in the presence of 100 µg of 3,4′,5,7-tetrahydroxyflavanone (1), 3′,4′,5,7-tetrahydroxyflavanone (2), 4′,5,7-trihydroxyflavanone (3), 3,4′,5-trihydroxy-7-methoxyflavanone (4), (+)-(2S)-4′,5,7-trihydroxy-6-methylflavanone (5), 4′,5,7-trihydroxy-6,8-dimethylflavanone (6), and 4′,5-dihydroxy-7-methoxyflavanone (7).

## 2.3. Anti-Pseudomonas Activity and Cytotoxicity of 3,4′,5,7-Tetrahydroxyflavanone

3,4′,5,7-tetrahydroxyflavanone (**1**) inhibited the growth of *P. aeruginosa*, with a minimum inhibitory concentration of 200 µg/mL (Table 3). This compound was bacteriostatic against *P. aeruginosa*. It significantly increased biofilm formation by *P. aeruginosa* at 48 h in comparison with the control, both in this experiment at masses of 200, 100, and 50 µg (Figure 5) and in the previous experiment at a mass of 100 µg (Figure 4). Further investigation is warranted to determine the mechanisms behind the biofilm-stimulating effect of this compound with *P. aeruginosa*. 3,4′,5,7-tetrahydroxyflavanone reduced adhesion of *P. aeruginosa* by 19%, 38%, and 35% at 200 µg, 100 µg, and 50 µg final mass, respectively (Table 3). This is an important result because adhesion to host tissue is an important step in bacterial survival and colonization [47]. The flavonoid induced no cytotoxic effects in Vero cell culture assay after 48 h (Table 3). Crude extracts from *C. torelliana* kino also had no cytotoxic effects on human colorectal epithelial adenocarcinoma (Caco-2) cells (Section 2.1, above). These results, therefore, confirm the potential of 3,4′,5,7-tetrahydroxyflavanone as a new antibacterial agent against *P. aeruginosa*. This is a significant finding because kino extracts have previously been found inactive against Gram-negative bacteria [16,17]. It was concluded that high molecular weight compounds, such as tannins, in the eucalypt kino could not penetrate the outer membrane of Gram-negative bacteria [17]. However, results in the current study suggest that compounds such as 3,4′,5,7-tetrahydroxyflavanone from the kino of *C. torelliana* can overcome this outer membrane barrier. This flavanonol has also been identified in kino from the distantly related species, *C. calophylla* and *C. gummifera* [48,49].

**Table 3.** Minimum inhibitory concentration (MIC) and reduction in adhesion of *Pseudomonas aeruginosa* in the presence of 3,4′,5,7-tetrahydroxy-flavanone from *Corymbia torelliana* kino.

| MIC (µg/mL) | Control Adhesion cfu (Mean ± S.E.) | Adhesion Difference (%) | | |
| --- | --- | --- | --- | --- |
| | | Final Mass (µg) | | |
| | | 200 | 100 | 50 |
| 200 | 3.86 ± 0.16 | −19 | −38 | −35 |

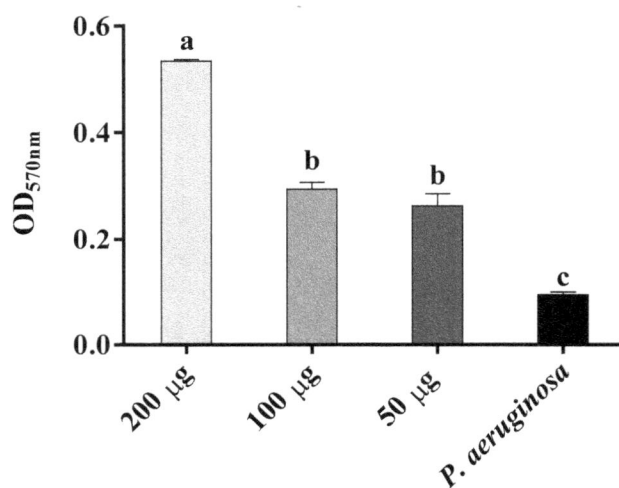

**Figure 5.** Biofilm formation of *Pseudomonas aeruginosa* in the presence of 200 µg, 100 µg, and 50 µg (final mass) of 3,4′,5,7-tetrahydroxyflavanone from kino exudate of *Corymbia torelliana*. Optical densities at 570 nm wavelength ($OD_{570nm}$) are presented as mean + S.E. Means with different letters are significantly different (ANOVA and Tukey's HSD test; $p < 0.05$; $n = 3$).

These results confirm that natural products from traditional medicine are promising leads to finding new sources of antibiotic drugs. Natural products have often played a key role in the formulation of new drugs [50] and there is particular interest in polyphenols such as flavonols, flavan-3-ols, and tannins for their potential antimicrobial effects [51]. Indigenous Australians have used kino exudates from eucalypt trees including *C. citriodora* to cure various ailments, and eucalypt kinos are well known for their polyphenol content [11]. The successful discovery of new sources of drugs based on traditional knowledge can also be guided by ecological cues [7]. *C. torelliana* is a geographically-restricted species that is closely related to the traditional medicinal species, *C. citriodora*, but which has a unique mutualism with stingless bees [21]. These bees are strongly attracted to the resin of *C. torelliana* fruits, which they use to construct their nests, and the bees disperse the seeds of this species [22–26]. Resin from the fruit of *C. torelliana* has also been shown to possess antimicrobial properties [27,28]. We were, therefore, guided by both traditional knowledge of the medicinal properties of this eucalypt group and ecological information on the attractiveness to bees of one particular species in this group, to identify potentially valuable antibacterial compounds.

## 3. Materials and Methods

### 3.1. Antimicrobial Activity and Cytotoxicity of Crude Extracts from Corymbia Trees

Fresh kino samples were obtained from *C. citriodora* subsp. *variegata* and *C. torelliana* × *C. citriodora* subsp. *variegata* (21 trees each) in a forestry plantation at Binjour (25°30′ S, 151°27′ E), Australia, established by the Queensland Department of Agriculture and Fisheries [52,53]. Samples were also collected from three *C. torelliana* trees on the Sunshine Coast (26°42′ S, 153°02′ E), Australia. *C. citriodora*, and hybrids between *C. citriodora* and *C. torelliana*, are grown extensively in forestry plantations [54–60] but *C. torelliana* is rarely grown in plantations, partly because of its invasive potential [21,25]. Therefore, the three samples of *C. torelliana* were obtained from isolated trees in amenity plantings. Kino samples were collected from naturally-occurring, freely-flowing trunk exudates into clean vials, transported to the laboratory on ice, and stored in the dark at −20 °C until testing. Collection of samples from older, crystallized exudates was avoided due to the effect of long-term sunlight exposure on the chemical composition of kino [61].

Crude extracts (1.0%, w/v) were prepared from the kino of each tree in ethanol (70%, v/v) at laboratory temperature and filtered through cotton wool to remove coarse debris. The kino extracts were then examined for their antimicrobial activity using well-diffusion methods [62,63]. Microbial suspensions were prepared by inoculating single colonies of the type culture strains, *P. aeruginosa* (ATCC 27853), *E. coli* (ATCC 25922), *S. typhimurium* (ATCC 13311), *S. aureus* (ATCC 25923), *B. cereus* (ATCC 11788), and a laboratory strain of *Candida albicans* into tryptone soya broth (TSB) (Oxoid, Australia). Cultures were incubated overnight at 37 °C on a rotary shaker (150 rpm). The bacterial suspensions were diluted using phosphate buffered saline (PBS, pH = 7.4) to approximately $1.0 \times 10^9$ colony forming units (cfu)/mL using a spectrophotometer at 600 nm wavelength. From each bacterial suspension, 2.5 mL was transferred to 247.5 mL of molten Mueller-Hinton agar at approximately 55°C, and thoroughly mixed to obtain a uniform concentration of $1.0 \times 10^7$ cfu/mL. Each of the six agar suspensions was poured into one sterilized acrylic plate (30 cm × 30 cm internal diameter and 1 cm depth) and allowed to set. Using a sterile cork borer of 4 mm external diameter, 49 holes were cut in each plate and the holes were inoculated with 40 μL of a 1.0% (w/v) solution of kino from each tree, providing a final mass of 400 μg of kino in each well. Plates were covered and incubated for 18 h at 37 °C, and the zone of inhibition for each kino extract was measured. A zone of inhibition >5 mm was considered as positive [64]. Ethanol (70%, v/v) was used as a negative control, and this provided no growth inhibition in any experiment. We also used *P. aeruginosa* (ATCC 27853) as the positive control for testing four clinical strains of *P. aeruginosa* (see below).

Based on the initial antimicrobial activity results, 40 μL of a 1.0% (w/v) solution of kino (final mass of 400 μg) from crude extract of *C. citriodora*, *C. torelliana* × *C. citriodora* or *C. torelliana* was also tested against four wild clinical strains of *P. aeruginosa*, representing different clonal groups. These strains were obtained from a study investigating the clonality of *P. aeruginosa* strains and their virulence properties [31]. The four clinical strains were also tested for their resistance to twelve antimicrobial agents according to Clinical Laboratory Standard Institute (CLSI) guidelines [33], and the results were compared with the antimicrobial activities of the kino extracts. The antimicrobial impregnated disks (Oxoid) included piperacillin (100 μg), ticarcillin (75 μg), piperacillin–tazobactam (100/10 μg), ticarcillin–clavulanic acid (75/10 μg), ceftazidime (30 μg), cefepime (30 μg), aztreonam (30 μg), imipenem (10 μg), gentamicin (10 μg), amikacin (30 μg), ciprofloxacin (5 μg) or norfloxacin (10 μg). Disks were placed on Mueller-Hinton agar that had been inoculated with bacterial suspension at a concentration of $1.0 \times 10^7$ cfu/mL. Plates were then incubated at 37 °C for 16–18 h, after which the diameter of the inhibition zone was measured. The strains were classified as susceptible (S), intermediate (I) or resistant (R) to the antibiotics according to CLSI guidelines [33].

MIC values of the extracts were measured using standard methods [65]. Fresh cultures of the microorganisms were prepared and added to Mueller-Hinton broth to give a final concentration of $1.0 \times 10^7$ cfu/mL. Five concentrations of kino samples (1.0%, 0.8%, 0.5%, 0.3%, and 0.1%; w/v) were prepared from three trees of each taxon and tested against each microorganism, giving final masses of 400 μg, 320 μg, 200 μg, 120 μg, and 40 μg of kino extract per tube. The tubes were incubated on a rotary shaker (150 rpm) at 37 °C for 18 h, and the growth of the microorganisms was recorded visually. A loopful of the last tube showing no growth was then subcultured on tryptone soy agar (TSA) to determine whether the kino was bactericidal or bacteriostatic. Plates were incubated at 37 °C for 18 h.

Biofilm formation in the presence of kino was tested in 96-well tissue culture plates. Suspensions of bacteria grown on TSA agar were prepared to a final concentration of $1.0 \times 10^6$ cfu/mL. As a control, an aliquot of 200 µL of each suspension was added to a well of the plate and incubated without shaking for 48 h at 37 °C. The highest concentration of prepared kino extract (i.e. 1.0%, w/v) from three trees of each taxon was used for biofilm formation assay. 100 µL of sample was added to 100 µL of TSB in each well, and bacterial growth was measured at 600 nm wavelength before staining. The growth medium was removed and plates were rinsed twice with PBS to rinse away non-adhering bacteria. The plates were then allowed to dry. Bacterial biofilm in the wells was stained with 220 µL of crystal violet (0.3%) for 10 min, excess stain was removed with tap water, and the dye was solubilised using 250 µL of acetone/ethanol (20/80, v/v) by shaking at 200 rpm for 10 min. Dissolved crystal violet was measured at 570 nm wavelength. The kino samples from each tree were tested in triplicate. The average optical density (OD) of all three wells at 570 nm wavelength was calculated.

Cytotoxicity of the crude extracts was tested against Vero cells (ATCC CCL-81) derived from African Green Monkey kidneys. Cells were maintained at 37 °C and 5% $CO_2$ in Eagle's minimal essential medium (EMEM) (Lonza, Australia) supplemented with 10% foetal bovine serum (FBS; Lonza) and 1% penicillin/streptomycin solution (Lonza). 200 µL of cell suspension was seeded into 96-well tissue culture plates and grown to 100% confluence in EMEM without antibiotics. Three concentrations of compound (200 µg, 100 µg, and 50 µg) were tested and all tests were performed in triplicate. Cells were visually examined for cytotoxic effects after 4, 24, and 48 h using an inverted phase contrast microscope (×400). The kino extracts were deemed cytotoxic if cell rounding and cell death (detachment from the bottom of wells) occurred in more than 50% of cells [38].

### 3.2. Antibacterial Activity of Flavonoids from Corymbia torelliana

Fresh kino samples from *C. torelliana* trees on the Sunshine Coast (see Section 3.1, above) were extracted in ethyl acetate/water (4/3; v/v). The extracts were stored at −20 °C until fractionation by preparative HPLC. The dry extract (100 mg) was dissolved in acetonitrile/water (1/1; v/v) for fractionation by preparative chromatography. Seven flavonoids were recovered and identified by spectroscopic and spectrometric methods including UV, 1D, and 2D NMR, and UPLC-HR-MS, as described previously [30]: 3,4′,5,7-tetrahydroxyflavanone (1), 3′,4′,5,7-tetrahydroxyflavanone (2), 4′,5,7- trihydroxyflavanone (3), 3,4′,5-trihydroxy-7methoxyflavanone (4), (+)-(2S)-4′,5,7-trihydroxy-6-methylflavanone (5), 4′,5,7-trihydroxy-6,8-dimethylflavanone (6), and 4′,5-dihydroxy-7-methoxyflavanone (7) (Figure 1). Two of these compounds, (1) and (4), are flavanonols, while the other compounds are flavanones.

Each flavonoid was prepared in ethanol (70%; v/v) at a final mass of 50 µg and tested for its antibacterial activity against *P. aeruginosa* (ATCC 27853) and *S. aureus* (ATCC 25923) using a well-diffusion method [63]. Bacterial suspensions were prepared in PBS ($1.0 \times 10^9$ cfu/mL) after overnight growth in TSB. Molten Mueller-Hinton agar was inoculated with bacterial suspension to give a concentration of $1.0 \times 10^7$ cfu/mL, and then allowed to set. Holes cut in the plate were inoculated with 50 µL of a 0.1% (w/v) solution of each flavonoid, providing a final mass of 50 µg in each well. This mass was chosen due to the concentration of pure compounds available after several rounds of extraction. Plates were covered and incubated for 18 h at 37 °C, and the zone of inhibition was measured. Ethanol (70%; v/v) provided no growth inhibition in any experiment. No differences in solubility in ethanol were observed among the seven flavonoids.

Anti-biofilm activity of the seven flavonoids was tested in 96-well tissue culture plates using *P. aeruginosa* (ATCC 27853) and *S. aureus* (ATCC 25923). Fresh bacterial suspensions ($1.0 \times 10^6$ cfu/mL) were prepared in PBS (pH 7.4), with their concentrations determined by observing the $OD_{600}$ value. This suspension was diluted 1:100 in sterile TSB, and 200 µL of suspension was inoculated into a sterile 96-well plate, avoiding the outermost wells to minimise the possibility of desiccation. 100 µL of each flavonoid solution (100 µg of compound) was added to 100 µL of TSB (diluted 1:50) inoculated with bacteria. The plates were processed and assessed for biofilm formation using the method described in

Section 3.1 (above). All tests were performed in triplicate wells. The mean ($\pm$standard error) zone of inhibition or optical density of the three wells was calculated for each sample.

### 3.3. Anti-Pseudomonas Activity and Cytotoxicity of 3,4′,5,7-tetrahydroxyflavanone

One of the most-abundant and active flavonoids, 3,4′,5,7-tetrahydroxyflavanone (1), was tested further against one of the most susceptible bacteria. *P. aeruginosa* (ATCC 27853) was grown and maintained in TSB. Cultures were incubated overnight at 37 °C on a rotary shaker (150 rpm). MICs were measured using methods described previously [65]. Fresh cultures of each bacterium were prepared and added to Mueller-Hinton broth to give a final concentration of c. $1.0 \times 10^7$ cfu/mL. Three concentrations of the flavonoid were prepared: 200 μg, 100 μg, and 50 μg per mL. Bacteria were grown at 100 rpm at 37 °C for 18 h, and the presence or absence of growth was determined visually. The last tube showing no growth was then subcultured on TSA plates to determine whether the MIC was bactericidal or bacteriostatic. These plates were incubated at 37 °C and the growth or lack of growth was observed after 18 h.

Antibiofilm activity of 3,4′,5,7-tetrahydroxyflavanone was tested in 96-well tissue culture plates against the same bacterial strains (above). Fresh bacterial suspensions ($1.0 \times 10^6$ cfu/mL) were prepared in sterile PBS (pH 7.4). This suspension was diluted 1:100 in sterile TSB and 200 μL was inoculated into a sterile tissue culture plate, avoiding the outermost wells. Each plate included positive controls (bacteria without any compounds) and negative controls (only TSB broth). 100 μL of each flavonoid solution (to reach 200 μg, 100 μg, and 50 μg final mass) was added to 100 μL of TSB (diluted 1:50) inoculated with bacterium. The plates were processed and assessed for biofilm formation using the method described in Section 3.1 (above). All tests were performed in triplicate and the average OD was calculated.

The ability of *P. aeruginosa* to adhere to Caco-2 cells (derived from human colon adenocarcinoma) was tested in the presence and the absence of 3,4′,5,7-tetrahydroxyflavanone. Cells were grown on glass coverslips (12 mm diameter, 1 mm thick) to 75% confluence in EMEM supplemented with 10% FBS and 1% penicillin/streptomycin in a 24-well culture plate (Nunc, Australia) at 37 °C in 5% $CO_2$. Cells were rinsed three times with 1 mL of EMEM to remove residual antibiotics, and the medium was replaced with antibiotic-free culture medium. Bacteria were grown in TSB at 110 rpm for 4 h at 37 °C. Bacterial suspensions were centrifuged at 3000 rpm for 10 min and the bacterial pellets were resuspended in sterile PBS (pH 7.4) and adjusted to a concentration of c. $1 \times 10^9$ cfu/mL. 100 μL of bacterial suspension was inoculated per well, and the plates were incubated at 37 °C in 5% $CO_2$ for 90 min. Non-adherent bacteria were removed by washing the cells three times with sterile PBS. Cells were fixed with 95% ethanol for 5 min, air dried, Gram-stained, and examined by light microscopy ($\times 1000$). Bacterial adhesion was assessed [66], with the percentage of adherent bacteria determined by the presence of bacteria on 100 randomly selected cells, and the degree of bacterial adhesion assessed by counting the number of attached bacteria on 25 randomly selected Caco-2 cells. Strains that adhered to <10% of Caco-2 cells were deemed non-adherent. These tests were performed in duplicate. Cytotoxicity of 3,4′,5,7-tetrahydroxyflavanone was tested in vitro against Vero cells as described above. 200 μL of cell suspension was seeded into 96-well tissue culture plates and grown to 100% confluence in medium without antibiotics. The flavonoid was tested at 200 μg, 100 μg and 50 μg final mass, with the tests performed in triplicate. Cells were examined for cytotoxic effects after 4, 24 and 48 h using and the cytotoxicity was interpreted as described above. These tests were performed in triplicate.

### 3.4. Data Analysis

Data were analysed by analysis of variance (ANOVA) followed by post-hoc Tukey's Honestly Significant Difference (HSD) test when the ANOVA detected significant differences ($p < 0.05$) among the means.

**Acknowledgments:** We thank Christina Neuman, Eva Hatje, Tim Smith, Tom Lewis, David Walton, Katie Roberts, Luke Verstraten, Nicholas Evans, Brooke Dwan, Tracey McMahon, and Bruce Randall for assistance. Motahareh Nobakht was supported by an Australian Postgraduate Award, an International Postgraduate Research Scholarship, and a top-up scholarship from Queensland Education and Training International.

**Author Contributions:** M.K., H.M.W., P.R.B., S.J.T. and M.N. conceived the study. M.N. and M.K. designed and performed the microbiology and cytotoxicity experiments with K.J.S. providing assistance. M.N., S.J.T. and H.M.W. designed and performed the plant sampling and statistical analyses. M.N. and P.R.B. designed and performed the flavonoid extraction and purification. All authors contributed to writing the paper.

**Conflicts of Interest:** The authors declare no conflict of interest.

# References

1.  Rappuoli, R. From Pasteur to genomics: Progress and challenges in infectious diseases. *Nat. Med.* **2004**, *10*, 1177–1185. [CrossRef] [PubMed]

2.  Schmidt, F.R. The challenge of multidrug resistance: Actual strategies in the development of novel antibacterials. *Appl. Microbiol. Biotechnol.* **2004**, *63*, 335–343. [CrossRef] [PubMed]

3.  Bell, B.G.; Schellevis, F.; Stobberingh, E.; Goossens, H.; Pringle, M. A systematic review and meta-analysis of the effects of antibiotic consumption on antibiotic resistance. *BMC Infect. Dis.* **2014**, *14*, 13. [CrossRef] [PubMed]

4.  Moloney, M.G. Natural products as a source for novel antibiotics. *Trends Pharmacol. Sci.* **2016**, *37*, 689–701. [CrossRef] [PubMed]

5.  Siller, G.; Rosen, R.; Freeman, M.; Welburn, P.; Katsamas, J.; Ogbourne, S.M. PEP005 (ingenol mebutate) gel for the topical treatment of superficial basal cell carcinoma: Results of a randomized phase IIa trial. *Australas. J. Dermatol.* **2010**, *51*, 99–105. [CrossRef] [PubMed]

6.  Russo, P.; Frustaci, A.; Fini, M.; Cesario, A. From traditional European medicine to discovery of new drug candidates for the treatment of dementia and Alzheimer's disease: Acetylcholinesterase inhibitors. *Curr. Med. Chem.* **2013**, *20*, 976–983. [CrossRef] [PubMed]

7.  Ogbourne, S.M.; Parsons, P.G. The value of nature's natural product library for the discovery of New Chemical Entities: The discovery of ingenol mebutate. *Fitoterapia* **2014**, *98*, 36–44. [CrossRef] [PubMed]

8.  Hamilton, K.D.; Brooks, P.R.; Ogbourne, S.M.; Russell, F.D. Natural products isolated from *Tetragonula carbonaria* cerumen modulate free radical-scavenging and 5-lipoxygenase activities in vitro. *BMC Complement. Altern. Med.* **2017**, *17*, 232. [CrossRef] [PubMed]

9.  Maiden, J. The gums, resins and other vegetable exudations of Australia. *J. R. Soc. NSW* **1901**, *1*, 161–212.

10. Penfold, A. *The Eucalypts*; Interscience Publishers: New York, NY, USA, 1961.

11. Locher, C.; Currie, L. Revisiting kinos—An Australian perspective. *J. Ethnopharmacol.* **2010**, *128*, 259–267. [CrossRef] [PubMed]

12. Reid, E.J.; Betts, T.J. *The Records of Western Australian Plants Used by Aboriginals as Medicinal Agents*; Western Australian Institute of Technology: Perth, Australia, 1977.

13. Packer, J.; Naz, T.; Yaegl Community Elders; Harrington, D.; Jamie, J.F.; Vemulpad, S.R. Antimicrobial activity of customary medicinal plants of the Yaegl Aboriginal community of northern New South Wales, Australia: A preliminary study. *BMC Res. Notes* **2015**, *8*, 276. [CrossRef] [PubMed]

14. Adeniyi, C.B.A.; Lawal, T.O.; Mahady, G.B. In vitro susceptibility of *Helicobacter pylori* to extracts of *Eucalyptus camaldulensis* and *Eucalyptus torelliana*. *Pharm. Biol.* **2009**, *47*, 99–102. [CrossRef] [PubMed]

15. Lawal, T.O.; Adeniyi, B.A.; Adegoke, A.O.; Franzblau, S.G.; Mahady, G.B. In vitro susceptibility of *Mycobacterium tuberculosis* to extracts of *Eucalyptus camaldulensis* and *Eucalyptus torelliana* and isolated compounds. *Pharm. Biol.* **2012**, *50*, 92–98. [CrossRef] [PubMed]

16. Satwalekar, S.S.; Gupta, T.R.; Narasimha, P.L. Chemical and antibacterial properties of kinos from Eucalyptus spp. Citriodorol—The antibiotic principle from the kino of *E. citriodora*. *J. Ind. Inst. Sci.* **1956**, *39*, 195–212.

17. Von Martius, S.; Hammer, K.A.; Locher, C. Chemical characteristics and antimicrobial effects of some *Eucalyptus* kinos. *J. Ethnopharmacol.* **2012**, *144*, 293–299. [CrossRef] [PubMed]

18. Ramos Alvarenga, R.F.; Wan, W.; Inui, T.; Franzblau, S.G.; Pauli, G.F.; Jaki, B.U. Airborne antituberculosis activity of *Eucalyptus citriodora* essential oil. *J. Nat. Prod.* **2014**, *77*, 603–610. [CrossRef] [PubMed]

19. Adeniyi, B.A.; Odufowoke, R.O.; Olaleye, S.B. Antibacterial and gastroprotective properties of *Eucalyptus torelliana* [Myrtaceae] crude extracts. *Int. J. Pharmacol.* **2006**, *2*, 362–365.

20. Lawal, T.O.; Adeniyi, B.A.; Idowu, O.S.; Moody, J.O. In vitro activities of *Eucalyptus camaldulensis* Dehnh. and *Eucalyptus torelliana* F. Muell. against non-tuberculous mycobacteria species. *Afr. J. Microbiol. Res.* **2011**, *5*, 3652–3657.

21. Wallace, H.M.; Trueman, S.J. Dispersal of *Eucalyptus torelliana* seeds by the resin-collecting stingless bee, *Trigona carbonaria*. *Oecologia* **1995**, *104*, 12–16. [CrossRef] [PubMed]

22. Wallace, H.M.; Howell, M.G.; Lee, D.J. Standard yet unusual mechanisms of long-distance dispersal: Seed dispersal of *Corymbia torelliana* by bees. *Divers. Distrib.* **2008**, *14*, 87–94. [CrossRef]

23. Wallace, H.; Lee, D.J. Resin-foraging by colonies of *Trigona sapiens* and *T. hockingsi* (Hymenoptera: Apidae, Meliponini) and consequent seed dispersal of *Corymbia torelliana* (Myrtaceae). *Apidologie* **2010**, *41*, 428–435. [CrossRef]

24. Leonhardt, S.D.; Wallace, H.M.; Schmitt, T. The cuticular profiles of Australian stingless bees are shaped by resin of the eucalypt tree *Corymbia torelliana*. *Austral Ecol.* **2011**, *36*, 537–543. [CrossRef]

25. Wallace, H.M.; Leonhardt, S.D. Do hybrid trees inherit invasive characteristics? Fruits of *Corymbia torelliana* × *C. citriodora* hybrids and potential for seed dispersal by bees. *PLoS ONE* **2015**, *10*, e0138868. [CrossRef] [PubMed]

26. Leonhardt, S.D.; Baumann, A.-M.; Wallace, H.M.; Brooks, P.; Schmitt, T. The chemistry of an unusual seed dispersal mutualism: Bees use a complex set of olfactory cues to find their partner. *Anim. Behav.* **2014**, *98*, 41–51. [CrossRef]

27. Drescher, N.; Wallace, H.M.; Katouli, M.; Massaro, C.F.; Leonhardt, S.D. Diversity matters: How bees benefit from different resin sources. *Oecologia* **2014**, *176*, 943–953. [CrossRef] [PubMed]

28. Massaro, C.F.; Smyth, T.J.; Smyth, W.F.; Heard, T.; Leonhardt, S.D.; Katouli, M.; Wallace, H.M.; Brooks, P. Phloroglucinols from anti-microbial deposit-resins of Australian stingless bees (*Tetragonula carbonaria*). *Phytother. Res.* **2015**, *29*, 48–58. [CrossRef] [PubMed]

29. Massaro, C.F.; Katouli, M.; Grkovic, T.; Vu, H.; Quinn, R.J.; Heard, T.A.; Carvalho, C.; Manley-Harris, M.; Wallace, H.M.; Brooks, P. Anti-staphylococcal activity of *C*-methyl flavanones from propolis of Australian stingless bees (*Tetragonula carbonaria*) and fruit resins of *Corymbia torelliana* (Myrtaceae). *Fitoterapia* **2014**, *95*, 247–257. [CrossRef] [PubMed]

30. Nobakht, M.; Grkovic, T.; Trueman, S.J.; Wallace, H.M.; Katouli, M.; Quinn, R.J.; Brooks, P.R. Chemical constituents of kino extract from *Corymbia torelliana*. *Molecules* **2014**, *19*, 17862–17871. [CrossRef] [PubMed]

31. Streeter, K.; Neuman, C.; Thompson, J.; Hatje, E.; Katouli, M. The characteristics of genetically related *Pseudomonas aeruginosa* from diverse sources and their interaction with human cell lines. *Can. J. Microbiol.* **2016**, *62*, 233–240. [CrossRef] [PubMed]

32. Tomasz, A. Mode of action of β-lactam antibiotics—A microbiologist's view. In *Antibiotics—Handbook of Experimental Pharmacology*; Demain, A.L., Solomon, A.N., Eds.; Springer: Berlin, Germany, 1983; pp. 15–96.

33. Clinical and Laboratory Standards Institute. *Performance Standards for Antimicrobial Susceptibility Testing; Twenty-Second Informational Supplement*; CLSI document M100-S22; Clinical and Laboratory Standards Institute: Wayne, PA, USA, 2012.

34. Polaquini, S.R.B.; Svidzinski, T.I.E.; Kemmelmeier, C.; Gasparetto, A. Effect of aqueous extract from Neem (*Azadirachta indica* A. Juss) on hydrophobicity, biofilm formation and adhesion in composite resin by *Candida albicans*. *Arch. Oral Biol.* **2006**, *51*, 482–490. [CrossRef] [PubMed]

35. Mittal, R.; Aggarwal, S.; Sharma, S.; Chhibber, S.; Harjai, K. Urinary tract infections caused by *Pseudomonas aeruginosa*: A minireview. *J. Infect. Public Health* **2009**, *2*, 101–111. [CrossRef] [PubMed]

36. Høiby, N.; Ciofu, N.; Bjarnshol, T. *Pseudomonas aeruginosa* biofilms in cystic fibrosis. *Future Microbiol.* **2010**, *5*, 1663–1674. [CrossRef] [PubMed]

37. Lynch, A.S.; Robertson, G.T. Bacterial and fungal biofilm infections. *Annu. Rev. Med.* **2008**, *59*, 415–428. [CrossRef] [PubMed]

38. Fiorentini, C.; Barbieri, E.; Falzano, L.; Mattaresse, P.; Baffone, W.; Pianetti, A.; Katouli, M.; Kühn, I.; Möllby, R.; Bruscolini, F.; et al. Occurrence, diversity and pathogenicity of mesophilic *Aeromonas* in estuarine waters of the Italian coast of the Adriatic Sea. *J. Appl. Microbiol.* **1998**, *85*, 501–511. [CrossRef] [PubMed]

39. Snowden, L.A.; Wernbacher, L.; Stenzel, D.; Tucker, J.; McKay, D.; O'Brien, M.; Katouli, M. Prevalence of environmental *Aeromonas* in South-East Queensland, Australia: A study of their interactions with human monolayer Caco-2 cells. *J. Appl. Microbiol.* **2006**, *101*, 964–975. [CrossRef] [PubMed]

40.  Hatje, E.; Neuman, C.; Katouli, M. Interaction of *Aeromonas* strains with lactic acid bacteria using Caco-2 cells. *Appl. Environ. Microbiol.* **2014**, *80*, 681–686. [CrossRef] [PubMed]

41.  Plésiat, P.; Nikaido, H. Outer membranes of Gram-negative bacteria are permeable to steroid probes. *Mol. Microbiol.* **1992**, *6*, 1323–1333. [CrossRef] [PubMed]

42.  Tsuchiya, H.; Sato, M.; Miyazaki, T.; Fujiwara, S.; Tanigaki, S.; Ohyama, M.; Tanaka, T.; Iinuma, M. Comparative study on the antibacterial activity of the phytochemical flavanones against methicillin-resistant *Staphylococcus aureus*. *J. Ethnopharmacol.* **1996**, *50*, 27–34. [CrossRef]

43.  Alcaraz, L.E.; Blanco, S.E.; Puig, O.N.; Tomas, F.; Ferretti, F.H. Antibacterial activity of flavonoids against methicillin-resistant *Staphylococcus aureus* strains. *J. Theor. Biol.* **2000**, *205*, 231–240. [CrossRef] [PubMed]

44.  Silva, J.F.M.; Souza, M.C.; Matta, S.R.; Andrade, M.R.; Vidal, F.V.N. Correlation analysis between phenolic levels of Brazilian propolis extracts and their antimicrobial and antioxidant activities. *Food Chem.* **2006**, *99*, 431–435. [CrossRef]

45.  Hoffman, L.R.; D'Argenio, D.A.; MacCoss, M.J.; Zhang, Z.; Jones, R.A.; Miller, S.I. Aminoglycoside antibiotics induce bacterial biofilm formation. *Nature* **2005**, *436*, 1171–1175. [CrossRef] [PubMed]

46.  Plyuta, V.; Zaitseva, J.; Lobakova, E.; Zagoskina, N.; Kuznetsov, A.; Khmel, I. Effect of plant phenolic compounds on biofilm formation by *Pseudomonas aeruginosa*. *APMIS* **2013**, *121*, 1073–1081. [CrossRef] [PubMed]

47.  Sousa, L.P.; Silva, A.F.; Calil, N.O.; Oliveira, M.G.; Silva, S.S.; Raposo, N.R.B. In vitro inhibition of *Pseudomonas aeruginosa* adhesion by xylitol. *Braz. Arch. Biol. Technol.* **2011**, *54*, 877–884. [CrossRef]

48.  Hillis, W. The chemistry of the Eucalypt kinos. Part I. chromatographic resolution. *Aust. J. Basic Appl. Sci.* **1951**, *3*, 385–397.

49.  Hillis, W. The chemistry of the Eucalypt kinos. Part II. Aromadendrin, kaempferol and ellagic acid. *Aust. J. Sci. Res.* **1952**, *2*, 379–386.

50.  Butler, M.S. The role of natural product chemistry in drug discovery. *J. Nat. Prod.* **2004**, *67*, 2141–2153. [CrossRef] [PubMed]

51.  Daglia, M. Polyphenols as antimicrobial agents. *Curr. Opin. Chem. Biol.* **2012**, *23*, 174–181. [CrossRef] [PubMed]

52.  Hayes, R.A.; Piggott, A.M.; Smith, T.E.; Nahrung, H.F. *Corymbia* phloem phenolics, tannins and terpenoids: Interactions with a cerambycid borer. *Chemoecology* **2014**, *24*, 95–103. [CrossRef]

53.  Nahrung, H.F.; Smith, T.E.; Wiegand, A.N.; Lawson, S.A.; Debuse, V.J. Host tree influences on longicorn beetle (Coleoptera: Cerambycidae) attack in subtropical *Corymbia* (Myrtales: Myrtaceae). *Environ. Entomol.* **2014**, *43*, 37–46. [CrossRef] [PubMed]

54.  Loumouamou, A.N.; Silou, T.; Mapola, G.; Chalcat, J.C.; Figuéredo, G. Yield and composition of essential oils from *Eucalyptus citriodora* × *Eucalyptus torelliana*, a hybrid species growing in Congo-Brazzaville. *J. Essent. Oils Res.* **2009**, *21*, 295–299. [CrossRef]

55.  Dickinson, G.R.; Wallace, H.M.; Lee, D.J. Reciprocal and advanced generation hybrids between *Corymbia citriodora* and *C. torelliana*: Forestry breeding and the risk of gene flow. *Ann. For. Sci.* **2013**, *70*, 1–10. [CrossRef]

56.  Trueman, S.J.; McMahon, T.V.; Bristow, M. Production of cuttings in response to stock plant temperature in the subtropical eucalypts, *Corymbia citriodora* and *Eucalyptus dunnii*. *New For.* **2013**, *44*, 265–279. [CrossRef]

57.  Trueman, S.J.; McMahon, T.V.; Bristow, M. Nutrient partitioning among the roots, hedge and cuttings of *Corymbia citriodora* stock plants. *J. Plant Nutr. Soil Sci.* **2013**, *13*, 977–989. [CrossRef]

58.  Trueman, S.J.; McMahon, T.V.; Bristow, M. Biomass partitioning in *Corymbia citriodora*, *Eucalyptus cloeziana* and *E. dunnii* stock plants in response to temperature. *J. Trop. For. Sci.* **2013**, *25*, 504–509.

59.  Lopes, E.D.; Laia, M.L.; Santos, A.S.; Soares, G.M.; Leite, R.W.P.; Martins, N.S. Influência do espaçamento de plantio na produção energética de clones de *Corymbia* e *Eucalyptus*. *Floresta* **2017**, *47*, 95–104. [CrossRef]

60.  Wendling, I.; Brooks, P.R.; Trueman, S.J. Topophysis in *Corymbia torelliana* × *C. citriodora* seedlings: Adventitious rooting capacity, stem anatomy, and auxin and abscisic acid concentrations. *New For.* **2015**, *46*, 107–120. [CrossRef]

61.  Maiden, J.H. Botany Bay of Eucalyptus kino. *Pharm. J. Trans.* **1889**, *3*, 221–321.

62.  Kudi, A.; Umoh, J.; Eduvie, L.; Gefu, J. Screening of some Nigerian medicinal plants for antibacterial activity. *J. Ethnopharmacol.* **1999**, *67*, 225–228. [CrossRef]

63. Boyanova, L.; Gergova, G.; Nikolov, R.; Derejian, S.; Lazarova, E.; Katsarova, N.; Mitov, I.; Krastev, Z. Activity of Bulgarian propolis against 94 *Helicobacter pylori* strains in vitro by agar-well diffusion, agar dilution and disc diffusion methods. *J. Med. Microbiol.* **2005**, *54*, 481–483. [CrossRef] [PubMed]

64. Palombo, E.A.; Semple, S. Antibacterial activity of traditional Australian medicinal plants. *J. Ethnopharmacol.* **2001**, *77*, 151–157. [CrossRef]

65. Wiegand, I.; Hilpert, K.; Hancock, R.E. Agar and broth dilution methods to determine the minimal inhibitory concentration (MIC) of antimicrobial substances. *Nat. Protoc.* **2008**, *3*, 163–175. [CrossRef] [PubMed]

66. Grey, P.A.; Kirov, S.M. Adherence to HEp-2 cells and enteropathogenic potential of *Aeromonas* spp. *Epidemiol. Infect.* **1993**, *110*, 279–287. [CrossRef] [PubMed]

# Permissions

# List of Contributors

**Christin Naumann and Bettina Hause**
Department of Cell and Metabolic Biology, Leibniz Institute of Plant Biochemistry, Weinberg 3, D-06120 Halle (Saale), Germany

**Markus Otto**
Department of Cell and Metabolic Biology, Leibniz Institute of Plant Biochemistry, Weinberg 3, D-06120 Halle (Saale), Germany
Department of Molecular Signal Processing, Leibniz Institute of Plant Biochemistry, Weinberg 3, D-06120 Halle (Saale), Germany

**Claus Wasternack**
Department of Molecular Signal Processing, Leibniz Institute of Plant Biochemistry, Weinberg 3, D-06120 Halle (Saale), Germany
Laboratory of Growth Regulators, Centre of the Region Haná for Biotechnological and Agricultural Research, Institute of Experimental Botany AS CR & Palacký University, Šlechtitelu° 11, CZ-78371 Olomouc, Czech Republic

**Wolfgang Brandt**
Department of Natural Product Chemistry, Leibniz Institute of Plant Biochemistry, Weinberg 3, D-06120 Halle (Saale), Germany

**Shinichi Enoki, Nozomi Fujimori, Chiho Yamaguchi, Tomoki Hattori and Shunji Suzuki**
Laboratory of Fruit Genetic Engineering, The Institute of Enology and Viticulture, University of Yamanashi, Yamanashi 400-0005, Japan

**Bettina Dudek, Anne-Christin Warskulat and Bernd Schneider**
Max Planck Institute for Chemical Ecology, Hans-Knöll-Straße 8, 07745 Jena, Germany

**Shanshan Li, Lin Liu and Shuying Fan**
College of Agronomy, Jiangxi Agricultural University, Nanchang 330045, China

**Chunpeng Wan and Chuying Chen**
Collaborative Innovation Center of Post-Harvest Key Technology and Quality Safety of Fruits and Vegetables in Jiangxi Province, Jiangxi Agricultural University, Nanchang 330045, China

**Simon E. Bull, Adrian Alder, Mathias Kohler and Wilhelm Gruissem**
Plant Biotechnology, Department of Biology, ETH Zürich, 8092 Zürich, Switzerland

**Hervé Vanderschuren**
Plant Biotechnology, Department of Biology, ETH Zürich, 8092 Zürich, Switzerland
Gembloux Agro-Bio Tech, University of Liège, 5030 Gembloux, Belgium

**Cristina Barsan**
Gembloux Agro-Bio Tech, University of Liège, 5030 Gembloux, Belgium

**Lars Hennig**
Department of Plant Biology and Linnean Centre for Plant Biology, PO Box 7080, The Swedish University of Agricultural Sciences, SE-750 07 Uppsala, Sweden

**Jean-Luc Wolfender**
School of Pharmaceutical Sciences, University of Lausanne, University of Geneva, quai Ernest-Ansermet 30, CH-1211 Geneva 4, Switzerland

**Adeline Chauvin**
School of Pharmaceutical Sciences, University of Lausanne, University of Geneva, quai Ernest-Ansermet 30, CH-1211 Geneva 4, Switzerland
Department of Plant Molecular Biology, University of Lausanne, CH-1015 Lausanne, Switzerland

**Aurore Lenglet and Edward E. Farmer**
Department of Plant Molecular Biology, University of Lausanne, CH-1015 Lausanne, Switzerland

**Nityananda Khanal and Anna Grisnich**
Department of Plant Sciences, University of Saskatchewan, Saskatoon, SK S7N 5A8, Canada

**Geoffrey E. Bray**
Department of Biochemistry, University of Saskatchewan, Saskatoon, SK S7N 5E5, Canada

**Gordon R. Gray**
Department of Plant Sciences, University of Saskatchewan, Saskatoon, SK S7N 5A8, Canada
Department of Biochemistry, University of Saskatchewan, Saskatoon, SK S7N 5E5, Canada

**Barbara A. Moffatt**
Department of Biology, University of Waterloo, Waterloo, ON N2L 3G1, Canada

**Tamalika Chakraborty**
Institute of Ethnobiology, School of Studies in Botany, Jiwaji University, Gwalior 474011, India

Chair of Site Classification and Vegetation Science, Institute of Forest Sciences, University of Freiburg, Tennenbacherstr. 4, D-79106 Freiburg, Germany

**Somidh Saha**
Resource Survey and Management Division, Forest Research Institute, PO New Forest, Dehra Dun 248006, India
Chair of Silviculture, Institute of Forest Sciences, University of Freiburg, Tennenbacherstr. 4, D-79106 Freiburg im Breisgau, Germany
Institute for Technology Assessment and Systems Analysis (ITAS), Karlsruhe Institute of Technology (KIT), Karlstr. 11, D-76133 Karlsruhe, Germany

**Narendra S. Bisht**
Resource Survey and Management Division, Forest Research Institute, PO New Forest, Dehra Dun 248006, India
Directorate of Extension, Indian Council of Forestry Research and Education, PO New Forest, Dehra Dun 248006, India

**Dil Thavarajah, Alex Abare and Indika Mapa**
Plant and Environmental Sciences, 270 Poole Agricultural Center, Clemson University, Clemson, SC 29634, USA

**Clarice J. Coyne**
USDA Agriculture Research Service, Western Regional Plant Introduction Station, Washington State University, Pullman, WA 99164-6434, USA

**Pushparajah Thavarajah**
BOV Solutions Inc., 1105 Garner Bagnal Blvd, Statesville, NC 28677, USA

**Shiv Kumar**
Biodiversity and Integrated Gene Management Program, International Centre for Agricultural Research in the Dry Areas (ICARDA), Rabat-Institute, Rabat, Morocco

**Vadim G. Lebedev, Nina P. Kovalenko and Konstantin A. Shestibratov**
Branch of Shemyakin and Ovchinnikov Institute of Bioorganic Chemistry of the Russian Academy of Sciences, Science avenue 6, Pushchino, Moscow Region 142290, Russia

**Gunbharpur Singh Gill, Riston Haugen, Steven L. and David H. Siemens**
Integrative Genomics Program, Black Hills State University, Spearfish, SD 57789, USA

**Matzner**
Biology Department, Augustana University, Sioux Falls, SD 57197, USA

**Abdelali Barakat**
Biology Department, University of South Dakota, Vermillion, SD 57069, USA

**Huiqin He**
College of Resources and Environmental Engineering, Yibin University, Yibin 644000, China

**Thomas Monaco**
US Department of Agriculture, Agricultural Research Service, Forage and Range Research Laboratory, Utah State University, Logan, UT 84322-6300, USA

**Huimin Zhao**
State Key Laboratory of Tree Genetics and Breeding, Northeast Forestry University, Harbin 150040, China

**Yucheng Wang**
State Key Laboratory of Tree Genetics and Breeding, Northeast Forestry University, Harbin 150040, China
Key Laboratory of Biogeography and Bioresource in Arid Land, Xinjiang Institute of Ecology and Geography, Chinese Academy of Sciences, Urumqi 830011, China

**Zilong Tan and Xuejing Wen**
Key Laboratory of Biogeography and Bioresource in Arid Land, Xinjiang Institute of Ecology and Geography, Chinese Academy of Sciences, Urumqi 830011, China

**María Pérez-Fernández, Elena Calvo-Magro, Irene Ramírez-Rojas and Laura Moreno-Gallardo**
Department of Physical, Chemical and Natural Systems, University Pablo de Olavide, Carretera de Utrera Km, Seville 141013, Spain

**Valentine Alexander**
Botany and Zoology Department, University of Stellenbosch, Private Bag X1, Matieland 7602, South Africa

**Franka Seiler, Jürgen Soll and Bettina Bölter**
Department Biologie I-Botanik, Ludwig-Maximilians-Universität, Großhadernerstr. 2-4, Planegg-Martinsried 82152, Germany

**Ann-Charlotte Wallenhammar**
Rural Economy and Agricultural Society, HS Konsult AB, SE-701 45 Örebro, Sweden

Precision Agriculture and Pedometrics, Department of Soil and Environment, Swedish University of Agricultural Sciences, P.O. Box 234, SE-523 23 Skara, Sweden

**Anders Jonsson**
Precision Agriculture and Pedometrics, Department of Soil and Environment, Swedish University of Agricultural Sciences, P.O. Box 234, SE-523 23 Skara, Sweden
Swedish Institute of Agricultural and Environmental Engineering, Green Tech Park, P.O. Box 63, SE-532 21 Skara, Sweden

**Albin Gunnarson**
Swedish Seed and Oilseed Growers, P.O. Box 53, SE-230 53 Alnarp, Sweden

**Fredrik Hansson**
Rural Economy and Agricultural Society, Borgeby Slottsväg 11, SE-237 91 Bjärred, Sweden

**Motahareh Nobakht, Stephen J. Trueman, Helen M. Wallace, Peter R. Brooks, Klrissa J. Streeter and Mohammad Katouli**
Centre for Genetics, Ecology and Physiology, University of the Sunshine Coast, Maroochydore DC, QLD 4558, Australia

# Index

**A**

Activity Regulation, 1
Adaptive (phenotypic) Plasticity, 55
Aerial Parts, 27-29, 31, 33
Antibiotic Resistance, 191, 202
Antimicrobial Activity, 191-196, 198-199, 202
Arabidopsis, 1-2, 6-8, 10-12, 16-17, 34-35, 37, 40-43, 49-55, 57, 59-75, 78-82, 117, 127-129, 131, 134-136, 150, 157-158, 170-171, 174, 179-181, 183-184
Arabidopsis Allene Oxide Cyclase Isoforms, 1
Arabidopsis Thaliana, 1, 8, 10-12, 16-17, 42, 52-55, 57, 75, 81-82, 129, 134-136, 158, 170-171, 179-180, 183
Ascorbate Acid, 149-150

**B**

Betula, 107, 111-115
Biofortificaiton, 96
Bioprospecting, 84, 94
Biotechnology, 34, 37, 40, 81, 104, 114
Breeding, 34-35, 37, 39-41, 96-97, 102, 104, 149, 171, 188, 190, 204

**C**

C Construction Costs, 159, 166
Caffeoylquinic Acid (CQA), 27
Caffeoylquinic Acids, 27-28
Carbon Isotope Ratio, 116, 120-121, 123, 127, 129, 131-133
Cassava, 34-41
Chlorophyll, 55-57, 59-61, 63, 72, 75-78, 82-83, 107, 110-113, 115, 152-153, 156, 170, 180-181
Chrysanthemum Coronarium L, 27, 32-33
Cold Acclimation, 55-57, 59, 61-66, 69, 71-76, 78-82
Cytotoxicity, 191, 193, 195, 198, 200-202

**D**

Defense, 12, 17, 42-43, 46-48, 50, 52-53, 116-118, 124-125, 127-135
Disturbance Age, 137
Drought, 82, 116-130, 132-136, 158

**E**

Eicosanoid, 42
Ethnobotany, 84, 93-94, 191
Ethnopharmacology, 84, 92-93, 95
Eucalyptus, 40, 191, 202-204
Eutrema Salsugineum, 55, 57, 75, 81-82

**F**

Fertile Flowering, 34
Flavonoids, 19-20, 24-27, 30, 184, 191, 196, 200-201, 204
Flc, 12-16, 117
Floral Transition, 12-15
Flower Pigmentation, 19
Flowering, 12, 14-17, 24, 34-37, 39-41, 84, 117, 134, 170-171, 183
Flowering Locus T, 34-35, 37
Forest Succession, 137

**G**

Gene Expression, 2, 42, 46, 49-50, 53, 57, 82, 110, 116, 123, 127-128, 130, 132-133, 135, 150, 155, 157-158, 170-171, 174, 176-182
Glasshouse Cassava, 34
Glucosinolates, 116, 122, 131, 135-136
Glutamine Synthetase, 107, 110, 113-114
Grafting, 34, 37, 39, 41
Grapevine, 12, 16

**H**

Herbivore, 42-43, 53, 116, 118, 121-125, 128-131, 135, 145
Herbivory, 10, 43, 116-119, 121-124, 126-130, 132-133, 135
Heteromerization, 1-2, 5-10

**I**

Inflorescence, 12, 16, 34-35, 37

**J**

Jasmonate, 1-2, 9-11, 42-43, 49-54, 136
Jasmonates (JAS), 1
Jasmonic Acid, 1, 42-43, 53-54, 117, 127

**K**

Kaempferol, 19, 21-25, 204

**L**

Lentil, 96-106
Lipoxygenases (LOXS), 42
Low Temperature, 55-56, 62, 65, 69-70, 72, 74-76, 78-80

**M**

Manihot Esculenta Crantz, 34, 40
Medicinal Plants, 84-85, 92-95, 202, 204-205
Mineral N, 159, 166
Mountain Plants, 84

# N

N2 Fixation, 159, 161, 163, 166, 168-169

Natural Products, 26, 191, 198, 202

Nitrogen Fertilization, 107, 110-112

Nudicaulins, 19-21, 23-26

# O

Oxylipin, 10-11, 42, 52-53

# P

Papaver Nudicaule, 19, 25-26

Pelargonidin, 19, 21, 23-26

Phenotype, 2, 35, 41, 170, 174, 179

Photoinhibition, 55-56, 62-63, 73, 77-78, 80, 82

Photosynthesis, 55-57, 59, 65, 68-69, 72-73, 78-83, 114, 130, 160, 162, 168-170, 174, 176, 179, 183

Photosynthetic Acclimation, 55, 57, 78-79, 82

Photosynthetic Organisms, 55-56

Plant Community Assembly, 137

Plasmodiophora Brassicae, 185-187, 190

Predictive Soil Tests, 185

Principal Component Analysis, 121-123, 137, 139-140, 144-145, 147

Protein Structure Analysis, 1

Pseudomonas Aeruginosa, 191, 194, 196, 198, 203-204

# Q

Qpcr Assay, 185, 189

# R

Recalcitrant Crops, 34

# S

Seed, 19, 34-35, 37, 39, 96-100, 102, 104-106, 137, 146, 185, 188, 190, 203

Selenium, 96-98, 103, 105-106

Site-directed Mutagenesis, 1, 3-4, 7

Species Colonization, 137

Stingless Bees, 191, 203

Syringe Agroinfiltration, 149-150, 153, 155-157

# T

Tetragonula, 191, 202-203

Traditional Knowledge, 84, 94, 198

Traditional Medicine, 84, 93, 191-192, 198

Transformation Efficiency, 149-151, 153, 155-157

Transgenic, 2, 10, 12, 16, 35, 37-39, 41, 107, 111-114, 150-151, 158

Transgenic Birch, 107, 112, 114

Tween-20, 149-150, 153, 155, 157

# V

Vitis Vinifera Glycosyl Hydrolase Family 17 (VvGHF17), 12

VvGHF17, 12-16

# W

Wild Germplasm, 96

Wounding, 2, 8, 10, 42-43, 46-48, 50-51, 53-54, 124

www.ingramcontent.com/pod-product-compliance
Lightning Source LLC
Chambersburg PA
CBHW080647200326
41458CB00013B/4763